Failure Analysis - Structural Health Monitoring of Structure and Infrastructure Components

Edited by Gobinath Ravindran, Vutukuru Mahesh and Moustafa Moufid Kassem

Published in London, United Kingdom

Failure Analysis - Structural Health Monitoring of Structure and Infrastructure Components
http://dx.doi.org/10.5772/intechopen.100773
Edited by Gobinath Ravindran, Vutukuru Mahesh and Moustafa Moufid Kassem

Contributors
Christof Sommitsch, Bernhard Sonderegger, Mohammad Ahmadi, Florian Riedlsperger, Felix Meixner, Bernhard Krenmayr, Josef Mergl, A. Toshimitsu Yokobori Jr, Go Ozeki, Gobinath Ravindran, Vutukuru Mahesh, A. Nikhil Kumar, Herda Yati Binti Katman, Samuel A. Ayowole Awe, Gonzalo Diego, Susana Merino, Ján Kováč, Kanhirodan Ravindranath, Abdulmuhsen Akbar, Bader Al-Wakaa, Zak Abdallah, Alireza Khalifeh, Cristina-Ileana Pascu, Raluca Malciu, Ilie Dumitru, Francisco Casanova-del-Angel, Alireda Aljaroudi, Ahmed Aljaroudi, Kenneth C. Crawford, Cameron R. Rusnak, Craig C. Menzemer, Ikwulono David D. Unobe, Andrew D. Sorensen, Moustafa Moufid Kassem, Ayman Mohammad Mansour

First published in London, United Kingdom, 2023 by IntechOpen
IntechOpen is the global imprint of INTECHOPEN LIMITED, registered in England and Wales, registration number: 11086078, 5 Princes Gate Court, London, SW7 2QJ, United Kingdom

British Library Cataloguing-in-Publication Data
A catalogue record for this book is available from the British Library

Additional hard and PDF copies can be obtained from orders@intechopen.com

Failure Analysis - Structural Health Monitoring of Structure and Infrastructure Components
Edited by Gobinath Ravindran, Vutukuru Mahesh and Moustafa Moufid Kassem
p. cm.
Print ISBN 978-1-80356-512-5
Online ISBN 978-1-80356-513-2
eBook (PDF) ISBN 978-1-80356-514-9

For EU product safety concerns:
IN TECH d.o.o., Prolaz Marije Krucifikse Kozulić 3, 51000 Rijeka, Croatia,
info@intechopen.com or visit our website at intechopen.com.

Meet the editors

Dr. Gobinath Ravindran currently works in the Department of Civil Engineering, SR University, Warangal, Telangana, India. He is also the Dean of Research and Ranking of the same University, which focuses on innovations related to high-performance and eco-friendly construction methods and materials. Dr. Ravindran has been working in the publishing industry for the last 14 years. He has published more than 200 research papers in peer-reviewed journals. He is also an editor for many journals. He possesses strong patent, industrial design drafting, and analysis skills in engineering and materials. Dr. Ravidran also has strong technical and content writing skills and volunteers to support budding researchers and scholars to promote their research..

Dr. V. Mahesh was awarded a Ph.D. in Mechanical Engineering from JNT University, Hyderabad, and an M.Tech in Computer Integrated Manufacturing from Regional Engineering College, presently the National Institute of Technology, Warangal. He has more than 25 years of teaching experience and successfully guided many students in their UG and PG projects. He has published more than 50 papers in various national and international journals and conferences. He is a reviewer of a number of reputed journals in the area of mechanical engineering. His areas of interest include Cognitive Science, Design Thinking, Social Entrepreneurship, Computer Integrated Manufacturing, and Composites. He is a Chartered Engineer and Member of The Institution of Engineers (INDIA). He is a life member of the Indian Society for Technical Education (ISTE) and the International Association of Engineers. Currently, he is working as a Professor of Mechanical Engineering at SR University, Warangal.

Dr. Moustafa Moufid Kassem is an assistant professor in the School of Civil Engineering, Universiti Sains Malaysia. He established the seismic design provision code for bridges in Malaysia and implemented it in 2022. He is collaborating with the Malaysian Public Work Department to create a novel seismic expansion joint as an innovative solution for irregular building types. Dr. Kassem holds associate memberships in both the American Society of Civil Engineers (ASCE) and the Institution of Civil Engineers (ICE), two prestigious engineering organizations in the United States and the United Kingdom. He specializes in earthquake engineering, risk assessment, seismic hazards, structural analysis, numerical simulation, and modeling. He has around forty research papers in structural and earthquake engineering to his credit.

Contents

Preface

There are numerous factors that contribute to the deterioration of structures and infrastructures over time, including fatigue failure induced by traffic loads, the effects of environmental conditions, and natural catastrophes such as earthquakes. Continuous monitoring of the structure is needed to provide an early warning of a potentially dangerous state and an assessment of the structure's changes over time, rather than only routine or critical event-based inspections. Thus, structural health monitoring (SHM) has recently become a popular area of research in structural engineering. SHM is a field that has grown tremendously over the past few decades. This book brings together the most recent studies and state-of-the-art methodologies to improve understanding of how SHM can be applied in the real world, as well as how it can be improved, by highlighting its practical, theoretical, and historical applications. The book also examines challenges in SHM and sheds light on the future trends of SHM techniques for different types of response measurement and their applications in civil engineering structures and infrastructures.

Dr. Gobinath Ravindran
Department of Civil Engineering,
SR University,
Warangal, India

Dr. Vutukuru Mahesh
Department of Mechanical Engineering,
SR University,
Warangal, India

Dr. Moustafa Moufid Kassem
School of Civil Engineering,
Universiti Sains Malaysia,
Penang, Malaysia

Chapter 1

The Quantitative Estimation of Mechanical Performance on the Creep Strength and Prediction of Creep Fracture Life for Creep Ductile Materials Based on QL^* Parameter

A. Toshimitsu Yokobori Jr and Go Ozeki

Abstract

Previously, we have proposed creep deformation law estimated by non-dimensional time representation to predict creep fracture and remnant life. Furthermore, using steady state creep rate coupled with crack growth rate law based on Q^* parameter, QL^* parameter was derived and it was found to enable us to discriminate creep ductility and predict creep fracture life. In this study, a quantitative estimation and a prediction methods of mechanical performance on creep strength (MPCS) and creep fracture life of the creep ductile materials including a weld joint notched specimen was noticed and the following studies were conducted. 1) The similarity law of creep deformation, 2) QL^*map, which discriminates creep ductility and predicts creep fracture life, 3) Derivation method of mechanical indicators, "Converted stress", and ΔQL^*, which quantify MPCS, 4) Example of the quantitative estimation by these parameters using P91 steel and its weld joint notched specimens. From these results, the concept of the converted stress and the ΔQL^*were found to enable us to conduct quantitative estimation of MPCS and prediction of creep fracture life, with the short experimental period, the small number of specimens, the reasonable accuracy and an economic efficiency, which is an engineering significance.

Keywords: QL^*, MPCS, converted stress, creep ductility, weld joint, creep fracture life, ΔQL^*

1. Introduction

Concerning the estimation of mechanical performance on creep and creep-fatigue interaction, the law on the relationship between applied stress and fracture life [1], the

Larson-Miller parameter [2], the Manson-Coffin law [3, 4] and the Ω method [5] were proposed for a smooth specimen. Furthermore, from the point of the application to actual engineering structure, the prediction of the crack growth life originated from a site of stress concentration is important and the establishment of the law of creep crack growth rate to predict creep crack growth life has been conducted [6, 7]. Especially, the establishment of the quantitative estimation method on the mechanical performance for a weld joint specimen under creep condition is important, because the effect of the morphology of heat affected zone (HAZ) line and the difference of material structure, between the weld metal, HAZ and the base metal, on the creep fracture life are significant factors to predict the creep fracture life [8].

Furthermore, from the point of actual application under recent social demand, such as the short experimental period and economical situation, it is necessary to establish the predicting method of creep fracture life by using small number of specimens and by conducting the short life creep test.

In this study, under these social demands mentioned above, by using a notched specimen of creep ductile material, the methods of estimating the quantitative mechanical performance on creep strength (MPCS) and of predicting creep fracture life of the weld joint by using small number of specimens and by conducting the short life creep test were proposed.

2. Theoretical foundation and background

2.1 Similarity law of creep deformation

For creep ductile materials, the time sequential characteristic of creep deformation plotted against non-dimensional time controlled by each fracture life, t_f was found to be independent of applied stress and temperature, that is, similarity law of creep deformation [9]. This behavior was illustrated as shown in **Figure 1a** and **b**. For such case, fracture life is possible to be predicted by the value of *RNOD* at the current

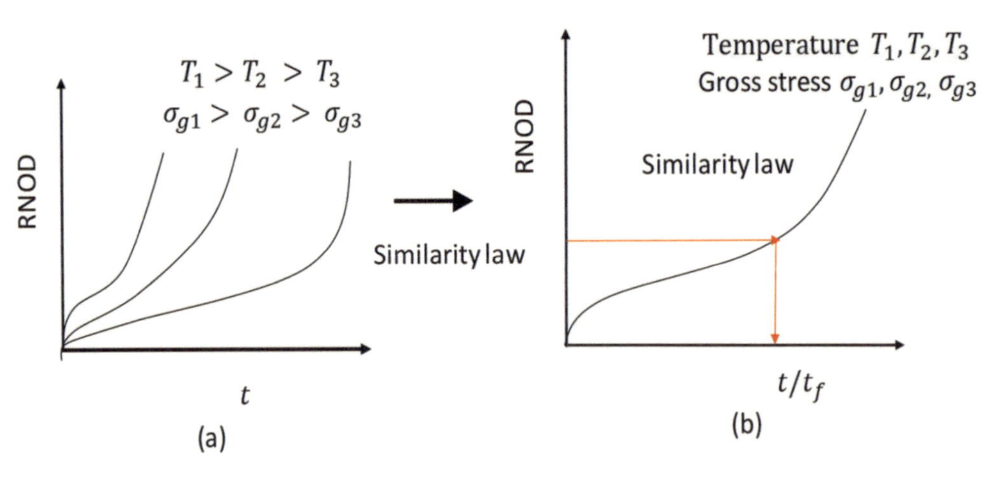

Figure 1.
Similarity law of creep deformation for a notched specimen. (a) The relationship between RNOD and loading time, (b) The relationship between RNOD and non-dimensional loading time controlled by each creep fracture life [9].

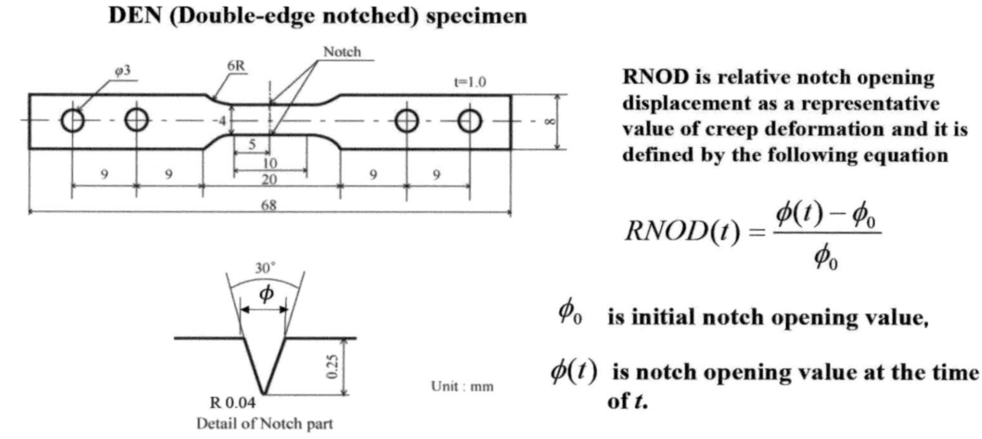

Figure 2.
Definition of relative notch opening displacement (RNOD) [10].

loading time, t as shown in **Figure 1b**. Where, as a measure of deformation for a notched specimen, the concept of the Relative Notch Opening Displacement (RNOD, a representative value of deformation of a DEN specimen) [10] given by Eq. (1) was used.

$$RNOD(t) = \frac{\phi(t) - \phi_0}{\phi_0} \tag{1}$$

where $\phi(t)$ is the notch opening value at the time of t and ϕ_0 is the initial notch opening value as shown in **Figure 2.**

2.2 Evaluation method of creep ductility (QL^* parameter)

The creep crack growth rate (CCGR) is written by Q^* parameter as shown in the Eq. (2) and (3) [6, 7].

$$\frac{da}{dt} = A^* \, exp\,(Q^*) = A^* \sigma_g^{m_g} \, exp\left(-\frac{\Delta H_g}{RT}\right) \tag{2}$$

$$Q^* = -\frac{\Delta H_g - m_g \ln \sigma_g}{RT} \tag{3}$$

By integrating Eq. (2) and (3), the life of creep crack growth is written by Eqns. (4) and (5),

$$t_f = \int_0^{t_f} \frac{da}{A^* \sigma_g^{m_g} \, exp\left(-\frac{\Delta H_g}{RT}\right)} = \frac{a_f - a_i}{A^* \sigma_g^{m_g} \, exp\left(-\frac{\Delta H_g}{RT}\right)} = \frac{C}{A^* \sigma_g^{m_g} \, exp\left(-\frac{\Delta H_g}{RT}\right)} \tag{4}$$

$$\frac{1}{t_f} = A_g \sigma_g^{m_g} \, exp\left(-\frac{\Delta H_g}{RT}\right) \tag{5}$$

where t_f is the life of crack growth and $A_g = \frac{A^*}{C}$.

The steady state creep strain rate is given by the Eq. (6).

$$\dot{\varepsilon}_s = A_c \sigma_g^{m_c} exp\left(-\frac{\Delta H_c}{RT}\right) \quad (6)$$

Dividing Eq. (6) by Eq. (5), QL^* parameter is given by Eq. (7) [11, 12].

$$QL^* = \dot{\varepsilon}_s{}^{\alpha} \bullet t_f = M\sigma_g^m exp\left(-\frac{\Delta H}{RT}\right) \quad (7)$$

Where A^*, A_g and A_c are constants. σ_g is gross stress (MPa), m_g and m_c are exponent of gross stress and ΔH_g, ΔH_c is activation energy of crack growth and creep strain (J/mol), respectively. R is gas constant (J/K · mol), T is absolute temperature (K). $m = m_c - m_g$, $\Delta H = \Delta H_c - \Delta H_g$, M is creep ductility. α is constant and it nearly equal to unity.

For creep ductile materials, since steady state creep rate well correlates with the inverse value of fracture life, values of m and ΔH almost equals to zero, respectively, that is, $m_c = m_g$ and $\Delta H_c = \Delta H_g$. For such case, Eq. (7) is written by Eq. (8), that is, the relationship between logarithmic values of $\dot{\varepsilon}_s$ and t_f is independent of gross stress and temperature.

$$QL^* = \dot{\varepsilon}_s{}^{\alpha} \bullet t_f = M \quad (8)$$

Based on the QL^* concept, the relationship between logarithmic values of $\dot{\varepsilon}_s$ and t_f for various materials are obtained as shown in **Figure 3** [11], that is, the QL^* map. The data band for each material shows the creep ductility as the value of QL^* (= M).

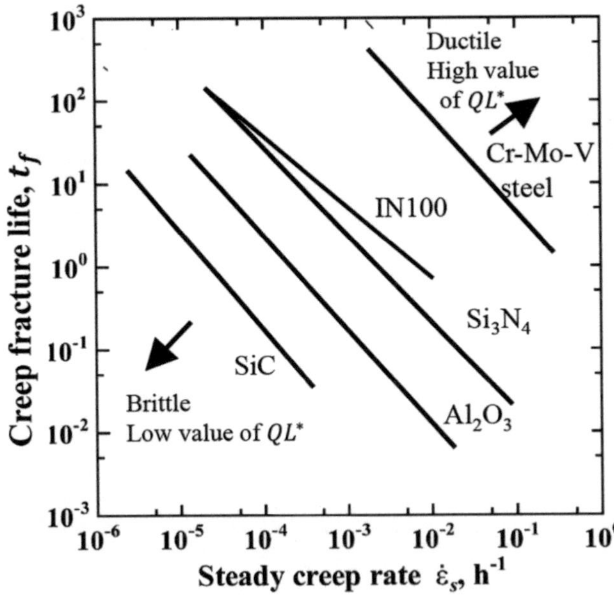

Figure 3.
QL^* map for creep ductile and brittle materials. The relationship between creep fracture life and steady state creep rate.

2.3 Derivation method of the converted stress and ΔQL^*

For the case of creep ductile materials such as Cr-Mo-V steel, the similarity law of creep deformation and Eq. (8) described in the Section 2.1 and 2.2 are valid [11, 12]. For such case, the concept of the converted stress is defined and valid to conduct a quantitative estimation of the MPCS and the prediction of creep fracture life.

This section shows the derivation method of the converted stress. When the same similarity law of creep deformation is valid both for A and B materials, the relationship between $ln\, t_f$ and $ln\, \dot{\varepsilon}_s$ in the QL^* map is unique as shown in **Figure 4** [12], where A material is a control material to make the QL^* line and B material is a target material to estimate MPCS.

The flow chart of the derivation method of the converted stress is shown in **Figure 5**.

When the similarity law of creep deformation is valid, Eq. (8) is unique both for A and B materials and it was written by Eq. (9),

$$t_f = M_1 \dot{\varepsilon}_s^{-\alpha}, \tag{9}$$

where M_1 is creep ductility for A and B materials.

Steady state creep rate for the A material is experimentally given by Eq. (10).

$$\dot{\varepsilon_{SA}} = K_A \sigma_A^{m_A} \tag{10}$$

Using Eqns. (9) and (10), t_{fA} is given by Eq. (11),

$$t_{fA} = M_1 K_A{}^{-\alpha} \sigma_A^{-\alpha m_A} \tag{11}$$

where t_{fA}, $\dot{\varepsilon_{SA}}$, K_A, σ_A and m_A is the creep fracture life, the steady state creep strain rate, the material constant, the applied stress and power coefficient value of σ_A, obtained by creep tests for the A material, respectively.

When the steady state creep rate, $\dot{\varepsilon_{SB}}$ for the B material is experimentally obtained by non-fracture test, substituting $\dot{\varepsilon_{SB}}$ in to Eq. (9), the predictive creep fracture life, t_{fB} for the B material is given by Eq. (12), because Eq. (9) is unique both for A and B materials.

$$t_{fB} = M_1 \dot{\varepsilon_{SB}}^{-\alpha} \tag{12}$$

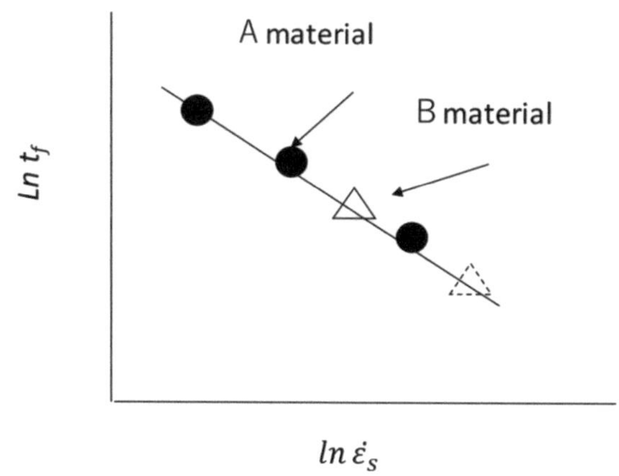

Figure 4.
QL^* line for A and B materials when the similarity law of creep deformation is valid both for A and B materials.

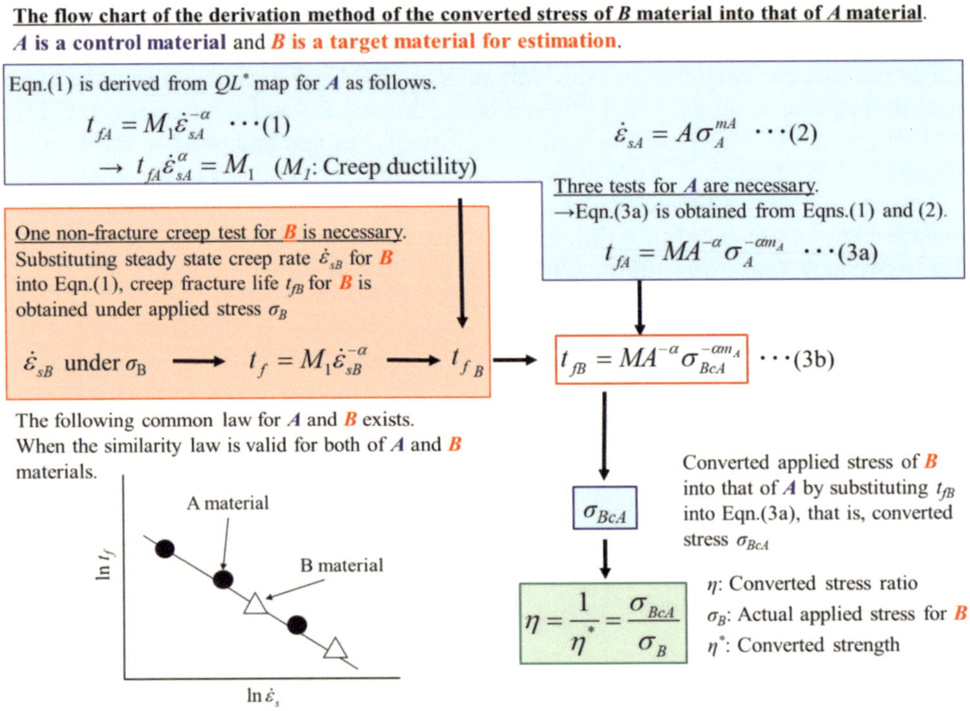

Figure 5.
Flow chart of derivation method of the converted stress.

Substituting t_{fB} given by Eq. (12) into t_{fA} in Eq. (11), the converted stress of the B material into that of the A material, σ_{BCA} is given by Eq. (13).

$$t_{fB} = M_1A^{-\alpha}\sigma_{BCA}^{-\alpha m_A} \tag{13}$$

where σ_{BCA} is the converted stress of the B material into that of the A material.

The converted stress of the B material into that of the A material, σ_{BCA} means applied stress which causes the equivalent steady state creep rate, $\dot{\varepsilon_{SB}}$ for the A material given by Eq. (10).

Furthermore, the converted stress ratio, η is defined and it is written by Eq. (14).

$$\eta = \frac{1}{\eta^*} = \frac{\sigma_{BCA}}{\sigma_B} \tag{14}$$

where σ_B is actual applied stress for the B material, η^* is the converted strength, which is the inverse value of the converted stress ratio, η. η is a quantitative indicator of the MPCS.

For the case of $\eta > 1.0$, $\sigma_B < \sigma_{BCA}$ being valid, it means creep strength of the B material is lower than that of the A material.

For the case of $\eta < 1.0$, $\sigma_B > \sigma_{BCA}$ being valid, it means creep strength of the B material is higher than that of the A material. The detailed derivation method is shown in **Figure 5**.

In addition, the concept of ΔQL^* is useful to discriminate the difference of the creep ductility caused by different materials or different local stress multi-axiality

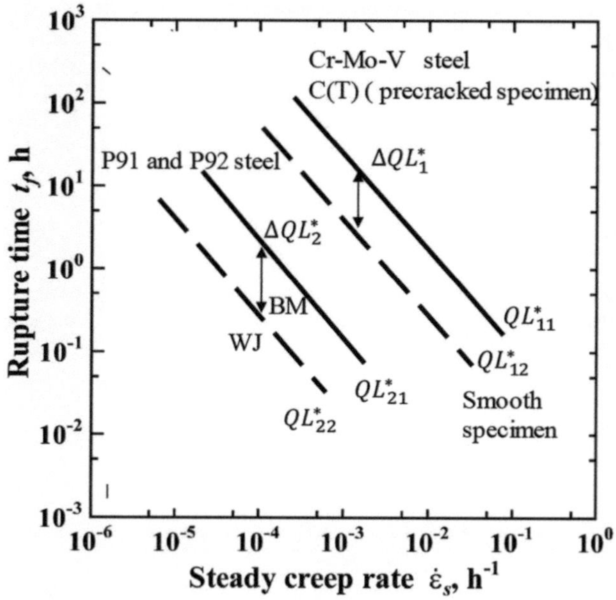

Figure 6.
The schematic illustration of the comparison of QL* *line between a notched specimen (C (T) specimen) and smooth specimen for Cr-Mo-v steel [12] and that between base metal and weld joint of C (T) specimen for P91 and P92 steels [14].*

characterized by $TF = \frac{3(\sigma_x+\sigma_y+\sigma_z)}{\sigma_{eq}} = \frac{3\sigma_P}{\sigma_{eq}}$ such as weld joint, that is, the structural brittleness [13]. σ_P is hydrostatic stress. σ_{eq} is equivalent stress.

Schematic illustration of experimental characteristics of ΔQL^* between a C (T) and a smooth specimens for Cr-Mo-V steel and a base metal and a weld joint of a notched specimen for P91 and P92 steels is shown in **Figure 6** [12, 14]. The difference in ΔQL^* between a C (T) and a smooth specimens is considered to be caused by the difference in compliance between a cracked specimen and a smooth specimen [12]. The difference in ΔQL^* between a base metal and a weld joint with a different weld metal will be caused by different local stress multi-axiality at the HAZ [8, 14]. For both cases, however, ΔQL^*, that is different creep ductility, exists, a parallel QL^* line appears [12, 14].

Using the concept of the converted stress and ΔQL^*, a quantitative estimation of MPCS and prediction of creep fracture can be conducted as shown in **Figures 5** and **6** mentioned in the Section 4.

3. The prediction of creep fracture life and the derivation of the converted stress to estimate the mechanical performance of creep strength of a double edge notched weld joint specimen for P91 steel based on the QL^* concept

3.1 Material and specimen

The material used for this study is P91 steel and matching weld metal of US-9Nb, which is a similar material as P91 steel. The chemical composition and mechanical properties are shown in **Tables 1** and 2.

C	Si	Mn	P	S	Ni
0.08–0.12	0.2–0.5	0.3–0.6	<0.02	<0.01	<0.40
Cr	Mo	V	Nb	Al	N
8.0–9.5	0.85–1.05	0.18–0.25	0.06–0.10	<0.04	0.03–0.07

Table 1.
Chemical compositions of P91 steel (wt%).

Tensile stress (MPa)	0.2% proof stress (MPa)
590	410

Table 2.
Mechanical properties of P91 steel.

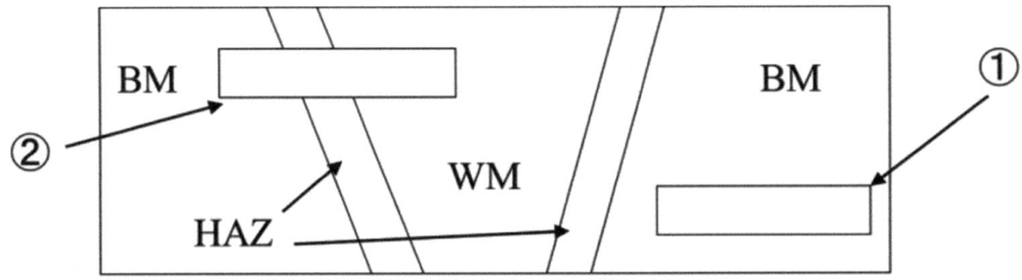

Figure 7.
Sampling site of specimen of base metal and weld joint. ① Base metal ② Weld joint.

The specimen used is a double-edge notched specimen (DEN) as shown in **Figure 2**. Sampling sites of specimen of base metal and weld joint are shown in **Figure 7**.

3.2 Experimental method

The machine system was designed and developed to enable automatic real-time observational experiments using a CCD microscope [15]. Now, CCD microscope was replaced to digital microscope manufactured by KEYENCE corporation. Using this apparatus, in situ observation of creep damage progression was conducted and the images of the damage region were quantified using a PC. The tests were conducted under high temperature vacuum conditions of 10^{-4} Pa. The creep damage region around the notch tip was found to be a dark region, composed of voids and micro-cracks originating along grain boundaries, as shown in previous results for SUS304 stainless steel [10, 15]. The dark region is defined as a creep damage region owing to the following reasons.

The specimens were heated using infrared rays (IR) under vacuum conditions, as shown in **Figure 8**. Creep damage is caused by micro-cracking along a grain boundary, which is composed of voids at the grain boundary [10, 15] that are considered to be caused by vacancy diffusion [16, 17]. In the damage region, a specimen surface becomes irregular due to micro-cracking along a grain boundary. In this region, diffused reflection of light by the lamp of IR was caused and it shows as the dark region.

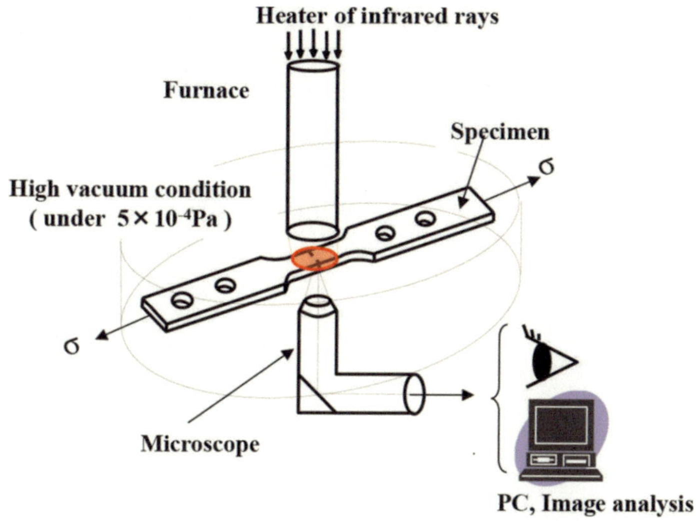

Figure 8.
Schematic illustration of in-situ observational creep-fatigue testing machine [15].

3.3 Experimental conditions and results

Previously, experimental results were published in Japanese [18], however, more detailed analyses are needed by more accurate analyses. In this section, updated results are written.

3.3.1 Similarity law of creep deformation

Experimental conditions, their results of steady state RNOD rate and creep fracture life are shown in **Table 3**. These results show that the fracture life of weld joint takes 3.5 ∼ 5% of that for the base metal.

	No.	Temp. (°C)	Stress (MPa)	Steady state RNOD rate (1/hr)	Creep fracture Life t_f, (hr)
Weld joint	W-1	650	135	5.65×10^{-2}	9.5
	W-2		113	9.00×10^{-3}	55.1
	W-3		113	2.00×10^{-2}	23.0 (predicted)
Base metal	B-1	650	200	1.13×10^{-1}	4.0 (predicted)
	B-2		135	1.62×10^{-3}	183.6
	B-3		113	1.44×10^{-4}	*1550.0

"Predicted" means that creep test was interrupted at the accelerated creep region and its final creep fracture was predicted from this point.
"" means that it almost covers total fracture life, however it is interrupt test.*

Table 3.
Experimental conditions, results of steady state RNOD rate and creep fracture life of DEN specimen (Base metal and weld joint for P91 steel).

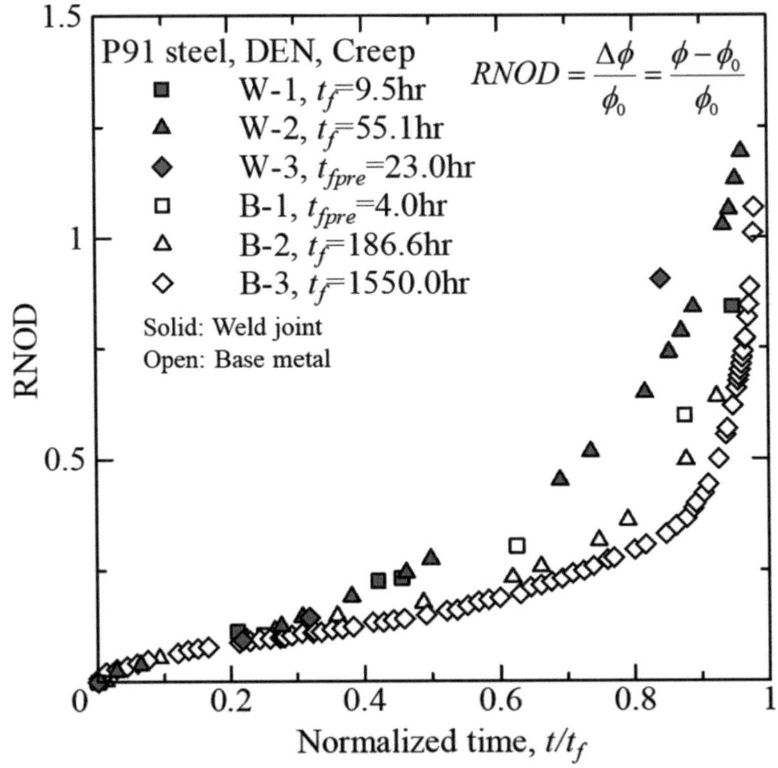

Figure 9.
Non-dimensional time sequential behavior of RNOD of the base metal and weld joint for P91 steel. t/t_f: non-dimensional creep fracture life of each specimen.

Non-dimensional time sequential characteristics of the RNOD curve (creep deformation) controlled by each fracture life are shown in **Figure 9**. For the base metal, the similarity law of RNOD curve caused, which is independent of applied stress. For the weld joint, the similarity law also caused for the case of $t/t_f < 0.5$.

3.3.2 Damage progression behavior of the notched specimen of the weld joint and the base metal

The time sequential behavior of damage progression around a notch tip of the weld joint specimen for P91 steel observed by the in situ observational testing machine under creep condition with a temperature of 650°C is shown in **Figure 10**. These results show that the creep damage preferentially originated at the HAZ of the base metal side, however this damage was not a dominant damage of final fracture. After that, another creep damage newly originated from both of the upper and the lower notch tips. When this damage area spread over the whole ligament from the upper notch tip to the lower notch tip, final fracture occurred, that is, creep fracture of a weld joint notched specimen is not creep crack growth dominant but it is creep damage dominant, which related to the similarity law of creep deformation as is mentioned in the Section 3.3.1.

The time sequential behavior of damage progression around a notch tip of the base metal observed by the in situ observational testing machine under creep condition with a temperature of 650°C for P91 steel is shown in **Figure 11**.

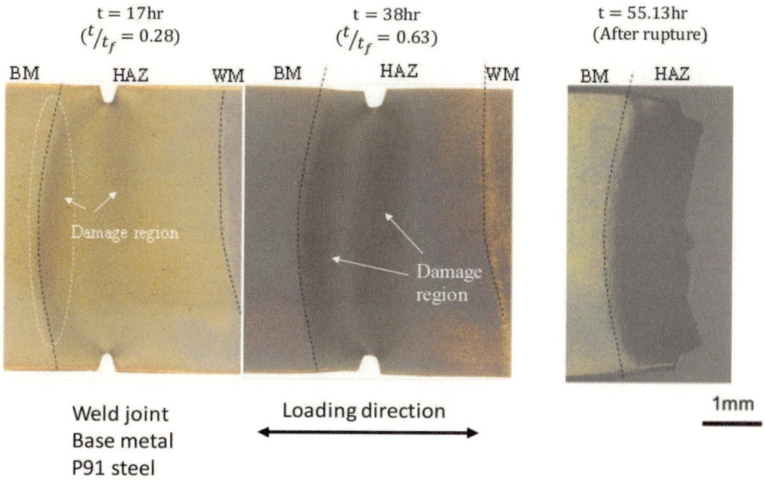

Figure 10.
The time sequential behavior of damage formation around a notch tip of the weld joint specimen under creep condition with a temperature of 650°C and 113 MPa for P91 steel.

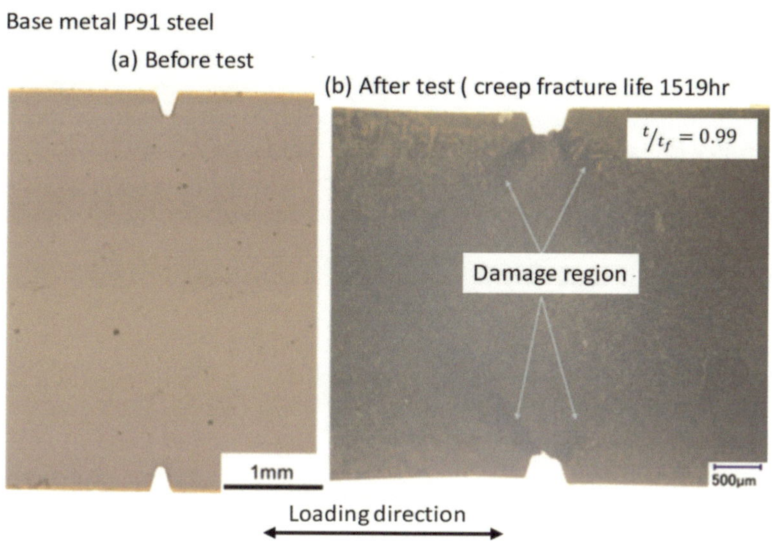

Figure 11.
The time sequential behavior of damage formation around a notch tip of the base metal under creep condition with a temperature of 650° C and 113 MPa for P91 steel.

For the base metal, creep damage originated from a notch tip in the direction of shearing stress. When the damage area spread over the specimen width, final fracture occurred. However, damage progression behavior is different from that of the weld joint, creep fracture mechanism is also creep damage dominant, which also related to the similarity law of creep deformation as is mentioned in the Section 3.3.1.

3.3.3 Experimental law of creep fracture life and creep strain rate

As shown in **Figure 12**, a unique linear logarithmic relationship between creep fracture life and steady state creep strain rate, that is, the unique QL^* relationship was

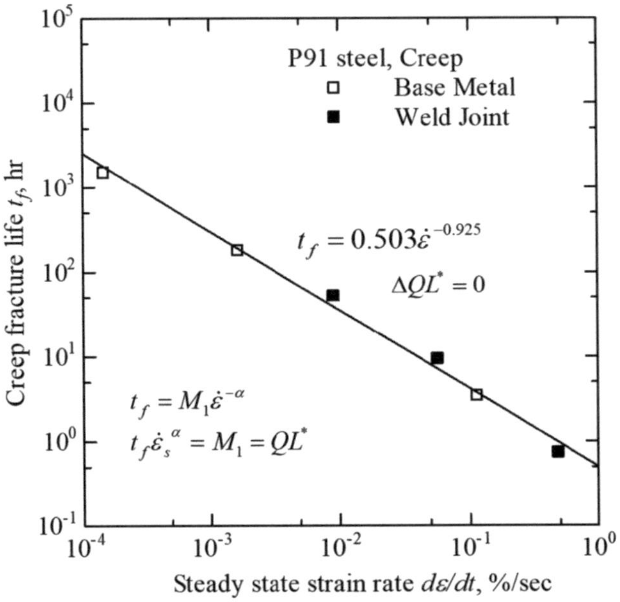

Figure 12.
The relationship between logarithmic value of t_f and $d\varepsilon/dt$.

found out both in base metal and weld joint, which means the weld joint is considered to have a similar property as that of base metal, that is, $\Delta QL^* = 0$.

As shown in **Figure 13**, the linear logarithmic relationship between the steady state creep strain rate and the applied stress was found out both for the base metal and the weld joint respectively.

From these experimental results, the following equations are obtained.

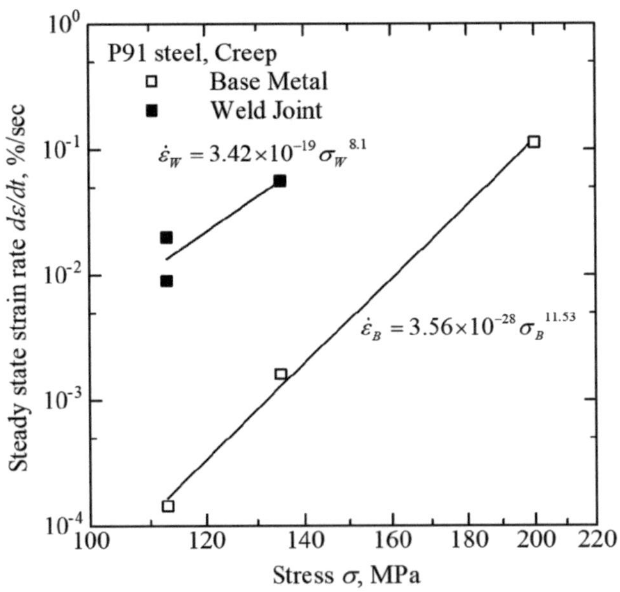

Figure 13.
The relationship between logarithmic value of $d\varepsilon/dt$ and applied stress, σ.

$$t_f = 0.503\dot{\varepsilon}^{-0.925} \tag{15}$$

$$\dot{\varepsilon}_B = 3.56 \times 10^{-28}{\sigma_B}^{11.53} \text{ (Base metal)}, \tag{16}$$

$$\dot{\varepsilon_W} = 3.42 \times 10^{-19}{\sigma_W}^{8.1} \text{ (Weld joint)}, \tag{17}$$

where t_f, $\dot{\varepsilon}_B$, σ_B, $\dot{\varepsilon_W}$ and σ_W is creep fracture life, steady state creep strain rate and applied stress for the base metal and weld metal respectively.

3.3.4 The derivation of the converted stress and the converted stress ratio

As shown in **Figure 12**, the QL^* relationship is written by Eq. (15) both of the base metal and the weld joint.

As shown in **Figure 13**, the linear logarithmic relationship between the steady state creep strain rate and the applied stress is given by Eq. (16) for the base metal.

Substituting Eq. (16) into Eq. (15), Eq. (18) is obtained for the base metal.

$$t_{fB} = 1.234 \times 10^{25}\sigma_B^{-10.67}, \tag{18}$$

where t_{fB} is the creep fracture life under applied stress, σ_B, for the base metal.

Using the experimental results of steady state creep strain rate, $\dot{\varepsilon}_W$ of the weld joint and substituting it into $\dot{\varepsilon}$ in Eq. (15), the fracture life, t_{fweld} of the weld joint specimen corresponding with $\dot{\varepsilon}_W$ is obtained.

Substituting t_{fweld} into t_{fB} in Eq. (18), the converted stress of the weld joint into that of the base metal, σ_{WCB}, with the same fracture life as t_{fweld} is obtained by Eq. (19).

$$\sigma_{WCB} = t_{fweld}^{-1/10.67} \times \left(1.234 \times 10^{25}\right)^{1/10.67} \tag{19}$$

The converted stress ratio, η, is defined by Eq. (20).

$$\eta = \frac{\sigma_{WCB}}{\sigma_W} \tag{20}$$

The accuracy of predictive creep fracture life of the weld joint derived from the QL^* line is about 71% $\sim$ 76% accuracy to the actual life as shown in **Table 4**.

The value of the converted stress ratio of the weld joint into that of the base metal, $\eta = \sigma_{WCB}/\sigma_W$ is 1.39 and 1.41 for the case of 135 MPa and 113 MPa of σ_W respectively as shown in **Table 4**. These results mean that the creep strength of the weld joint is lower than that of base metal of 39% and 41%, respectively. The scattering of the converted stress ratio is 0.7%. On the other hand, the value of the ratio of creep fracture life of the weld joint to that of the base metal is 0.052 and 0.035 respectively. The scattering of the creep fracture ratio is 20%.

From these results, the converted stress ratio is more accurate indicator of estimating the MPCS than the fracture life ratio from the point of the data scattering.

For creep ductile materials, the prediction of the creep fracture life of the target materials and the quantitative comparative estimation of the MPCS of the target materials to the control material are found to be possible measuring the steady state creep rate of the target materials by the short time range experiments.

	Applied stress (MPa)	Creep fracture life (hr) of the base metal, t_{fBM} and the weld joint, t_{fWJ}	Converted stress (MPa), σ_{WCB}	Converted stress ratio, $\eta\left(=\frac{\sigma_{WCB}}{\sigma_W}\right)$, σ_{WCB}: converted stress σ_W: actual applied stress for the weld joint η: converted stress ratio	Ratio of fracture life of weld joint to that of base metal, $\frac{t_{fWJ}}{t_{fBM}}$
Base metal	135	183.6	135	1.0	1.0
Weld joint		9.5 7.2 (predicted by QL^*)	187	1.39	0.052
Base metal	113	1519.3	113	1.0	1.0
Weld joint		55.13 39.3 (predicted by QL^*)	160	1.41	0.035

Table 4.
Converted stress and the ratio of converted stress to actual applied stress of the weld joint.

4. Discussions

4.1 Application of this theory to the case of C (T) specimens of the weld joint for P91 and P92 steels

The QL^* map of the base metal and weld joint for the ASME grade P92 steel and ASME grade P91 steel are shown in **Figure 14** [14]. Weld metal is 12Cr steel and 9Cr steel, respectively.

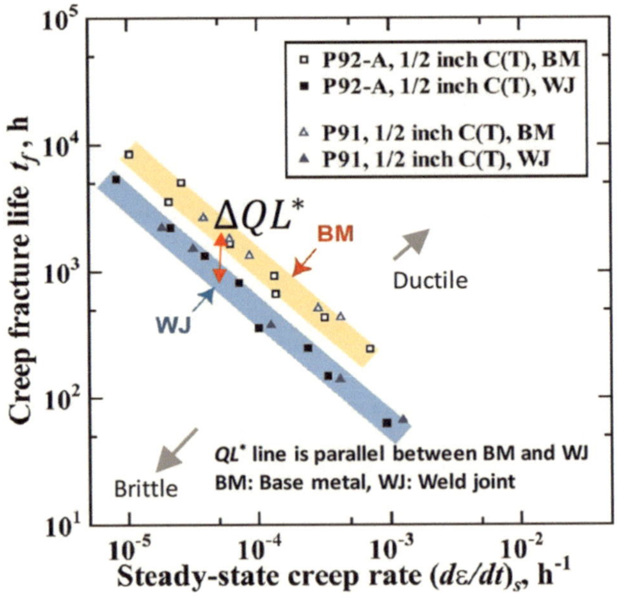

Figure 14.
QL^*map of the base metal and weld joint for the P92 and P91 steels [14].

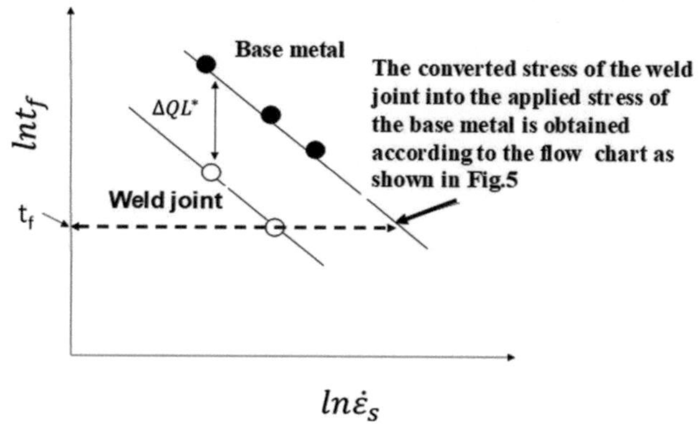

Figure 15.
The prediction and derivation methods of creep fracture life and the converted stress of the weld joint into that of the base metal.

These results showed that the QL^* line of the weld joint is parallel to and lower of ΔQL^* than that of the base metal. Therefore, one creep fracture test of the short range life is necessary to determine the value of ΔQL^*, after that, as shown in **Figure 15**, the converted stress of the weld joint in to the applied stress of base metal, σ_{WCB} and its ratio, η are feasible to obtain according to the flow chart given by **Figure 5**.

4.2 Application of this theory to the case of a smooth specimen and a notched specimen

The QL^* map of a double edge notched and a smooth specimen for Cr-Mo-V steel is shown in **Figure 16**.

These results showed that the QL^* line of the smooth specimen is parallel to and lower of ΔQL^* than that of the notched specimen [12]. Therefore, one creep fracture test of the short range life is necessary to determine the value of ΔQL^*, after that, as shown in **Figure 17**, the converted stress of the smooth specimen in to the applied stress of the notched specimen and its ratio, η are feasible to obtain according to the flow chart given by **Figure 5**.

5. Summary

The experimental law on the relationship between the steady state creep rate and creep fracture life given by Eq. (21) was well known as the Monkman – Grant law [19], however, Yokobori, et al. showed this law is only valid under limited case, and the data band is essentially different depending on the creep ductility as shown in **Figure 3** [11].

Furthermore, however the similar equation as Eq. (21) was found to be valid for creep ductile materials and it was correctly written by Eq. (8).

$$t_f \times \dot{\varepsilon}_s = Const. \tag{21}$$

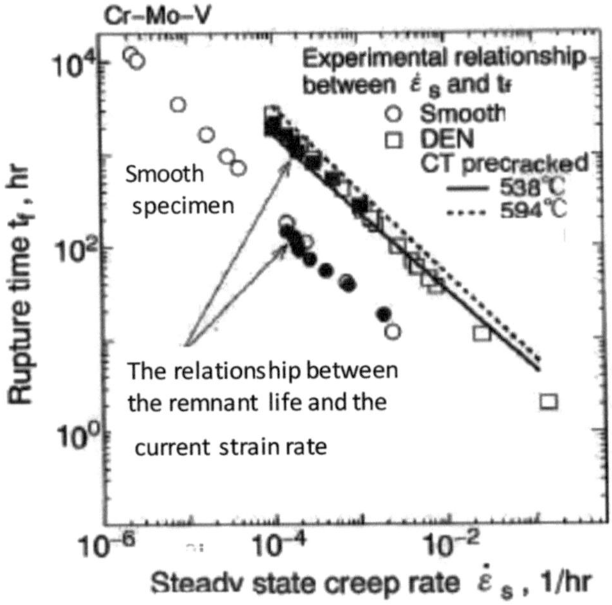

Figure 16.
*QL*map of a notched specimen and a smooth specimen for the Cr-Mo-V steel [12]. DEN: a double edge notched specimen of a plate specimen. CT: a C (T)specimen. Open mark: steady state creep rate. Solid mark: remnant life of creep fracture for the current creep rate. Experimental data of solid mark exist on the same QL* line and the QL* line of the notched specimen is parallel to that of smooth specimen.*

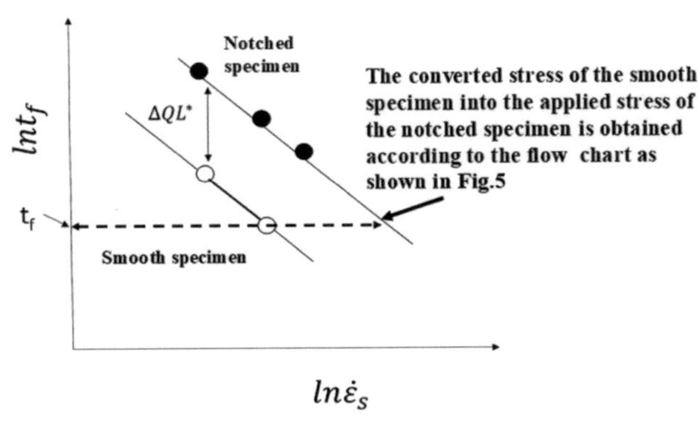

Figure 17.
The prediction method of creep fracture life and the converted stress of the applied stress of a smooth specimen into that of a notched specimen.

In this study, for creep ductile materials, QL^* line was found to be written by Eq. (8) and its parallel lines of a notched and a smooth specimen for the base metal and the weld joint.

On the basis of these results mentioned above, the prediction of the creep fracture life and the quantitative estimation of the MPCS were found to be possible using the QL^* map and the concept of converted stress, with the short experimental period, the small number of specimens, the reasonable accuracy and an economic efficiency. It is significant from engineering point.

The experimental data obtained by the commissioned research with Tohoku Electric Power Co.Inc. from April 2016 to March 2019 were used in the section of 3.3 in this proof [20].

Acknowledgements

Parts of this work was supported by Council for Science, Technology and Innovation (CSTI), Cross-ministerial Strategic Innovation Promotion Program (SIP), "Materials Integration for revolutionary design system of structural materials" (Funding agency: JST).

Conflict of interest

The author declares that there is no conflict of interest associated with this study.

Author details

A. Toshimitsu Yokobori Jr* and Go Ozeki
Advanced Comprehensive Research Organization, Teikyo University, Tokyo, Japan

*Address all correspondence to: toshi.yokobori@med.teikyo-u.ac.jp

References

[1] Tabuchi M, Takahashi Y. Evaluation of creep strength reduction factors for welded joints of modified 9Cr-1Mo steel. Journal of Pressure Vessel Technology. 2012;**134**:31401-311∼6

[2] Larson FR, Miller J. Time-dependent relationship for rupture and creep stresses. Trans of ASME. 1952;**74**:765-771

[3] Manson SS. Behaviour of materials under conditions of thermal stress. Heat Transfer Symposium, NACA-TN-2933. 1953

[4] Coffin LF Jr. A study of the effects of cyclic thermal stresses on a ductile metal. Transactions of the ASME. 1954;**76**: 931-950

[5] Prager M. Development of the MPC OMEGA method for life assessment in the creep range. Journal of Pressure Vessel Technology. 1995;**117**:95-103

[6] Yokobori AT Jr, Yokobori T. New concept to crack growth at high temperature creep and creep-fatigue. In: Salama K et al., editors. Advances in Fracture Research, Proc. of the ICF7. Vol. 2. Pergamon Press; 1989. pp. 1723-1735

[7] Yokobori AT Jr, Uesugi T, Yokobori T, Fuji A, Kitagawa M, Yamaya I, et al. Estimation of creep crack growth rate in IN-100 based on the Q* parameter concept. Journal of Materials Science. 1998;**33**:1555-1562

[8] Yokobori AT Jr, Sugiura R, Sato K, Shigeyama H, Tabuchi M. The effect of vacancy diffusion and concentration on creep crack initiation in weldment of a C (T) specimen for high Cr steel. Strength, Fracture and Complexity. 2018;**11**(2–3): 107-120

[9] Yokobori AT Jr, Yokobori T, Tabuchi M. The master curve and constitutive equation for creep deformation and fracture for Cr-Mo-V steel throughout smooth, notched and precracked specimens. Journal of Materials Science. 1996;**31**:4767-4773

[10] Yokobori AT Jr, Yokobori T, Kuriyama T, Kako T. Relative notch opening displacement concept for crack initiation in temperature time dependent fracture. In: Proc. of the 6th International Conference on Fracture. Vol. 3. Pergamon Press; 1984. pp. 2181-2190

[11] Yokobori AT Jr, Yokobori T, Yamazaki K. A characterization of high temperature creep fracture life for ceramics. Journal of Materials Science Letters. 1996;**15**:2002-2007

[12] Yokobori AT Jr, Prager M. Proposal of a new concept on creep fracture remnant life for a pre-cracked specimen. Materials at High Temperatures. 1999; **16**(3):137-141

[13] Yokobori AT Jr, Sugiura R, Tabuchi M, Fuji A, Adachi T, Yokobori T. The effect of multi-axial stress component on creep crack growth rate concerning structural brittleness. In: Proc. of 11th Int. Conf. on Fract. CDrom; 2005

[14] Shigeyama H. Tohoku University. 2014

[15] Yokobori AT Jr, Kuriyama T, Kaji Y. Damage progression behaviour under high temperature creep and fatigue conditions, advances in fracture research. In: Proc. of ICF 10, in the content of Special Lecture of CDrom, Honolulu 2001. Elsevier Science; 2001 CDrom

[16] Yokobori AT Jr, Abe K, Tsukidate H, Ohmi T, Sugiura R, Ishikawa H. Micro

mechanics based on vacancy diffusion coupled with damage mechanics related to creep deformation and prediction of creep fracture life. Materials at High Temperatures. 2011;**28**(2):126-136

[17] Yokobori AT, Jr., Sugiura R, Ohmi T, Ainsworth R. A review of time-dependent fracture life law (or model) based on a proposed multi-scale analysis. Strength, Fracture and Complexity, An International Journal. 2014;**8**:205-218

[18] Yokobori AT Jr, Ozeki G. The proposal of quantitative estimation parameter of mechanical performance on the creep strength of weld joint notched specimen for 9Cr steel (P91), Journal of the Strength and Fracture of Materials. 2021;**55**:1-7 (in Japanese)

[19] Monkman FC, Grant NJ. An empirical relationship between rupture life and minimum creep rate in creep-rupture tests. In: Proc. of the ASTM. Vol. 56. 1956. pp. 593-620

[20] Ozeki G, Yokobori AT Jr, Endo T, Sato F, Ohmi T, Sato K. Creep damage formation behavior and crack growth life for notched specimen of P91 steel including weld joint, 2018 International Conference on Materials Strength and Applied Mechanics (MSAM2018), paper ID1193

Chapter 2

Perspective Chapter: Fatigue of Materials

Alireza Khalifeh

Abstract

This chapter deals with the fatigue fracture of the materials under cyclic loadings. Components of structures and machines may be subjected to cyclic loads and the resulting cyclic stress that can lead to microscopic physical damage and fracture of the materials involved. It has been seen at a stress well below the ultimate strength, this microscopic damage can accumulate under action of cyclic loadings until it develops into a crack that leads to final separation of the component. In addition, the material inherently has cracks and other microscopic defects that grow due to cyclic loads and lead to fracture of machine or structure parts. The failures are more often sudden, unpredictable and catastrophic which may occur after a short period of design life. The main objective in writing this chapter is to present scientific findings and relevant engineering practice involving materials fatigue failures.

Keywords: failures analysis, fatigue fracture, cyclic loadings, machinery equipment, structures

1. Introduction

It has been found that a metal subjected to cyclic stress will fail at a stress level much lower than that of a single application load. Fractures occurring under cyclic loadings are known as fatigue fractures [1, 2]. Indeed, one of the main reasons for unpredictable and premature material failures in service is the application of cyclic loads and the occurrence of fatigue [3–8].

Two event that caused a lot of human and financial losses due to fatigue were observed during the 1994 Northridge and 1995 Kobe earthquakes. Investigations have shown that cyclic loading of earthquakes alongside presents of high strain rates, notch and poor material properties were responsible for these premature failures in steel structures [9–11]. It should be noted that earthquake loads in the form of low cycle fatigue (LCF) and extremely low cycle fatigue (ELCF) caused the failure of steel structures [12–15].

Machinery equipment's such as compressors, turbines and pumps are more prone to this type of damage. Numerous destructions in these devices have been reported due to incorrect design or manufacturing defects and have caused loss of production and financial resources [16–19]. Failure of a Ti6Al4V alloy compressor impeller used in a petrochemical plant is shown in **Figure 1**. Investigations revealed fatigue has been responsible in the failure of compressor impeller. Stress concentration in the blade root causes the formation of fatigue cracks and final failure of the part [16].

Figure 1.
Fracture of a gas compressor turbine blade [16].

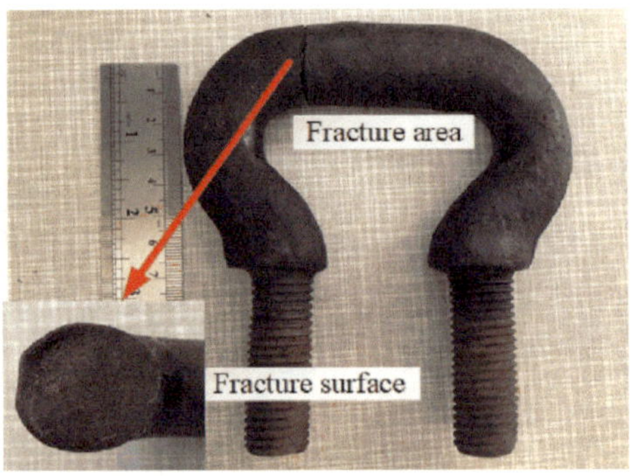

Figure 2.
Fatigue failure of U-bolt of an elevator.

Another practical example of fatigue failure is shown in **Figure 2**. This failure occurred in an AISI4140 steel material as a result of not considering the metallurgical parameters in the construction of U-bolts for a lift. Experience showed that surface modification technique is a suitable strategy for extending the life of U-bolts under cyclic loadings. The technique consisted of heating, quenching, tempering and transforming the initial ferritic/pearlitic microstructure to tempered martensite with a higher surface hardness. The idea was taken from the fact that surface hardening process produced a reduction in grain size, retained austenite level, compressive residual stress, and as a result significantly improves the fatigue limit of the low alloy steels [20, 21]. Thermo-chemical surface treatment such as carburizing and nitriding can also improve the fatigue properties of these steels [20, 22–24].

This type of failure is insidious because it led the equipment's to failure and plants to shut downs without any warning. Three main factors are necessary for this type of failure [2]:

1. Adequate tensile stress

2. Fluctuation of applied stress

3. Adequate cyclic loads

Other factors such as stress concentration, overload, temperature, metallurgical structure, surface finishing, and residual stresses accelerate the occurrence of these type of fractures [2, 16, 17, 25–29].

The purpose of this chapter is to present the fatigue failure of materials and the methods of minimizing such damages for safety, durability and reliability of the products. To achieve this, the mechanical aspects of fatigue are explained first. Then, fatigue damage mechanism and fatigue futures are discussed. In the next step, the author is placed a focus on the types of fatigue failure and their characteristics by stating several practical examples. Finally, it has been dealing with factors that affecting material fatigue properties. It should be noted that the material in this chapter is based on our interaction with fatigue damages of components in the industry as well as based on that were taught university.

2. Mechanical aspects of fatigue fracture

Analysis of stress and strains for cyclic loading is needed for dealing with engineering situation such as vibratory loading which lead the component to fatigue fracture. In some practical applications the material operates at a maximum and minimum stress levels that are constant. This is known as constant amplitude stressing and is shown in **Figure 3**.

The stress range is the difference between the maximum and minimum stress values, $\Delta\sigma = \sigma_{max} - \sigma_{min}$, the mean stress, σ_m is the average of maximum and

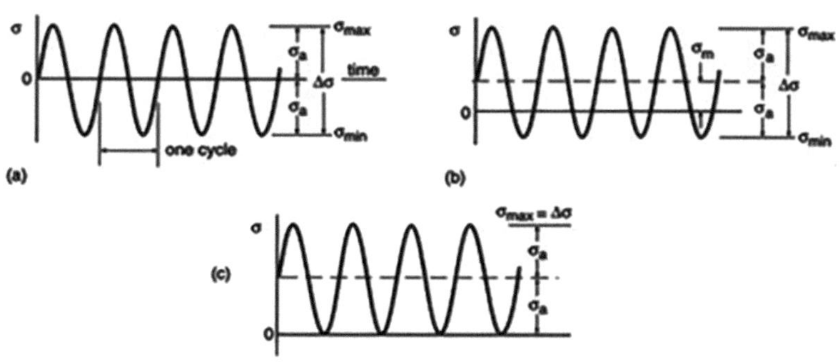

Figure 3.
Various type of cyclic loadings. a) Completely reversed stressing, (b) nonzero mean stress, (c) zero to tension stressing.

minimum of stress values that may be zero (**Figure 3a**). Half range is named stress amplitude. Mathematical expressions are as follows [1]:

$$\sigma_a = \frac{\sigma_{max} - \sigma_{min}}{2} \tag{1}$$

$$\sigma_m = \frac{\sigma_{max} + \sigma_{min}}{2} \tag{2}$$

The stress ratio R and amplitude ratio A is defined as:

$$\sigma_a = \frac{\sigma_{min}}{\sigma_{max}} \tag{3}$$

$$A = \frac{\sigma_a}{\sigma_m} \tag{4}$$

2.1 S-N curves

Stresses (S) versus life (N) is an engineering representative of fatigue behavior of materials. When we do cyclic test on a sample at a stress level, S, the sample will be failed after N cycle. If the test repeated at higher stress level the cyclic to failure will be smaller. A typical plot of S-N curve in a rotating bending test of an aluminum alloy in logarithm scale is shown in **Figure 4**.

S-N data in Log–Log scale is usually in form of a straight line. An equation can be fitted on these data's is [1]:

$$\sigma_a = AN_f^B \tag{5}$$

It should be noted that Eq. 5 describes the linear part of the S-N curve and is known Baskin equation. The nonlinear part is called fatigue limit or endurance limit, Se which is seen in S-N curve of some materials like plain carbon and low alloy steels.

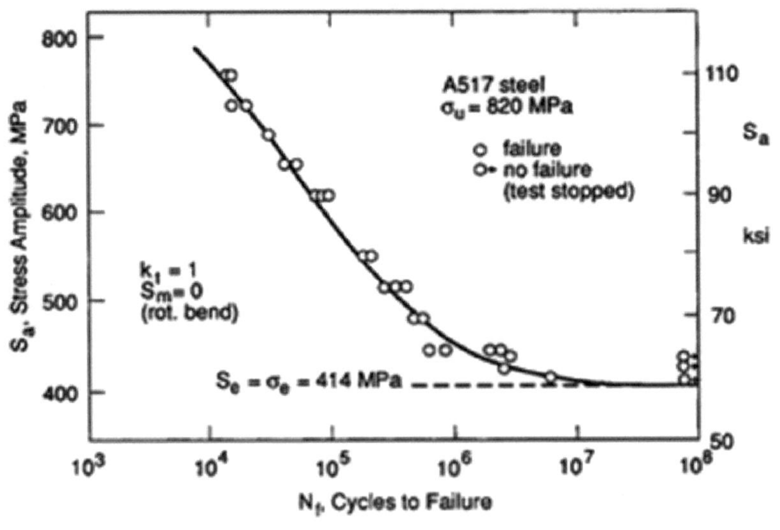

Figure 4.
S-N curve for specimens of steel. Fatigue limit can be seen at about Se = 414 MPa [30].

This is a stress level that fatigue failure does not occur under ordinary conditions or the cycle number to failure is unlimited. It should be noted in practical applications irregular loads versus time histories are more commonly encountered [1]. Examples for these conditions are given in **Figure 5**.

For such situations that the amplitude of loading is variable, the Palmgren-Miner rule predicts fatigue life of the component [2]:

$$\frac{n1}{N1} + \frac{n2}{N2} + \frac{n3}{N3} + \ldots + \frac{nk}{Nk} = 1 \tag{6}$$

Where n1, n2, n3 ... and nk are the number of work cycles at each of the different stress levels and N2, N2, N3 ... and Nk are the life of the part at each similar stress levels.

According to this equation, the total life of the part is estimated from the sum of the percentage of lives spent by each stress cycle.

2.2 Fracture mechanic

The presence of cracks significantly reduces the strength and longevity of a component due to increasing the probability of occurring brittle fracture [32–34]. Cracks may be produced during the manufacturing process or other inherent flaws that convert to crack and grow until its rich critical sizes for brittle fracture. Paris equation describes the crack growth behavior of a material under cyclic loadings [1]:

$$\frac{da}{dN} = C(\Delta K)^m \tag{7}$$

$$\Delta K = F\Delta\sigma\sqrt{\pi\alpha} \tag{8}$$

$$\Delta\sigma = \sigma_{max} - \sigma_{min} \tag{9}$$

Where $\frac{da}{dN}$ cyclic crack growth rate, ΔK stress intensity range, C and m are constants.

In fact this equation is log–log plot of $\frac{da}{dN}$ versus ΔK as shown in **Figure 6**. In this diagram there is a vertical part denoted ΔK_{th}, which named the fatigue crack growth threshold. This quantity is a lower limiting value that below of that the crack growth does not occur. At high growth rate the curve again become steep due to rapid unstable crack growth [1].

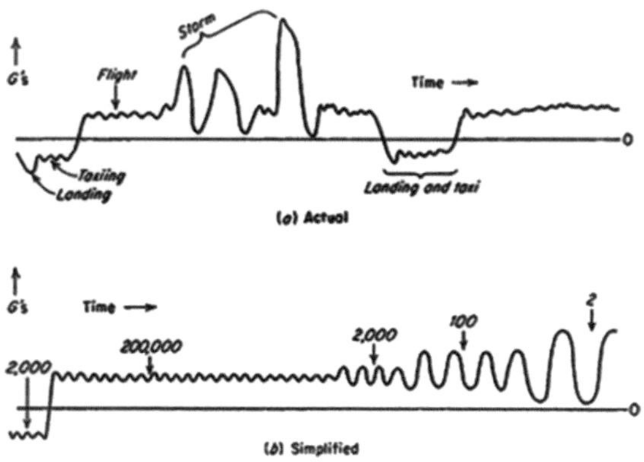

Figure 5.
Loads for one flight of a fixed-wing aircraft (a), and a simplified form of this loading [31].

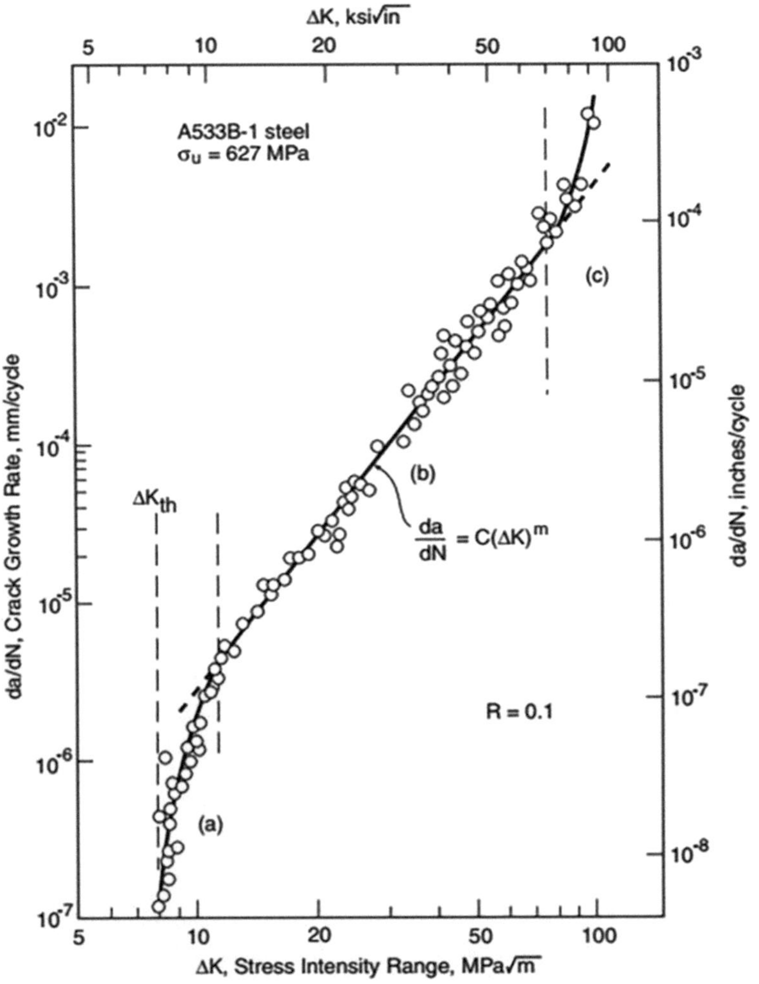

Figure 6.
Fatigue crack growth rate for a ductile pressure vessel steel. (a) Threshold intensity factor, ΔK_{th}, (b) intermediate region which shows with power equation, (c) rapid crack growth [35].

3. Fatigue damage mechanism

Fatigue is a damage processes of components caused by cyclic loads. The process involves four stages [1, 36]:

1. Crack nucleation 2. Short crack growth 3. Long crack growth 4. Final separation

The first step is crack nucleation. It has been observed that crack of fatigue damage starts at near high stress concentration sites such as slip bands, inclusions, porosities or other manufacturing discontinuity. Localized shear plans that usually occurs at surface or within grain boundaries is another location for nucleation of fatigue cracks. This stage of fatigue cracking may be relieved with proper annealing treatment.

The next step in fatigue damage process is the crack growth. This stage is the deepening of the initial crack on the planes with maximum shear stress and it is often called crack growth stage I. This stage is greatly affected by microstructure characteristic such as grain size, slip mode and stress level because the crack size is in the order of the microstructure.

Step 3 is the crack growth stage II. At this stage, the crack created in the previous stage grows in the direction perpendicular to the planes with high tensile stress. This stage is less affected by the microstructure because the plastic zone in crack tip is much larger than the material grain size.

Stage 4 is final separation. This stage is when the crack length reaches a critical value and the remaining cross-section cannot withstand the applied load.

Fatigue studies show that fatigue cracks usually start from a free surface. If these cracks start from the inside, the nucleation site of the crack is usually a carburized or similar surface layer [2].

4. Fatigue features

In macroscopic scale fatigue failure is seen with a brittle appearance and without any gross deformation in the failure. The fatigue failure surface usually consists of a smooth area due to crack growth and surface wear on each other, and a rough area formed when the load is unbearable [36]. Another characteristic of fatigue failure is the beach marks, fine and arch-shaped lines, that starts from the place of crack initiation and progresses to the area of ductile fracture as it is seen in **Figure 7**. They are also known as macroscopic striations.

It should be noted that striations are not always formed on the fracture surfaces. Inert environments, high strength materials for examples steels with hardness above 30 HRC, aqueous environments or high temperature air, creep fatigue condition, Low ΔK value for example stainless steel at $\Delta K < 30MPa\sqrt{m}$ Mode II and III fracture [36]. **Figure 8** shows a fatigue fracture of an AISI 316 L stainless steel in absence of striations.

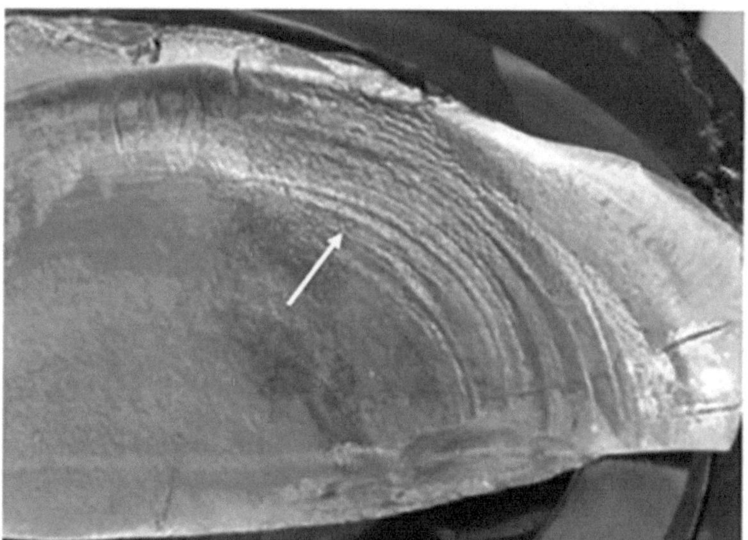

Figure 7.
Beach marks on fractured surface. The arrow indicates the crack direction propagation [37].

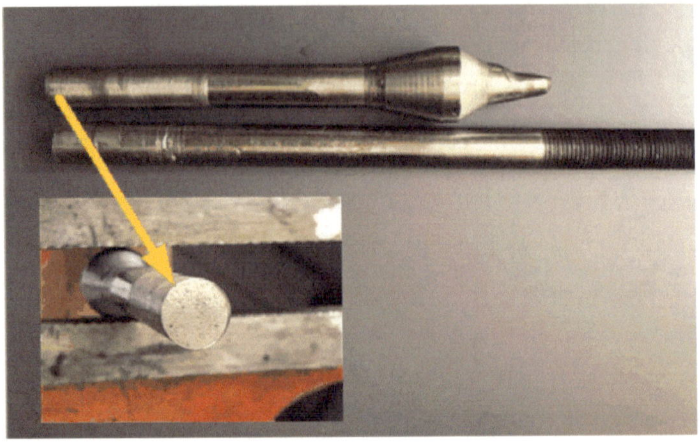

Figure 8.
Surface fatigue failure of a duplex stainless steel due to high vibration in services.

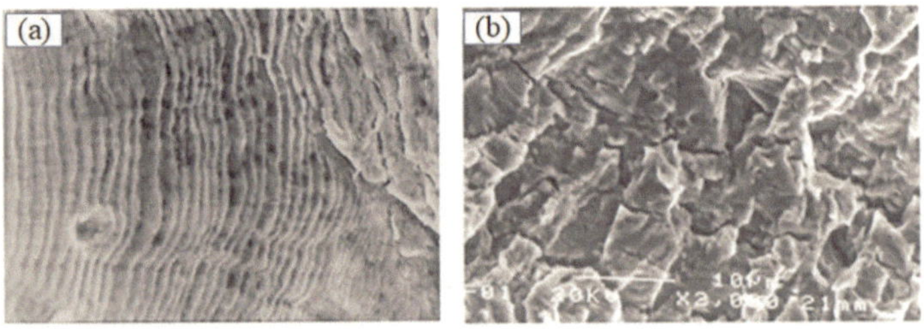

Figure 9.
Microscopic striation in a fatigue fracture surface of a ductile material (a), brittle type of striations (b) [37].

The microscopic evaluation of fracture surface is generally carried out with a scanning electron microscopy (SEM). The most important characteristic of fatigue fracture that is manifested in failure fracture surface is the presence of striations. Seriations are small groves extended perpendicular to the crack growth direction as seen in **Figure 9**. In general, striations indicate the growth rate of cracks in microscopic scales where each striation correspond to one load cycle. It is important to note that striation in brittle materials are different from ductile materials [36]. In ductile state the striation profile is wavy and smooth while brittle striations are irregular or saw tooth-like as seen in **Figure 9a, b**.

5. Types of fatigue

Fatigue failure of parts and components can be categorized to high cyclic, low-cycle fatigue, extremely low cycle fatigue, corrosion fatigue, and thermal fatigue. Here the features and parameters that control each process are discussed:

5.1 High cycle fatigue

High cycle fatigue is characterized by high number of cycles to failure and little plastic deformation. This type of failure occurs with a brittle appearance. **Figure 1** shows the occurrence of a typical high cycle fatigue failure in the stem of gas compressor turbine blade, due to high vibration and cyclic stress. In this case failure usually occurs at a stress concentration point such as a sharp corner or groove or a metallurgical stress concentration point such as an impurity [2].

The controlling parameter in this state is stress and this type of fatigue is called control stress fatigue. Failure evaluation of structures with this mechanism is done by testing the samples at different stress levels (N) and the number of cycles leading to failure (N) is obtained in this way.

In this case, the fatigue life of this type of fatigue can be approximated by the Baskin eq. [2]:

$$\sigma_a = AN_f^B \tag{10}$$

Where σ_a is the stress amplitude, N is the number of cycles to failure, A and B are the material constants.

5.2 Low cycle fatigue

Low-cycle fatigue is for situations where failure occurs in less than 10^2–10^4 cycles [13, 38]. This type of fatigue occurs at relatively high stresses and a small number of cycles. Steam reactors and power generators are more prone to this type of failure [39–41]. In these cases, cyclic stresses usually have a thermal origin and the material fails from fatigue due to thermal contraction and expansion. Special laboratory methods have been developed to study the cyclic behavior of materials [1]. Standard ASTM E 606 provides details of the study of the cyclic behavior of materials. These tests are usually performed in a constant strain range [42].

It is widely accepted that in this situations, generalized deformations, such as strain, displacement and rotation) are more representative than stress, force and moment. **Figure 10** shows a stress–strain loop of a strain control cyclic test under a constant strain cycle in a fatigue test. The dimensions of this loop are described by its width, which indicates the amplitude of the strain $\Delta\varepsilon$, and its height, which is the amplitude of the stress $\Delta\sigma$. The strain amplitude consists of two components, elastic and plastic strain. The common method showing low fatigue cycle data is plotting of the plastic strain amplitude, $\Delta\varepsilon_p$, in terms of cycle number N. This representation in log–log scale is in a form of straight line as it is seen in **Figure 11**. The relation that can be fitted to the data in this diagram is as follows [1]:

$$\frac{\Delta\varepsilon_p}{2} = \varepsilon_f'(2N_f)^c \tag{11}$$

where $\frac{\Delta\varepsilon_p}{2}$ is the plastic strain amplitude, ε_f' is the fatigue ductility coefficient, c is the fatigue ductility exponent and 2Nf is the number of reversals to failure. The relationship is often called Coffin-Manson [1].

Another classic example of LCF is the fracture of steel structures under earthquake loadings [12, 13, 43]. In these cases, as well structural deformation substitute in fatigue strength curve to establish the fatigue deformability curve of the structural connections. **Figure 12** presents structural deformation of a beam to column steel structure during a seismic loading. Here φ is structural deformation parameter and somewhat represent the rotation intensity of the rigid connection. Drawing the

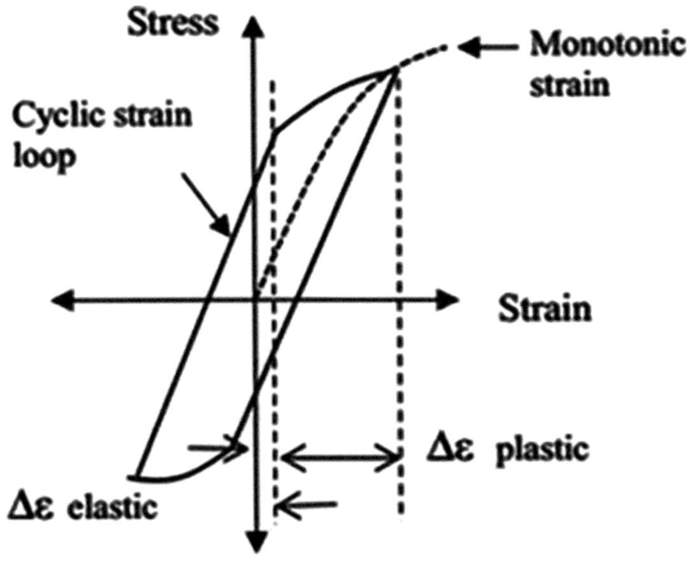

Figure 10.
Stable stress–strain hysteresis loop.

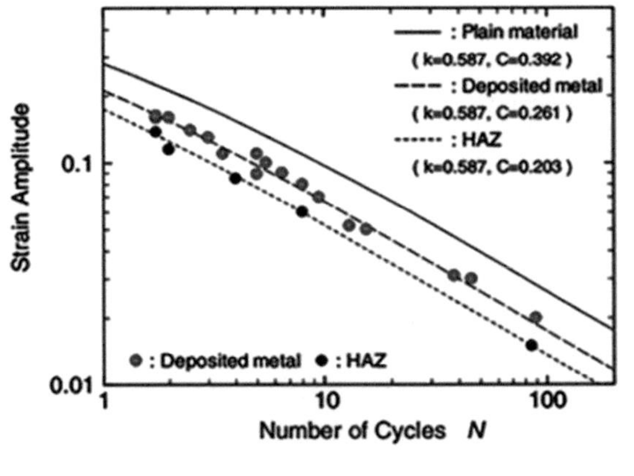

Figure 11.
Low cycle fatigue strength of a plain carbon steel and its weldment [14].

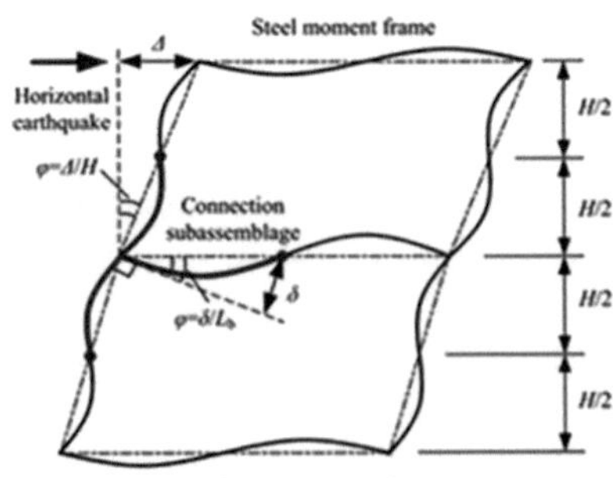

Figure 12.
Schematic presentation of deformation of steel frame under horizontal earthquake loadings [12].

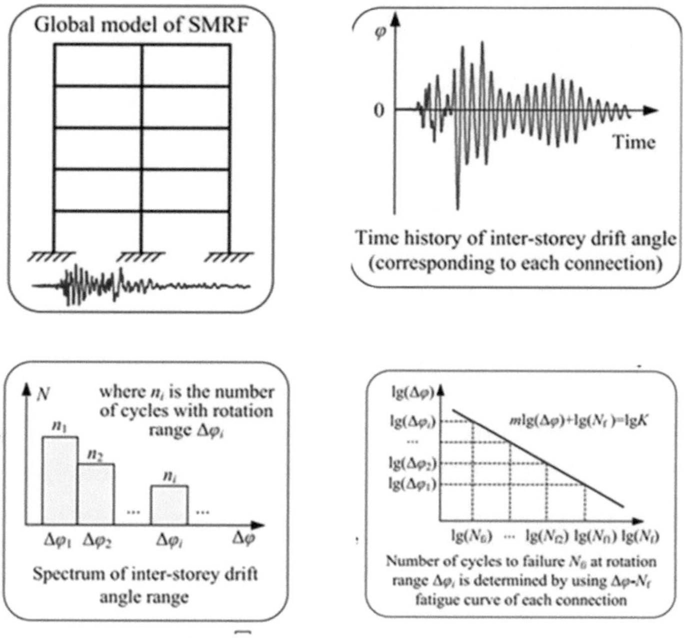

Figure 13.
Extremely low-cycle fatigue behavior of steel structures under earthquake loadings [12].

logarithmic curve of changes $\Delta\varphi$ in the number of cycles to failure N_f is a straight line (**Figure 13**) that can be expressed by the following eq. (12):

$$\Delta\varphi^m N_f = K \tag{12}$$

Where m is the slope of the fatigue curve and K is constant.

5.3 Extremely low-cycle fatigue

Extremely low-cycle fatigue is a fatigue failure characterized by large plastic strains (several times of yield strains) and a number of cycles to failure less than 10^2 [14, 44]. This type of fatigue failure is observed under extreme seismic conditions, structural members, particularly those acting as dissipative elements [44]. A typical example of this failure mode is the failure of structures during the 1994 Northridge earthquake in USA and 1995 Kobe earthquakes in Japan. Extensive failures during these two events led to many casualties and financial losses [9, 45]. Since this type of failure in a large volume causes the destruction of industrial buildings and structures, we study specifically and discuss the governing relationships.

ELCF is quite different from conventional high cycle fatigue where stresses are below the yield strength or low cycle fatigue where strains are in the order of the yield strains. In this type of failure, the level of deformation is much greater than the yield stress and the so-called control strain fatigue conditions prevail. It has been shown that when the number of cycles to failure N_f falls below approximately 200, estimation of fatigue life using the Coffin- Manson model will be associated with inaccuracy due to changes in the damage mechanism. As the strain amplitudes increases from LCF regime to the ELCF regime, the failure mode varies from fatigue fracture to accumulation of ductile damage, due to changes in damage mechanism. A series of modifications were made to the Coffin- Manson model by Tateishi et al. which also accounted the Ductile Damage. According to this model the total damage of material is the sum of the ductile damage fraction and fatigue damage fraction. Eq. 13 describes this mechanism [46]:

$$\frac{\Delta\varepsilon_p}{2} = \varepsilon_f'(2N_f)^c\, C_m \tag{13}$$

$$C_m = \begin{cases} \left(\dfrac{\varepsilon_f - \Delta\varepsilon_{max}}{\varepsilon_f - \varepsilon_u}\right) if\, \Delta\varepsilon_{max} > \varepsilon_u \\ 1.0\; if\, \Delta\varepsilon_{max} \leq \varepsilon_u \end{cases} \tag{14}$$

Where ε_f' and c are the Coffin- Manson constants. $\Delta\varepsilon_{max}$ maximum plastic strain range, ε_u ultimate strain in monotonic tensile test and C_m a factor linked to the ductile damage fraction.

5.4 Corrosion fatigue

High reactivity of fracture surfaces along aggressive micro-environment in the crack cavity lead to strong interaction of the corrosion and cyclic plastic deformation and rupture of the material which is called corrosion fatigue [47–49]. When fatigue corrosion occurs, corrosion strongly accelerates the rate at which fatigue cracks spread. In corrosion fatigue fracture surfaces may contain brittle striations on large facets or surfaces similar to what we see in quasi-cleavage fracture. A typical corrosion fatigue fracture at the macroscopic and microscopic scales are shown in **Figures 14** and **15**, respectively [37].

The fracture mechanism during a corrosion fatigue can be summarize as follow:

- Initiation of fatigue cracks due to mechanical stresses
- Penetration of the corrosive solution into the crack tip
- Reaction of solution with the material at the crack tip

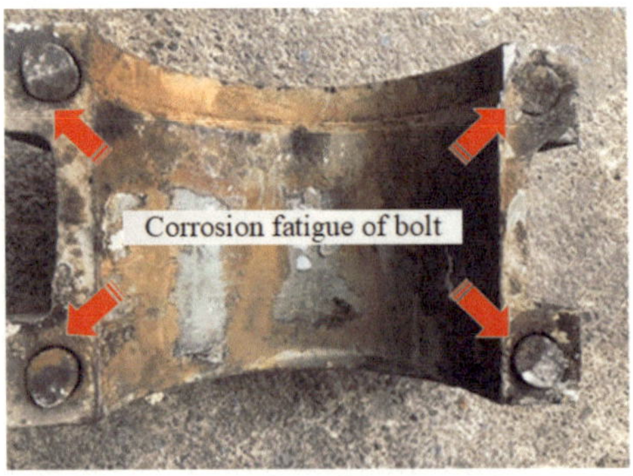

Figure 14.
Macroscopic view of a typical corrosion-fatigue fracture surfaces.

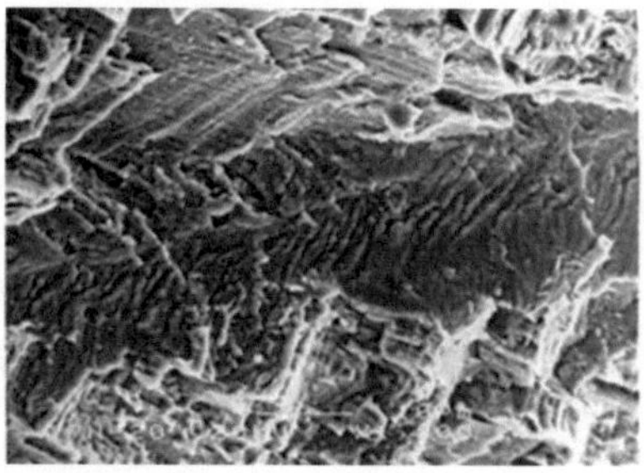

Figure 15.
Microscopic view of a typical corrosion-fatigue fracture surfaces [37].

- Passive layer rupture during cyclic strain at the crack tip
- Production of corrosion products that affect the effective stress factor
- This type of fatigue failure can occur in a high cycle or low cycle fatigue mode.

5.5 Thermal fatigue

Components may fail due to thermal stresses generated during cooling and heating at high temperatures. This is called thermal fatigue [50]. This type of failure can occur in a situation where no mechanical stress presents. In other words, the stresses that lead to the fracture of the part here have only a heat source. Thermal stresses occur when a constraint prevents dimensional changes due to variation in temperature. For a bar fixed on both sides, the heat stress due to ΔT is calculated from the following equation [50, 51]:

$$\sigma = \alpha E \Delta T \tag{15}$$

Where α is thermal expansion coefficient and E is elastic modulus.

Thermal fatigue can be categorized in the low cycle fatigue state due to the low number of cycles; it causes the destruction of the part. Austenitic stainless steels are susceptible to this type of failure due to their low thermal conductivity and high thermal expansion [2].

6. Factors affecting material fatigue properties

The fatigue behavior of the material is very sensitive to design and structure. Three very important factors that affected fatigue properties are the stress concentration, the residual stresses and material selection.

6.1 Effects of microstructure and material properties

The microstructure significantly affects the fatigue properties [52]. It was found that any changes in the microstructure altering the fatigue behavior especially in the case of high cycle fatigues. Decreasing in grain sizes and increasing in density of dislocation also noticeably improved the fatigue lives. In brass alloys an increase in fatigue lives observed by cold working and increasing in the dislocation density [1, 2]. The analyses carried out after Northridge earthquake on material consumables showed that the fracture toughness levels of some of electrode materials were very poor and this has been a strong reason for the decrease in fatigue properties of metal structures during these events [53].

In metals, reducing the size of inclusions and impurities significantly increases the fatigue properties. It has been well accepted that second-phase particles in the microstructures play a major role in the fracture of steels and failure resistance can be improved through changes in the volume fraction and morphology of these particles [54–57]. These particles are the centers of stress concentration and cause a decrease in the fatigue properties of the material [58]. Heat treatment is an effective factor in affecting the microstructure and improving fatigue properties.

6.2 Effects of surface

The source of all fatigue failures is the surface of the components. There is much evidence that fatigue properties are highly sensitive to surface conditions. Surface factors that affect the fatigue behavior consists of surface roughness, changes in the surface properties, and residual stress.

Surface roughness: Tests performed on metal samples have shown that the smoother the surface of the parts, the longer their fatigue life in the test [59]. This is due to the fact that local superficial scratches are the stress concentration points and the onset of fatigue cracking.

Surface properties: Because fatigue failure is highly dependent on surface conditions, any factor that affects the surface strength also affects its fatigue properties [60, 61]. For example, surface heat treatment of carbonation and nitration, which increase the surface hardness, improve the fatigue properties. On the other hand, carbonation operation, which reduces the surface hardness of the part, reduces the fatigue properties.

Surface residual stress: Residual stress is a type of stress that remain in a part after manufacturing processes even without supplementary thermal gradient and external loads. In welded parts due to local heating during welding, complex thermal stress produces during welding which led to residual stress and distortion in component [62–64]. Residual stresses are also created by deformation of formwork and fabrication. The residual stresses are combined with the applied stresses and in the tensile state reduce the fatigue life of the part during dynamic loadings [2]. It is important to note that these type of stresses in the compressed state can increase the fatigue life of components and structures. In fact, there are commercial methods such as ball bearing and surface rolling that produce compressive residual stress and are used to improve fatigue properties [65–69].

6.3 Notch effects

The manufacturing defects is a factor that produced stress concentration point and reducing the fatigue properties of a material [25, 70, 71]. Investigations on fracture of steel structures in Kobe and Northridge earthquakes have clarified the fatigue brittle fractures triggered by the crack-like defects in the weld metal [53]. In addition, device components usually have stress concentration areas such as fillets, grooves, keyways, and holes which called stress raisers. These areas generically termed notches for brevity and usually reduces the resistance of the equipment to fatigue failures.

DOI: http://dx.doi.org/10.5772/intechopen.107400

Figure 16 provides an example of a notch in a machinery equipment, in particular, the attachment of blade to shroud in a CO2 compressor. Despite carful design to minimize the severity of the notch, a fatigue crack led the equipment to premature failures. Another example is given in **Figure 17**. This is a fracture in beam to column steel structure under seismic loadings during Northridge earthquake where the source of fracture is a notch in welded part (lack of fusion) [72]. Stress raisers may also be due to metallurgical defects such as porosity, impurities, and defects due to crushing and surface decarbonization due to working at high temperatures [71, 73].

Stress intensity factor, K_t is a parameter that characterizes the degree of severity of a notch or stress concentration point [1]:

$$K_t = \frac{\sigma}{S} \tag{16}$$

Where σ is local notch stress and S is the nominal stress.

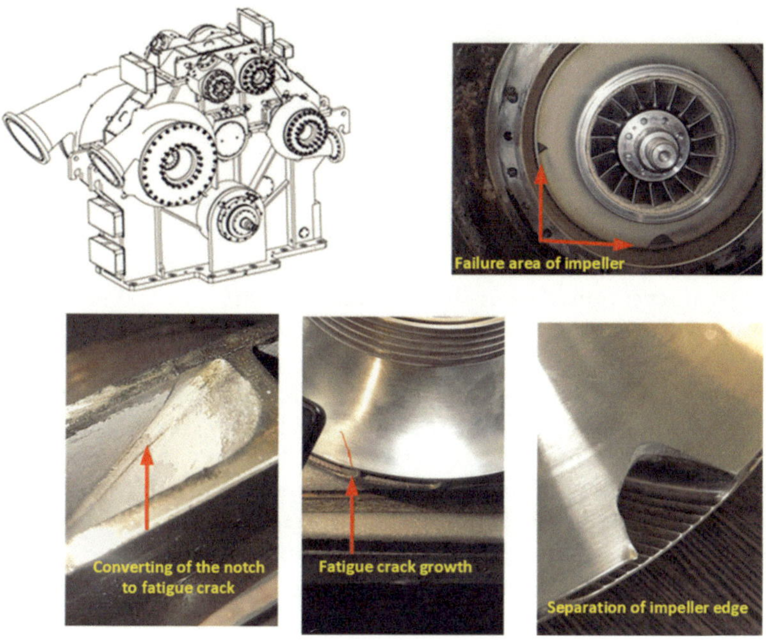

Figure 16.
Fatigue failure of impeller of a compressor due to presents of a notch.

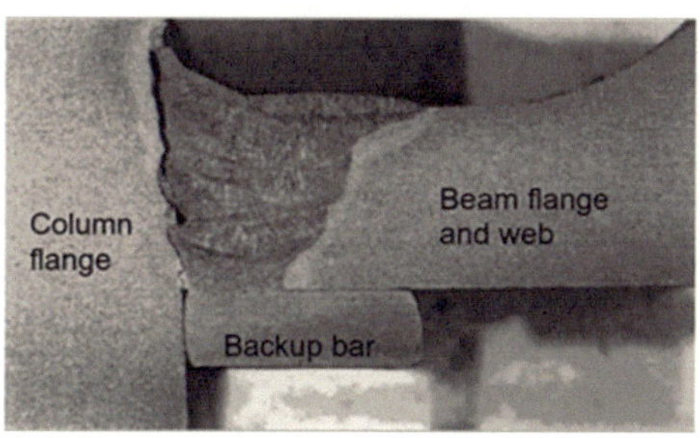

Figure 17.
Effects of notch on fracture of steel structures under Northridge earthquake [72].

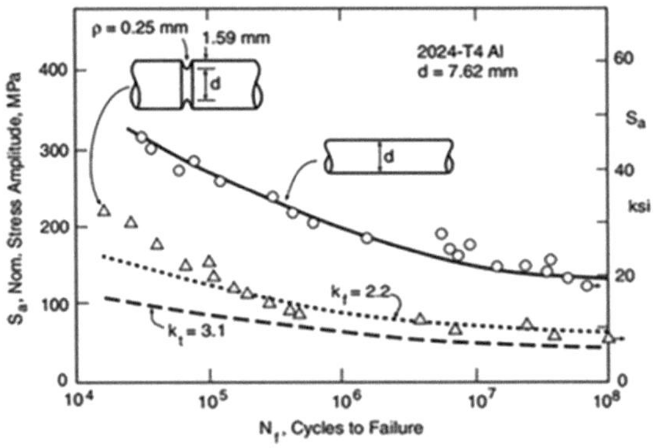

Figure 18.
Effects of notch on S-N behavior of an aluminum alloy (Kt is estimate of fatigue life and Kf is data's obtain by test) [74].

On a plot of S versus life Nf, the fatigue life decreases in proportion to K_t factor as presented in **Figure 18**.

7. Conclusion

Cyclic loads may lead the machines and structural components to premature failure that is called fatigue. Concern about fatigue failure is due to the fact that it occurs at a stress level much lower than the ultimate strength and in a completely unpredictable way. Macroscopically fatigue failure is seen with a brittle appearance and without any gross deformation in the fracture area. Fatigue failure can be occurred in form of high cycle, low cycle, and extremely low cycle fatigue. There are metallurgical and mechanical parameters that affect the occurrence of fatigue failures. Hostile environment causes corrosion fatigue and decreases the operation life of the components. Presents of notch causes stress concentrations points and accelerated the fatigue failures. Residual stress in the tensile form reduces the fatigue life while in the form of compressive stress increases the life of components.

Author details

Alireza Khalifeh
Shiraz Petrochemical Complex, Shiraz, Iran

*Address all correspondence to: ar.khalifeh@spc.co.ir

References

[1] Lewinsohn CA. Mechanical Behavior of Materials by Norman E. Dowling. UK: Pearson; 2013

[2] Dieter GE, Bacon D. Mechanical Metallurgy. Vol. 3. New York: McGraw-Hill; 1976

[3] Zhang C et al. Fatigue failure of welded details in steel bridge pylons. Engineering Failure Analysis. 2021;**127**: 105530

[4] Lourenço N et al. Fatigue failure of a compressor blade. Engineering Failure Analysis. 2008;**15**(8):1150-1154

[5] Ebara R. Corrosion fatigue phenomena learned from failure analysis. Engineering Failure Analysis. 2006;**13**(3):516-525

[6] Bhaumik S, Sujata M, Venkataswamy M. Fatigue failure of aircraft components. Engineering Failure Analysis. 2008;**15**(6):675-694

[7] Gilbert JL et al. Intergranular corrosion-fatigue failure of cobalt-alloy femoral stems. A failure analysis of two implants. J Bone Joint Surg Am. 1994;**76**: 110-115

[8] Ren P-R et al. High-cycle fatigue failure analysis of cast Al-Si alloy engine cylinder head. Engineering Failure Analysis. 2021;**127**:105546

[9] Miller DK. Lessons learned from the Northridge earthquake. Engineering Structures. 1998;**20**(4–6):249-260

[10] Nakashima M, Inoue K, Tada M. Classification of damage to steel buildings observed in the 1995 Hyogoken-Nanbu earthquake. Engineering Structures. 1998;**20**(4–6): 271-281

[11] Khalifeh A et al. Investigating of the tensile mechanical properties of structural steels at high strain rates. Materials Science and Engineering: A. 2018;**712**:232-239

[12] Zhou H, et al. Seismic low-cycle fatigue evaluation of welded beam-to-column connections in steel moment frames through global–local analysis. International Journal of Fatigue. 2014;**64**:97-113

[13] Lashkari B. Seismic risk evaluation of steel structures based on low-cycle fatigue. Reliability Engineering & System Safety. 1988;**20**(4):297-302

[14] Tateishi K, Hanji T, Minami K. A prediction model for extremely low cycle fatigue strength of structural steel. International Journal of Fatigue. 2007; **29**(5):887-896

[15] Jia L-J, Ge H. Ultra-Low-Cycle Fatigue Failure of Metal Structures under Strong Earthquakes. Singapore: Springer; 2019

[16] Kermanpur A et al. Failure analysis of Ti6Al4V gas turbine compressor blades. Engineering Failure Analysis. 2008;**15**(8):1052-1064

[17] Shahrivar A, Abdolmaleki A. Failure of a screw compressor shaft. Engineering Failure Analysis. 2006;**13**(4):698-704

[18] Ul-Hamid A et al. Failure analysis of an impeller blade. Materials and Corrosion. 2015;**66**(3):286-295

[19] Sivaprasad S et al. Investigation on the failure of air compressor. Engineering Failure Analysis. 2010; **17**(1):150-157

[20] Xiao N et al. High cycle fatigue behavior of a low carbon alloy steel: The

influence of vacuum carburizing treatment. Engineering Failure Analysis. 2020;**109**:104215

[21] Prabhudev K. Handbook of Heat Treatment of Steels. India: Tata McGraw-Hill Education; 1988

[22] Genel K, Demirkol M, Capa M. Effect of ion nitriding on fatigue behaviour of AISI 4140 steel. Materials Science and Engineering: A. 2000;**279** (1–2):207-216

[23] Kim H-J, Kweon Y-G. High cycle fatigue behavior of gas-carburized medium carbon Cr-Mo steel. Metallurgical and Materials Transactions A. 1996;**27**(9):2557-2563

[24] Limodin N, Verreman Y. Fatigue strength improvement of a 4140 steel by gas nitriding: Influence of notch severity. Materials Science and Engineering: A. 2006;**435**:460-467

[25] Zerbst U et al. Defects as a root cause of fatigue failure of metallic components. I: Basic aspects. Engineering Failure Analysis. 2019;**97**: 777-792

[26] Zhu X et al. Failure analysis of a failed connecting rod cap and connecting bolts of a reciprocating compressor. Engineering Failure Analysis. 2017;**74**: 218-227

[27] Webster G, Ezeilo A. Residual stress distributions and their influence on fatigue lifetimes. International Journal of Fatigue. 2001;**23**:375-383

[28] James M et al. Residual stresses and fatigue performance. Engineering Failure Analysis. 2007;**14**(2):384-395

[29] Torres M, Voorwald H. An evaluation of shot peening, residual stress and stress relaxation on the fatigue life of AISI 4340 steel. International Journal of Fatigue. 2002;**24**(8):877-886

[30] Brockenbrough RL, Johnston BG. Steel design manual.Technical Rep. ADUSS 27-3400-01. US, Monroeville, PA. 1968

[31] Sines G. Behavior of metals under complex static and alternating stresses. Metal fatigue. 1959;**1**:145-169

[32] Lin J. Durability and damage tolerance analysis methods for lightweight aircraft structures: Review and prospects. International Journal of Lightweight Materials and Manufacture. 2022;**5**:224-250

[33] Sankaran K, Perez R, Jata K. Effects of pitting corrosion on the fatigue behavior of aluminum alloy 7075-T6: Modeling and experimental studies. Materials Science and Engineering: A. 2001;**297**(1–2):223-229

[34] Chapetti MD. Fracture mechanics for fatigue design of metallic components and small defect assessment. International Journal of Fatigue. 2022;**154**:106550

[35] Paris P et al. Extensive study of low fatigue crack growth rates in A533 and A508 steels. ASTM STP. 1972;**513**: 141-176

[36] González-Velázquez JL. Fatigue Fracture. In: Fractography and Failure Analysis. Switzerland: Springer; 2018. pp. 71-95

[37] González-Velázquez JL. Fractography and Failure Analysis. Vol. 24. Switzerland: Springer; 2018

[38] Tateishi K, Hanji T. Low cycle fatigue strength of butt-welded steel joint by means of new testing system with image technique. International

Journal of Fatigue. 2004;**26**(12): 1349-1356

[39] Mathew M, Laha K, Sandhya R. Creep and low cycle fatigue behaviour of fast reactor structural materials. Procedia Engineering. 2013;**55**:17-26

[40] Rao CV et al. Low cycle fatigue, deformation and fracture behaviour of Inconel 617 alloy. Materials Science and Engineering: A. 2019;**765**:138286

[41] Golański G, Mroziński S. Low cycle fatigue and cyclic softening behaviour of martensitic cast steel. Engineering Failure Analysis. 2013;**35**:692-702

[42] ASTM E. 606-92. Standard practice for strain-controlled fatigue testing. In: Annual Book of ASTM Standards. Baltimore, MD, USA: ASTM International; 2002. p. 569

[43] Zambrano A, Foti D. Damage indices evaluation for seismic resistant structures subjected to low-cycle fatigue phenomena. International Journal of Mechanical Sciences. 2014;**78**:106-117

[44] Nip K et al. Extremely low cycle fatigue tests on structural carbon steel and stainless steel. Journal of Constructional Steel Research. 2010;**66**(1):96-110

[45] Kuwamura H. Fracture of steel during an earthquake—State-of-the-art in Japan. Engineering Structures. 1998; **20**(4–6):310-322

[46] Bayoumi M, Abd El Latif A. Characterization of cyclic plastic bending of austenitic AISI 304 stainless steel. Engineering Fracture Mechanics. 1995;**51**(6):1049-1058

[47] Gangloff RP. Corrosion fatigue crack propagation in metals. In: International Conference on Environment Induced Cracking of Metals. USA: NASA; 1990

[48] Fleck C, Eifler D. Corrosion, fatigue and corrosion fatigue behaviour of metal implant materials, especially titanium alloys. International Journal of Fatigue. 2010;**32**(6):929-935

[49] McEvily AJ, Wei RP. Fracture Mechanics and Corrosion Fatigue. In: Proceedings of the Conference on Corrosion Fatigue. Stores, CN: N.A.C.E, 1971:381-395

[50] Weronski A. Thermal Fatigue of Metals. New York: CRC Press; 1991

[51] Khalifeh A. Stress corrosion cracking damages. In: Failure Analysis. UK: IntechOpen; 2019

[52] Sanders R, Starke E. The effect of intermediate thermomechanical treatments on the fatigue properties of a 7050 aluminum alloy. Metallurgical Transactions A. 1978;**9**(8):1087-1100

[53] Kermajani M et al. Effect of weld metal toughness on fracture behavior under ultra-low cycle fatigue loading (earthquake). Materials Science and Engineering: A. 2016;**668**:30-37

[54] Zerbst U et al. Defects as a root cause of fatigue failure of metallic components. II: Non-metallic inclusions. Engineering Failure Analysis. 2019;**98**:228-239

[55] Zhang J et al. Inclusion size evaluation and fatigue strength analysis of 35CrMo alloy railway axle steel. Materials Science and Engineering: A. 2013;**562**:211-217

[56] Schmiedt A et al. Evaluation of maximum non-metallic inclusion sizes in engineering steels by fitting a generalized extreme value distribution based on vectors of largest observations. Acta Materialia. 2015;**95**:1-9

[57] Fang X-Y, Zhao Y-X, Liu H-W. Study on fatigue failure mechanism at

various temperatures of a high-speed railway wheel steel. Materials Science and Engineering: A. 2017;**696**:299-314

[58] Yoshida H, Nagumo M. Microstructures controlling the ductile crack growth resistance of low carbon steels. Metallurgical and Materials Transactions A. 1998;**29**(1):279-287

[59] Pegues J et al. Surface roughness effects on the fatigue strength of additively manufactured Ti-6Al-4V. International Journal of Fatigue. 2018; **116**:543-552

[60] Fathallah R et al. Effect of surface properties on high cycle fatigue behaviour of shot peened ductile steel. Materials Science and Technology. 2003; **19**(8):1050-1056

[61] Ocaña J et al. Laser shock processing: An emerging technique for the enhancement of surface properties and fatigue life of high-strength metal alloys. International Journal of Microstructure and Materials Properties. 2013;**8**(1/2):38-52

[62] Khalifeh A. Stress corrosion cracking behavior of materials. Engineering Failure Analysis. 2020;**10**:55-75

[63] Thumser R, Bergmann JW, Vormwald M. Residual stress fields and fatigue analysis of autofrettaged parts. International Journal of Pressure Vessels and Piping. 2002;**79**(2):113-117

[64] Garcia C et al. Fatigue crack growth in residual stress fields. International Journal of Fatigue. 2016;**87**:326-338

[65] Acevedo R et al. Residual stress analysis of additive manufacturing of metallic parts using ultrasonic waves: State of the art review. Journal of Materials Research and Technology. 2020;**9**(4):9457-9477

[66] Masubuchi K. Analysis of Welded Structures: Residual Stresses, Distortion, and their Consequences. Vol. 33. USA: Elsevier; 2013

[67] Dong P, Brust F. Welding residual stresses and effects on fracture in pressure vessel and piping components: A millennium review and beyond. Journal of Pressure Vessel Technology. 2000;**122**(3):329-338

[68] Feng Z. Processes and Mechanisms of Welding Residual Stress and Distortion. UK: Woodhead Publishing Limited; 2005

[69] Khalifeh A et al. Stress corrosion cracking of a circulation water heater tubesheet. Engineering Failure Analysis. 2017;**78**:55-66

[70] Taylor D, O'Donnell M. Notch geometry effects in fatigue: A conservative design approach. Engineering Failure Analysis. 1994;**1**(4):275-287

[71] Vincent M et al. Fatigue from defect under multiaxial loading: Defect stress gradient (DSG) approach using ellipsoidal equivalent inclusion method. International Journal of Fatigue. 2014;**59**: 176-187

[72] Wang Y et al. Fracture behavior analyses of welded beam-to-column connections based on elastic and inelastic fracture mechanics. International Journal of Steel Structures. 2010;**10**(3):253-265

[73] Kakiuchi T et al. Prediction of fatigue limit in additively manufactured Ti-6Al-4V alloy at elevated temperature. International Journal of Fatigue. 2019;**126**:55-61

[74] MacGregor CW, Grossman N. Effects of Cyclic Loading on Mechanical Behavior of 24S-T4 and 75S-T6 Aluminum Alloys and SAE 4130 Steel. USA: NASA-TN-2812; 1952

Chapter 3

Microstructurally Based Modeling of Creep Deformation and Damage in Martensitic Steels

Christof Sommitsch, Bernhard Sonderegger, Mohammad Ahmadi, Florian Riedlsperger, Felix Meixner, Josef Mergl and Bernhard Krenmayr

Abstract

This chapter deals with modeling the microstructural evolution, creep deformation, and pore formation in creep-resistant martensitic 9–12% Cr steels. Apart from the stress and temperature exposure of the material, the input parameters for the models are as-received microstructure and one single-creep experiment of moderate duration. The models provide predictive results on deformation rates and microstructure degradation over a wide stress range. Due to their link to the underlying fundamental physical processes such as classical nucleation theory, Gibbs energy dissipation, climb, and glide of dislocations, etc., the models are applicable to any martensitic steel with similar microstructure to the presented case study. Note that we section the chapter into part 1: creep deformation and part 2: pore formation.

Keywords: microstructure, dislocations, precipitates, subgrains, pores, deformation rate, damage

1. Introduction

When trying to simulate deformation rates during creep, one can basically follow two different approaches: (i) phenomenologically based—and or (ii) physically inspired models. In phenomenological models, the deformation (rate) is usually stated as analytical function with system parameters such as temperature and stress as input [1–4]. These kinds of models are easily and quickly employed; however, the approaches carry some disadvantages as well: (1) they give little or no insight into the actual underlying physical processes governing the creep rate, and (2) the model parameters cannot be usually determined independently from the creep experiment the model is actually aiming to predict. For these reasons, we choose to focus on a physically based model instead [5], which is a reviewed, corrected, and extended version of the seminal work of Ghoniem [6]. In addition to avoiding the mentioned drawbacks of phenomenological approaches, our physical model has the advantage of including a variety of microstructural elements such as dislocations and subgrain boundaries (SGBs), and their interactions. This allows for getting a deeper understanding of the creep process and opens the opportunity of rating individual material badges by taking their as-received microstructure as starting condition for a creep simulation. Finally, we end up with an assembly of rate equations for the microstructural elements along with some side equations modeling the physical processes. In summary, the model gives us insights into the specific reasons why a material badge features good or bad creep behavior, as long as its microstructure can be considered homogeneous. The model also has the potential to rate the impact of individual microstructural phenomena.

Our model includes the microstructure by mean values of specific microstructural elements (e.g. dislocation density, grain boundary precipitates, etc.) instead of a spatially resolved features. This allows for a simpler construction of a "representative volume element," which speeds up computation and is more easily compared to microstructural investigations.

In our work, we focus on martensitic 9–12% Cr steels. We select the material P91 to demonstrate the validity of the model due to its widespread use and industrial significance. Nevertheless, the concept can be adapted to other material groups by including their specific microstructural elements and their interactions.

Please note that all symbols used in the equations are explained in **Table 6** at the end of the chapter.

2. Creep model: our microstructurally based approach

The aim of our creep model is to predict the creep rate and microstructural evolution, based on the initial microstructure and the system parameters stress and temperature. A very simple, yet useful approximation of the creep rate $\dot{\varepsilon}$ has been introduced by Orowan [7] and later extended by Yadav [8] to include damage. While Yadav considers damage due to cavities D_{cav} and precipitates D_{ppt}, damage in our model is based on cavities alone, since the effect of precipitates is dealt with other equations in the framework in a physical manner. The resulting modified Orowan equation calculates the creep rate from the physical inputs of mobile dislocation density ρ_m and an effective dislocation velocity v_{eff}; see Eq. (1). Latter term has been introduced by Riedlsperger [5] incorporating glide and climb processes. Since both quantities, ρ_m and v_{eff}, result from interactions within the microstructure, we take a closer look at those, see **Figure 1**. Within **Figure 1**, we indicate following interactions:

a. Frank—read sources (generation of mobile dislocations ρ_m)

b. Emission from static dislocations $\rho_s$into mobile dislocations ρ_m

c. Immobilization of mobile dislocations ρ_m—generation of static dislocations ρ_s

d. Recovery by climb processes for mobile (d_1) and static dislocations (d_2)

e. Spontaneous annihilation of mobile (e_1) and static dislocations (e_2)

f. Subgrain boundaries produced from static dislocations

g. Subgrain growth minus Zener—pinning of boundary precipitates

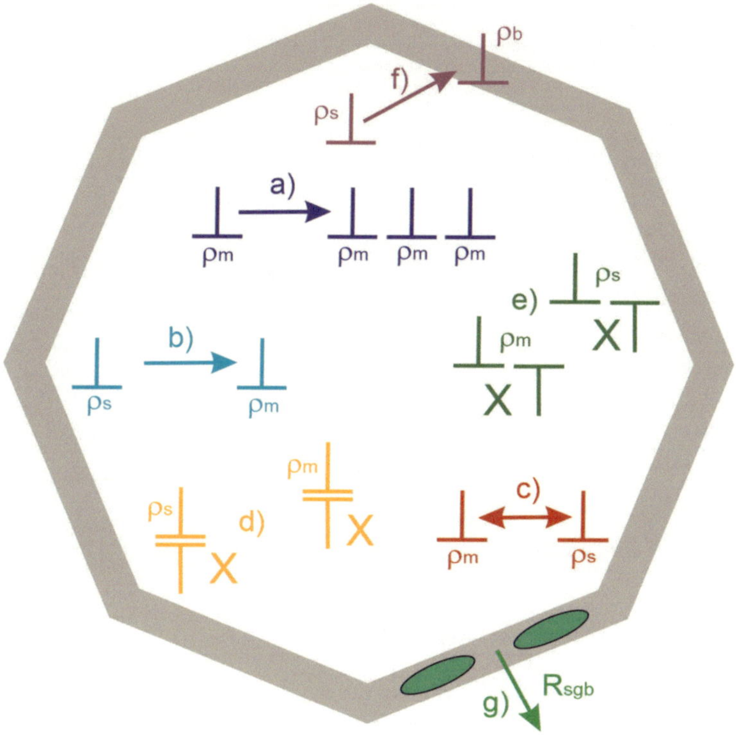

Figure 1.
Microstructural interactions within a subgrain. See text for individual interactions. "X" signifies annihilation of dislocations and arrows mark transformations.

These interactions (a–g) are also integrated into the rate equations for the microstructural evolution of mobile dislocations ρ_m, static dislocations ρ_s, dislocations within subgrain boundaries ρ_b, and subgrain boundaries R_{sgb}; see Eqs. (2)–(5) in **Table 1**. **Table 1** also indicates the original source for each equation.

In addition to the "rate equations" of the microstructural evolution of the material, **Table 1** also assembles the framework of the underlying physical phenomena. Within this paragraph, we only give a brief overview. Detailed discussions can be found in the cited sources.

The effective subgrain growth pressure P_{eff}, Eq. (6), is rewritten, but it is equivalent to Ghoniem's work [6] and includes dissipation of grain boundary energy as well as Zener pinning from precipitates. The mobility of the subgrains, M_{sgb}, Eqs (7) and (8), is governed mostly by diffusion coefficients and the boundary misorientation [6]. Dislocation spacing within a subgrain boundary, h_{b}, Eq. (9), has been deducted by geometrical means [6]. The effective velocity of mobile dislocations v_{eff} includes glide processes as well as climbing of dislocations over precipitates within the subgrain interior [5]. The glide velocity of mobile dislocations $v_g$considers forward and potential backward movements according to their jump probabilities, which are linked to mechanical and thermal activation [9]. The corrected applied stress σ_{app}, as shown in Eq. (12), considers the reduced cross section of a creep sample due to its poisson ratio [5]. Since the climb of mobile dislocations depends on local diffusion, the climb velocity, v_c, as shown in Eq. (13) [6], is split into a lattice diffusion share, v_{cl}, as shown in Eq. (14) [5], and a pipe diffusion share, v_{cp}, as shown in Eq. (16) [5], with additional terms regarding the

Creep model equations	Eq.
Creep strain rate (modified Orowan equation, see text):	(1)
$\frac{d\varepsilon}{dt} = \frac{b \bullet \rho_m \bullet v_{eff}}{M \bullet (1 - D_{cav})}$	
Mobile dislocation density rate [5]:	
$\frac{d\rho_m}{dt} = v_{eff} \bullet \left\{ \underbrace{\rho_m^{3/2}}_{a} + \underbrace{\frac{\beta \rho_s R_{sgb}}{h_b^2}}_{b} - \underbrace{\frac{\rho_m}{2R_{sgb}}}_{c} - \underbrace{8\rho_m^{3/2} \frac{v_c}{v_{eff}}}_{d1} - \underbrace{\delta_{anh}(\rho_m + \rho_s)\rho_m}_{e1} \right\}$	(2)
Static dislocation density rate [6]:	
$\frac{d\rho_s}{dt} = v_{eff} \bullet \left\{ \underbrace{\frac{\rho_m}{2 \bullet R_{sgb}}}_{c} - \underbrace{8 \bullet \frac{v_c}{v_{eff}} \bullet \frac{\rho_s}{h_b}}_{d2} - \underbrace{\delta_{anh} \bullet \rho_s \bullet \rho_m}_{e2} \right\}$	(3)
Boundary dislocation density rate [6]:	
$\frac{d\rho_b}{dt} = \underbrace{8 \bullet (1 - 2\zeta) \bullet \rho_s \bullet \frac{v_c}{h_b}}_{f} - \underbrace{M_{sgb} \bullet P_{eff}}_{g} \bullet \frac{\rho_b}{R_{sgb}}$	(4)
Subgrain growth rate (h: nucleation of subgrains) [5]:	
$\frac{dR_{sgb}}{dt} = \underbrace{M_{sgb} \bullet P_{eff}}_{g} - \underbrace{G \bullet \eta_V \bullet K_c^2 \bullet \left(\sqrt{\rho_m + \rho_s} - \frac{K_c}{2R_{sgb}} \right) \bullet \frac{D_s \Omega}{2k_B T}}_{h}$	(5)
Effective subgrain growth pressure (only precipitates at boundaries) [6]:	
$P_{eff} = \frac{4}{3} \bullet G \bullet b^2 \bullet \rho_b \bullet \left[1 - \frac{\pi}{2} \bullet \left(\sum_{i=1}^{m} r_{p,i}^2 \bullet N_{V,i} \right) \bullet R_{sgb} \right]$	(6)
Subgrain boundary mobility – case 1, if $P_{eff} \geq 0$ [6]:	
$M_{sgb} = \frac{2 \bullet \pi \bullet \eta_v \bullet D_s \bullet \Omega}{b \bullet k_B \bullet T \bullet \theta_i f_i} + \frac{2 \bullet \pi \bullet b \bullet D_{vp} \bullet \Omega \bullet \theta_i f_i}{h_b^2 \bullet k_B \bullet T}$	(7)
Subgrain boundary mobility – case 2, if $P_{eff} < 0$ [6]:	
$M_{sgb} = \frac{D_s \bullet \Omega}{2 \bullet \pi \bullet k_B \bullet T \bullet \sum_{i=1}^{m} r_{p,i}^4 \bullet N_{V,i} \bullet \theta_i f_i}$	(8)
Mean dislocation spacing within the subgrain boundary [6]:	
$h_b = \frac{1}{(\rho_b + \rho_s) \bullet R_{sgb}}$	(9)
Effective dislocation velocity (only precipitates in grain interior) [5]:	
$\frac{1}{v_{eff}} = \frac{1}{v_g} + \sum_i \frac{\pi}{2} \bullet N_{V,i} \bullet r_{p,i}^3 \bullet \frac{1}{v_c}$	(10)
Dislocation glide velocity [9]:	
$v_g = a_1 \bullet exp\left\{ -\frac{Q}{k_B \bullet T} \right\} \bullet exp\left\{ -\frac{\sigma_i \bullet V_r}{k_B \bullet T} \right\} \bullet 2 \bullet sinh\left\{ \frac{\sigma_{app} \bullet V_r}{k_B \bullet T} \right\}$	(11)
Corrected applied stress [5]:	
$\sigma_{app} = \sigma_{app,0} \bullet (1 + \nu \bullet \varepsilon)$	(12)
Dislocation climb velocity [6]:	
$v_c = v_{cl} + v_{cp}$	(13)
Dislocation climb velocity-lattice diffusion share [5]:	
$v_{cl} = \frac{2 \bullet \pi \bullet \eta_v \bullet D_s}{\left[1 - \eta_v \bullet \ln\left(L_\alpha \bullet \sqrt{\rho_t} \right) \right] \bullet b} \bullet exp\left\{ -\frac{\sigma_i \bullet \Omega}{k_B \bullet T} \right\} \bullet 2 \bullet sinh\left\{ \frac{\sigma_{app} \bullet \Omega}{k_B \bullet T} \right\}$	(14)

Creep model equations	Eq.
Parameter for elastic interactions between dislocations/defects [6]:	
$L_\alpha = \frac{(1+\nu)\cdot G\cdot b\cdot 0.5\cdot\Omega}{3\cdot\pi\cdot(1-\nu)\cdot k_B\cdot T}$	(15)
Dislocation climb velocity-pipe diffusion share [5]:	
$\upsilon_{cp} = \frac{2\cdot\pi\cdot b\cdot D_{vp}}{L_p^2}\cdot exp\left\{-\frac{\sigma_i\cdot\Omega}{k_B\cdot T}\right\}\cdot 2\cdot sinh\left\{\frac{\sigma_{app}\cdot\Omega}{k_B\cdot T}\right\}$	(16)
Distance of core-vacancy diffusion before evaporation into lattice [10]:	
$L_p = \sqrt{2}\cdot a_g\cdot exp\left\{\frac{\Delta W}{2\cdot k_B\cdot T}\right\}$	(17)
Internal stress [11]	
$\sigma_i = \alpha\cdot M\cdot G\cdot b\cdot\sqrt{\rho_m + c_s\cdot\rho_s}$	(18)
Damage parameter for cavities, see text:	
$\dot{D}_{cav} = \mathrm{A}\cdot\varepsilon\cdot\dot{\varepsilon}^{1/2}$	(19)
Particle coarsening:	
$r_i^3(t) - r_{i,t=0}^3 = \mathrm{k}_i\cdot\mathrm{t}$	(20)
Reduction of number density of particles due to coarsening:	
$\frac{N_i(t)}{N_{i,t=0}} = \frac{r_{i,t=0}^3}{r_i^3(t)}$	(21)

Table 1.
Creep model equations and their primary literature sources.

vacancy-dislocation interactions from Hirth and Lothe [10]; see Eq. (17). The internal stress σ_i, which is an important input for the glide and climb velocities and considers interactions of mobile-mobile and mobile-static dislocations, is taken from Basirat [11]; see Eq. (18). Finally, the phenomenological factor "cavitation damage" D_{cav}, as shown in Eq. (19), has been introduced by Basirat [11] as well but slightly adapted here for a better agreement with the observed master creep curve.

3. Creep model: experimental input and simulation setup

The input parameters needed to simulate creep over a range of stresses can be divided into three groups/types:

i. General material data from the material group;

ii. Microstructural data in as-received condition of the material badge of interest;

iii. Parameters which have to be adapted to one single master-creep-curve.

Group (i) is applicable to a wider range of materials and has already been collected in [5]. **Table 2** gives an overview on the findings.

Group (ii) is the parameter for a specific material badge. This group accounts for the different creep behavior stemming from specific processing routines, e.g.

Input	Value	Details	Source
a_g	2.866×10^{-10} m	Fe-bcc	[12]
α	0.02	9Cr-1Mo steel at 650°C	[11]
b	2.48×10^{-10} m	Fe-bcc	[13]
c_s	0.3	9Cr-1Mo steel at 650°C	[11]
δ_{anh}	5b	Edge dislocations	[11]
D_s	2×10^{-19} m^2/s	Fe in Fe-1.5Mo at 650°C	[14]
D_{vp}	4.75×10^{-19} m^2/s	P91	[5]
η_V	2×10^{-4}	Ferritic steels	[6]
G	61.9 GPa	9–12% Cr steels at 650°C	[15]
k_B	1.38065×10^{-23} J/K	Fundamental constant	[16]
K_c	2.1	P91	[5]
M	3	Crystals	[11]
Ω	1.177×10^{-29} m	Volume/atom	[5]
Q	4.01×10^{-19} J	HT-9 at 550°C	[6]
$\theta_i f_i$	3°=0.0524 rad	9–12% Cr steels	[17]
V_r	35Ω	P91 at 650°C	[5]
ν	0.317	9–12% Cr steels at 650°C	[15]
ΔW	1.26×10^{-19} J	P91 at 650°C	[5]
ζ	0.034	9–12% Cr steels	[18]

Table 2.
Input parameters for the material group. Input parameters type (i).

chemical composition and heat treatment. Result from the processing routine is the as-received microstructure, which also acts as an input for our simulation. In our case, we directly measured the subgrain size R_{sgb} and the boundary dislocation density $\rho_b$via electron backscatter diffraction (EBSD) [5]. The typical as-received mobile dislocation density $\rho_m$was taken from Panait [19]. Simulations reveal that ρ_m quickly converges toward generic values during creep, so the results are not very sensitive to the initial value of this parameter. The same is true forρ_s, which has been adopted from [11]. The precipitate data have to be split into particles at subgrain boundaries (SGBs) and particles within subgrains (VN, NbC, and AlN). Both have been calculated by the thermodynamic software MatCalc [20] with the chemical composition of the heat, the actual heat treatment, and matrix data [5]. The results of the particle simulation, coarsening of particles from as-received condition onward, have been mimicked by a simple coarsening law, as shown in Eqs. (20) and (21), which is now used as input for the creep simulation. **Table 3** summarizes the microstructure-specific input data for the as-received condition, which are the starting values of our creep simulation.

After these parameters have been set, only the variables a_1, A, and β remain (group (iii) of input parameters), whereas A is of phenomenological nature and thus cannot be measured directly as a matter of principle, and the parameters a_1 and β have physical interpretations but are extremely inconvenient for direct assessment. We therefore have to set up one single (!) creep experiment at set temperature T and stress $\sigma_{app,0}$ to fit these parameters.

Input	Value	Note	Source
R_{sgb}	4.0×10^{-7} m	EBSD	[5]
ρ_b	5.9×10^{14} m^{-2}	EBSD	[5]
ρ_m	4.5×10^{14} m^{-2}	TEM	[19]
ρ_s	4.5×10^{13} m^{-2}	Estimate 1/10 of ρ_m	[11]
r_i (SGB)	1.49×10^{-7} m	At boundaries	[5]
r_i (VN)	4.17×10^{-8} m	Grain interior	[5]
r_i (NbC)	7.55×10^{-8} m	Grain interior	[5]
r_i (AlN)	4.43×10^{-8} m	Grain interior	[5]
N_i (SGB)	9.52×10^{18} m^{-3}	At boundaries	[5]
N_i (VN)	1.09×10^{20} m^{-3}	Grain interior	[5]
N_i (NbC)	4.04×10^{18} m^{-3}	Grain interior	[5]
N_i (AlN)	8.85×10^{18} m^{-3}	Grain interior	[5]
k_i (SGB)	6.90×10^{-28} s^{-1}	At boundaries	[5]
k_i (VN)	$2.21 \cdot 10^{-30}$ s^{-1}	Grain interior	[5]
k_i (NbC)	0 s^{-1}	Grain interior	[5]
k_i (AlN)	1.19×10^{-29} s^{-1}	Grain interior	[5]

Table 3.
Microstructure-specific input data for the as-received state. Input parameters type (ii).

We conducted/performed an instrumented creep test at 650°C and a nominal stress of 70 MPa which led to a rupture time of 8740 h [5]. The simulated creep curve has been validated against the experimental results and the missing parameters a_1, A, and β were optimized using a least-squares fit of the time-dependent creep deformation. **Table 4** summarizes the results which are also used for all other creep simulations presented in the following text.

Since the creep model already contains the impact of stress and temperature explicitly within its network of equations, the found input parameters are stress- and temperature-independent. We can thus use the same input for other stresses to produce time-to-rupture (TTR) diagrams. To do so, we have left all input parameters unchanged as indicated in **Tables 2–4** and generated creep simulations in the stress range of 50–120 MPa. The next section comprises the result of the simulated master creep curve including the calculated microstructural evolution, as well as the changes of the creep behavior with altering stresses, leading to the construction of the TTR diagram.

Parameter	Value
a_1	3.725 m/s
A	0.035
β	0.07

Table 4.
Parameters found by fitting against one creep curve. Input parameters type (iii).

4. Creep model: simulation results and discussion

First simulated result is the master-creep curve at 650°C and 70 MPa, indicating the creep deformation and deformation rate. **Figures 2** and **3** demonstrate the agreement between the simulated and the experimental result of the creep deformation: the simulated creep curve is very close to the experiment including primary, secondary, and tertiary creep stage, and also the final fracture of the sample. The experimental minimum of the creep strain rate is about $2.5 \times 10^{-6}\ h^{-1}$ in the range between 1.000 and 3.000 h, whereas the simulated result is $3.0 \times 10^{-6}\ h^{-1}$ at 900 h, suggesting good agreement as well.

Regarding the microstructural evolution, the simulation predicts a quick recovery of the mobile dislocation density ρ_m from $4.5 \times 10^{14}\ m^{-2}$ to $1.5 \times 10^{13}\ m^{-2}$ within the first 500 h of creep, which is exactly mirroring the continuous decrease in the creep rate during the primary creep regime. According to the (modified) Orowan equation (Eq. (1)), two potential reasons can be responsible for the creep rate: the mobile dislocation density and the effective velocity of the dislocations. Our model demonstrates the dominating role of the dislocation density in this regard. After reaching its minimum, ρ_m increases slowly up to about $2 \times 10^{13}\ m^{-2}$ at the time of fracture. This level agrees well with several literature data on similar material under comparable conditions, indicating dislocation densities at time of fracture of $2 \times 10^{13}\ m^{-2}$ [21] or $2.7 \times 10^{13}\ m^{-2}$ to $3.5 \times 10^{13}\ m^{-2}$ [22].

ρ_b and R_{sgb} were verified using EBSD on the fractured sample of the master-creep experiment, showing very good agreement with the simulated data, whereas simulation indicates a subgrain radius of 0.95 μm at the time of fracture and experimental verifications reveal a mean value of 0.7 μm [5]. The experimental value of boundary dislocation density of $3.4 \times 10^{14}\ m^{-2}$ also compares very well to the simulated result of $3.0 \times 10^{14}\ m^{-2}$ (**Figure 4**) [5].

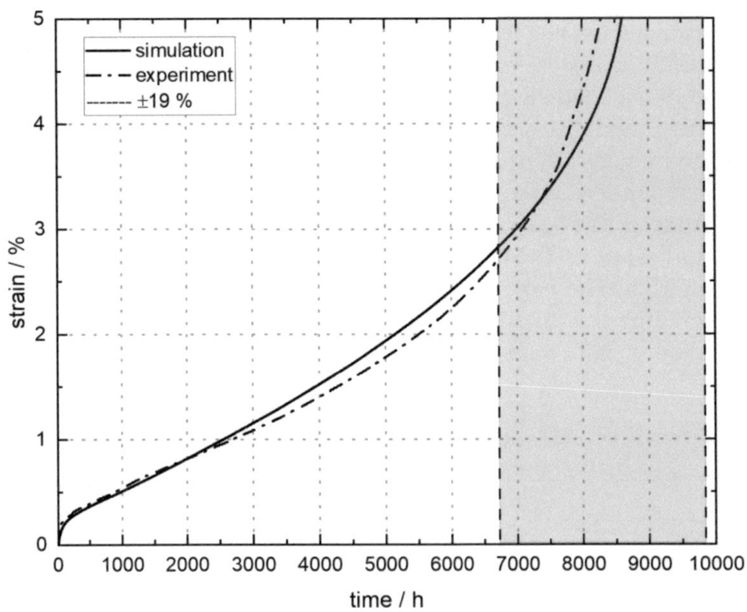

Figure 2.
Experimental and simulated master-creep curve at 650°C, 70 MPa.

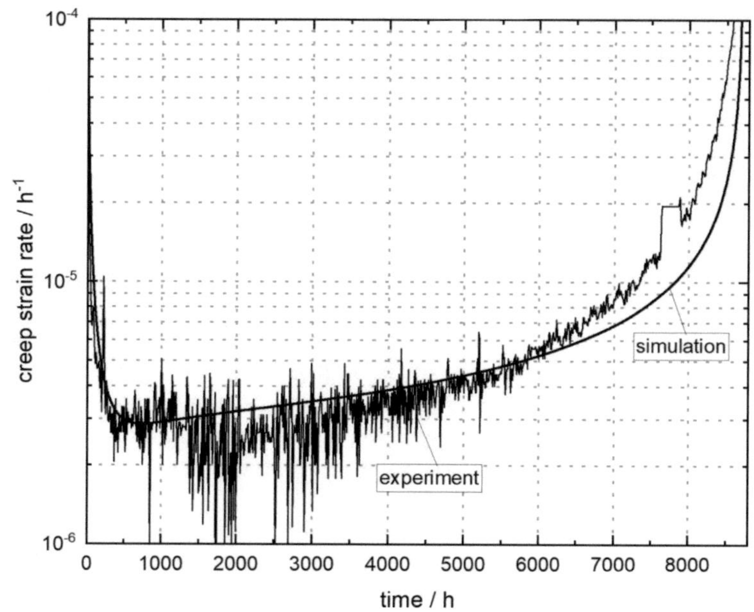

Figure 3.
Experimental and simulated creep-strain rates at 650°C, 70 MPa.

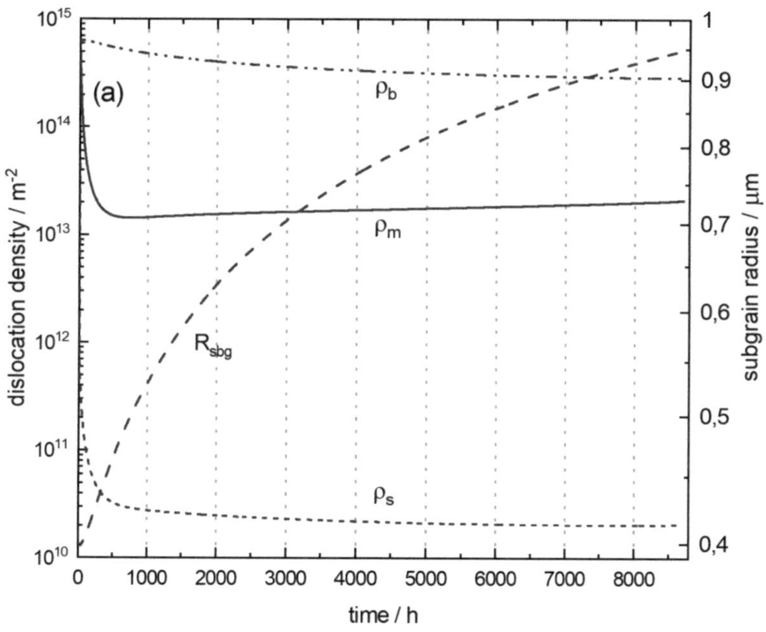

Figure 4.
Simulated microstructural evolution at 650°C, 70 MPa.

In summary, the simulation results of the test case of the master-creep experiment are good enough to apply the model to multiple stresses. We carried out creep simulation in the stress range of 50–120 MPa in steps of 10 MPa, with a resulting creep curve and accompanying microstructural evolution in each simulation. One side result, the rupture time t_R, could then be used for reconstructing a time-to-rupture (TTR) diagram. In the simulation, the sample ruptures when the damage parameter D_{cav} reaches a level of 1, which is basically a result of extensive creep strain and/or strain rate [5]. This typically occurs at a strain between 3% (low stresses) and 10% (high stresses). Please note that the fracture elongation is not an input, but a result of the simulation. Please also note that the simulation does not consider local necking but only deals with sample areas of uniform cross sections. **Figure 5** shows the individual simulated creep curves within the investigated stress range.

Please note that the creep curves appear to look different from **Figure 2** because the logarithmic time-scale is used in order to simultaneously show all results. Each of the creep curves feature primary, secondary, and tertiary creep regimes. **Figure 6** now finally shows the constructed TTR stemming from the simulated creep data and compares them against the standard literature data from European Creep Collaborative Committee (ECCC) [23], ASME [24], and NIMS [25].

The agreement is excellent, and the simulated curve lies right between the data from the three standard literature sources for creep rupture data of P91. Once again, it is important to mention that all model input data (except for the system stress) were identical for all creep simulations. This detail is very important, because the simulation allows for a predictive extrapolation from a single-creep experiment carried out for 8.740 h to up to six times longer creep times. As it appears, the simulation also allows for extrapolating to shorter running times by a factor of about 40 in our case. We thus motivate to use and test our model for even shorter reference experiments in order to extrapolate for long running times.

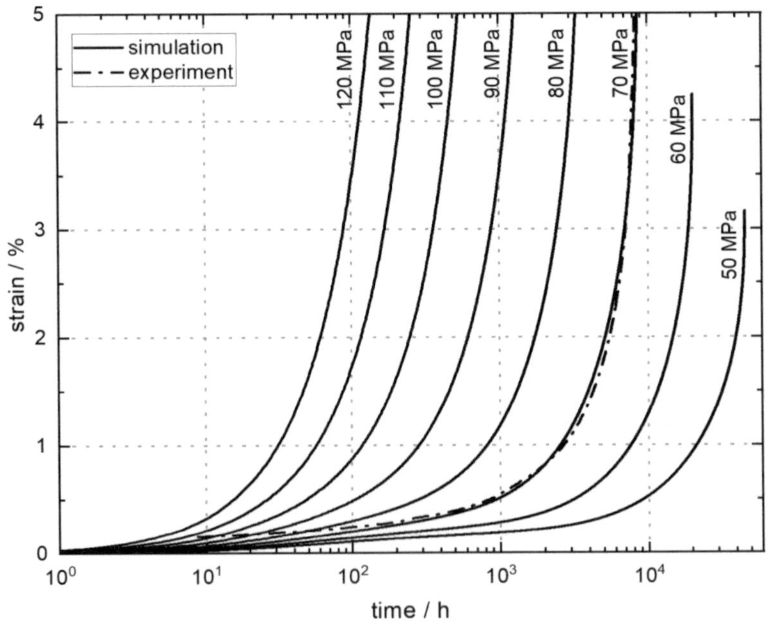

Figure 5.
Simulated microstructural evolution at 650°C, 70 MPa.

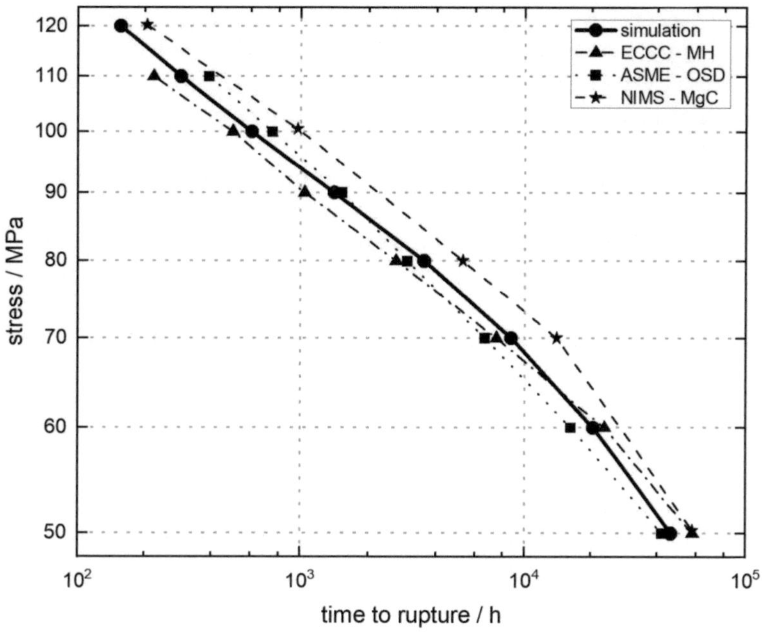

Figure 6.
Simulated TTR diagram compared against standard literature data.

5. Pore formation model: introduction

It has been long established [26] and is now well accepted [27, 28] that failure during creep loading results mainly due to intergranular rupture. Cavities nucleate predominantly at grain boundaries, grow during creep exposure, and coalesce to form microcracks. In tertiary creep, these cracks are so numerous that they significantly weaken the microstructure, and the remaining available cross section is put under more stress which further promotes damage and accelerates the strain rate.

In some cases, the remaining creep life can be directly correlated with the degree of cavitation [29, 30].

6. Pore formation model: nucleation and growth

It is not well established by what mechanism cavities nucleate [31]; however, the linear relationship between cavity nucleation rate, first observed by Needham et al [32], still holds true to this day [33]. Grain boundary sliding, as necessitated to maintain contact between the grains when they elongate during diffusional creep, is one proposed mechanism [34]. This sliding generates cavities at ledges that are pulled apart at the grain boundary.

We propose a model based on the physics of diffusion and fluctuational theory, known as classical nucleation theory (CNT). CNT was formulated at the beginning of the twentieth-century by the works of Volmer and Weber [35], Becker and Döring [36], Frenkel [37], and Zeldovich [38]. It has been prominent and successful in modeling the nucleation of new phases, precipitates, and similar phenomena.

Balluffi [39] was the first to explain the nucleation of holes (cavities) by vacancy supersaturation. However, Raj and Ashby [40] were the first to consider the mechanical stress as the driving force for nucleation, a theory which was further developed by Hirth and Nix [41] and Riedel [28] and forms the basis for our nucleation model.

While CNT generally speaks of nuclei, which may form new phases, we specify these as clusters of vacancies which may form cavities.

The free energy change on formation of such a cluster in the bulk encompasses the pressure-volume work done by the external stress, σ, and the energy required by the newly formed surface between the cluster and the matrix, excluding the elastic energy [42]. Eq. (22) shows the relation between the free energy change, ΔF, and the cluster radius, r, with the specific surface energy, γ:

$$\Delta F = -\frac{4}{3}\pi r^3\,\sigma + 4\pi r^2\,\gamma \tag{22}$$

Plotting this relation over the cluster radius, as shown in **Figure** 7, we see that the free energy reaches a maximum at a certain cluster size, r^*, designated as the critical radius. This energy barrier needs to be overcome for a cluster to become a stable cavity which will then continue to grow as its total free energy decreases.

From Eq. (22), we derive the critical radius and critical (maximum) free energy as follows:

$$r^* = \frac{2\gamma}{\sigma} \tag{23}$$

$$\Delta F^* = \frac{16}{3}\pi\frac{\gamma^3}{\sigma^2} \tag{24}$$

Clusters below the critical size are naturally/always present in the microstructure due to thermal fluctuations [35, 43]. Their concentration is determined by the number

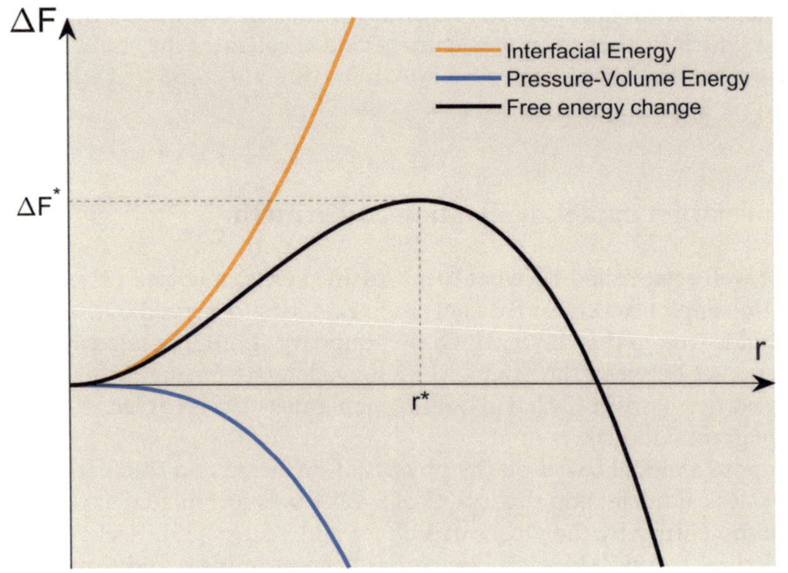

Figure 7.
Free energy change vs. radius for a spherical cluster of vacancies.

of possible nucleation sites, N_S, and an Arrhenius term comprised of the free energy change over the atomic thermal energy, kT. These are the first two terms in Eq. (25) for the nucleation rate, and they quantify the number of critical cavities available in quasi-static equilibrium. The next term, β^*, is the vacancy attachment rate and Z, is the Zeldovich factor:

$$I = N_s \, exp\left(-\frac{\Delta F^*}{kT}\right)\beta^* \, Z \tag{25}$$

Some vacancies, which exist throughout the microstructure and are more prevalent at higher temperatures, may find themselves on the surface of a critical cluster and only one atomic jump away from joining it. The number of these vacancies jumping toward the critical cluster per unit time is described by the vacancy attachment rate, β^*. In Eq. (26), this is shown to depend on the diffusion coefficient, D, the concentration of vacancies, X_V, and the surface area of a critical cluster, A^*, with a being the average interatomic distance:

$$\beta^* = \frac{D}{a^2} X_v \frac{A^*}{a^2} \tag{26}$$

The Zeldovich factor is explained by its namesake [38] and other literature [28, 44] to reduce the nucleation rate, since steady-state nucleation artificially removes supercritical clusters and because slightly supercritical clusters are still more likely to dissolve rather than grow. It is defined in Eq. (27) with n^* signifying the number of vacancies in a critical cluster:

$$Z = \frac{1}{n^*}\sqrt{\frac{\Delta F^*}{3\pi kT}} \tag{27}$$

While the nucleation rate is only directly proportional to most of the physical parameters in Eq. (25), the Arrhenius term dominates. Small changes in the height of

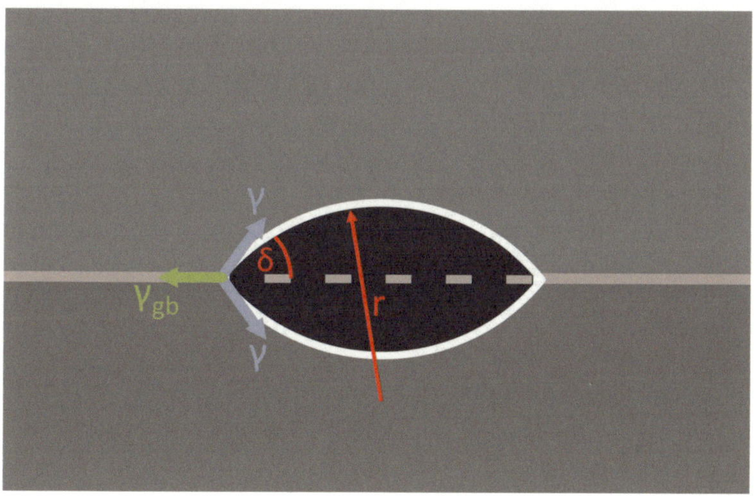

Figure 8.
Cluster formed between two grains.

the nucleation barrier lead to large variations in the equilibrium number of critical cavities available for nucleation and therefore the final nucleation rate. Smaller critical clusters are more likely to nucleate, such as in the case for clusters formed on grain boundaries as shown in **Figure 8**. The dihedral angle, δ, formed between the surface and the grain boundary as a result of the competing grain boundary surface energy, γ_{gb}, and the cluster surface energy, γ, reduce the volume and surface area [45] of a critical cluster, even though the curvature, r^*, is unchanged. Removing the previous grain boundary area (dashed gray line in **Figure 8**) also reduces the critical free energy:

$$\delta = acos\left(\frac{\gamma_{gb}}{2\,\gamma}\right) \tag{28}$$

Nucleation is further boosted by the quicker diffusion of vacancies along the grain boundaries and the effect of converting the multiaxial stress state to an average stress on the grain boundary [46]. Also, real defects, such as dislocations, interacting with the grain boundary supply additional vacancies which can effectively increase the driving force by several gigapascals [47]. These effects predict cavity nucleation almost exclusively at grain boundaries and do not require extreme threshold stresses for nucleation, which is an enduring criticism of classical nucleation of cavities [48].

Finally, a theory based on generalized broken bonds (GNBBs) [49] is used to calculate the free surface energy from the energy of vacancy formation and a correction is applied [50] when dealing with nanosized critical clusters.

Diffusional cavity growth is less controversial and commonly assumed to follow the rate of radial growth in Eq. (29) by Hull and Rimmer [51]. Its resemblance to the Svoboda, Fischer, Fratzl, Kozeschnik (SFFK) model [20] used in precipitate growth simulations further strengthens its prestige:

$$\dot{r} = \frac{D\,X_v a^3}{kT\,r}\left(\sigma - \frac{2\gamma}{r}\right) \tag{29}$$

The sintering stress, $2\gamma/r$, that opposes the driving stress, σ, also explains the shrinking of subcritical clusters predicted by CNT.

Var.	Description	Unit	Source
γ	Free surface energy	$J\,m^{-2}$	[49]
σ	External stress	Pa	Experiment
N_s	Number of nucleation sites	—	Experiment and [53]
T	Temperature	K	Experiment
σ_D	Driving force of defects	Pa	[47]
γ_{gb}	Grain boundary energy	$J\,m^{-2}$	[54]
Q_v	Vacancy formation energy	$J\,mol^{-1}$	[55]
D	Diffusion coefficient	$m^2\,s^{-1}$	[56]
a	Interatomic spacing	m	[57]
d_g	Grain diameter	m	Experiment

Table 5.
Model parameters and physical constants.

7. Pore formation model: model implementation

The equations for nucleation and growth are integrated into a Kampmann-Wagner framework [52] at a constant temperature and external stress state. At regularly spaced time intervals, a class of newly formed cavities with a population derived from Eq. (25) is formed, each with a radius slightly (20%) above the critical radius from Eq. (23).

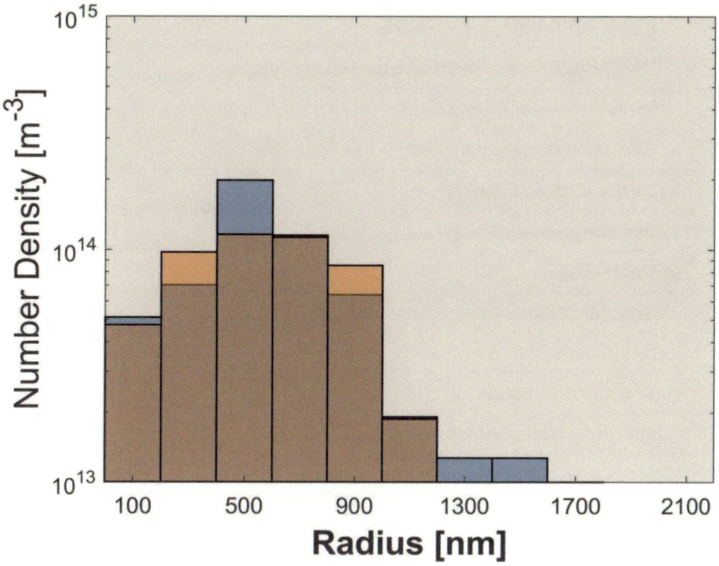

Figure 9.
Comparison between simulated (red) and measured (blue) histograms of cavities in Nickel-based alloy 625 after 5500 h at 700°C and 183 MPa.

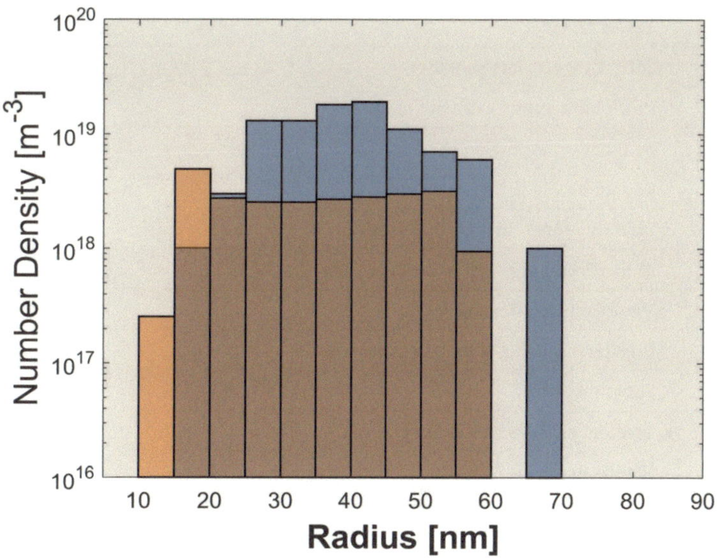

Figure 10.
Comparison between simulated (red) and measured (blue) histograms of cavities in P23 steel after 9000 h at 600°C and 90 MPa.

Var.	Description	Unit
a_1	Parameter for glide velocity	[m/s]
a_g	Size of unit cell	[m]
A	Material-specific constant	[–]
α	Dislocation interaction factor	[–]
b	Burgers vector ½ <111> for bcc	[m]
β	Parameter for density of sources	[–]
c_s	Weighting factor for statics in the internal stress	[–]
δ_{anh}	Dislocation annihilation length	[m]
D_{cav}	Cavitation damage parameter	[–]
D_s	Lattice diffusion coefficient	[m^2/s]
D_{vp}	Pipe diffusion coefficient	[m^2/s]
ε	Creep strain	[–]
η_v	Transfer coefficient for defects into jogs	[–]
G	Shear modulus	[Pa]
γ	Free surface energy	[J m^{-2}]
γ_{GB}	Specific grain boundary energy	[J m^{-2}]
h_b	Mean dislocation spacing within the subgrain wall	[m]
k_B	Boltzmann constant	[J/K]
K_c	Holt constant	[–]
k_i	Precipitate coarsening parameter of precipitate class i	[s^{-1}]
L_α	Parameter for elastic interactions between dislocation/defects	[m]
L_p	Diffusion path of core-vacancy before evaporation into lattice	[m]
M	Taylor factor	[–]
M_{sgb}	Subgrain boundary mobility	[m/Pa s]
N_i	Precipitate number density of class i	[m^{-3}]
N_s	Number of nucleation sites	–
ν	Poisson's ratio	[–]
P_{eff}	Effective subgrain growth pressure	[Pa]
Q	Activation energy for dislocation glide	[J]
Q_v	Vacancy formation energy	[J mol^{-1}]
r_i	Precipitate radius of particle class i	[m]
R_{sgb}	Subgrain radius	[m]
ρ_b	Boundary dislocation density	[m^{-2}]
ρ_m	Mobile dislocation density	[m^{-2}]
ρ_s	Static dislocation density	[m^{-2}]
ρ_t	Total dislocation density	[m^{-2}]
$\sigma_{app,0}$	Applied initial stress	[Pa]
σ_{app}	Corrected applied stress	[Pa]

Var.	Description	Unit
σ_i	Internal stress	[Pa]
σ_{D}	Driving force of defects	[Pa]
t	Time	[s]
t_R	Rupture time	[s]
T	Temperature	[K]
$\theta_i f_i$	Mean misorientation of subgrains	[rad]
v_c	Climb velocity	[m/s]
v_{cl}	Lattice diffusion share of v_c	[m/s]
v_{cp}	Pipe diffusion share of v_c	[m/s]
v_{eff}	Effective velocity of mobile dislocations	[m/s]
v_g	Glide velocity	[m/s]
V_r	Apparent activation volume for thermally activated glide	[m^3]
ΔW	Activation energy difference between pipe and lattice diffusion	[J]
Ω	Atomic volume (unit cell size divided by two atoms in bcc)	[m^3]
ζ	Fraction of statics annihilating when transforming to boundaries	[–]

Table 6.
List of symbols and variables used within the creep model.

During the intervals, the respective radii of all classes grow according to Eq. (29). As the available nucleation sites are used, the nucleation rate diminishes. The simulation ends when all nucleation sites are consumed and there is no more uncavitated grain boundary area. The number of nucleation sites at grain boundaries is calculated [53] from average grain diameters, assuming all grains to be tetrakaidekahedral (**Table** 5).

8. Pore formation model: results and comparison with experiments

Figures 9 and **10** compare simulated results of nucleated cavities with experimental results obtained from secondary electron microscopy and density measurements. Case studies comparing our model to experimental investigations have been published [42, 58].

9. Conclusion

We have introduced a complex physically based creep model and demonstrated its capabilities in the case of the martensitic steel P91. The model is capable of simulating the creep deformation as well as the microstructural evolution during creep. As soon as some final parameters have been set, based on a single-creep experiment, those parameters can be used for simulating the creep behavior over a wide range of stresses allowing for extrapolating the creep behavior. Current results suggest an extrapolation of the creep lifetime by a factor of at least 6 over a reference experiment.

Furthermore, we have introduced a physically based model for the formation of creep pores due to vacancy diffusion, which is also showing very good agreement with experiments.

Acknowledgements

The authors gratefully acknowledge funding from Austrian Science Fund (FWF) within project "Software development on dislocation creep in alloys" (P-31374).

Author details

Christof Sommitsch[1*], Bernhard Sonderegger[2], Mohammad Ahmadi[1], Florian Riedlsperger[1], Felix Meixner[1], Josef Mergl[2] and Bernhard Krenmayr[2]

1 Institute of Materials Science, Joining and Forming, Graz University of Technology, Graz, Austria

2 Institute for Engineering Materials – Metals and Alloys, Johannes Kepler University Linz, Linz, Austria

*Address all correspondence to: christof.sommitsch@tugraz.at

References

[1] Larson F, Miller J. A time-temperature relationship for rupture and creep stresses. ASTM Proceedings. 1952; **74**:765-775

[2] Monkman F, Grant N. An empirical relationship between rupture life and minimum creep rate in creep rupture tests. ASTM Proceedings. 1956;**56**:593-620

[3] Norton F. The Creep of Steel at High Remperatures. New York: McGraw-Hill; 1929

[4] Holdsworth S, Askins M, Baker A, Gariboldi E, Holmström S, Klenk A, et al. Factors influencing creep model equation selection. International Journal of Pressure Vessels and Piping. 2008;**85**: 80-88

[5] Riedlsperger F, Krenmayr B, Zuderstorfer G, Fercher B, Niederl B, Schmid J, et al. Application of an advanced mean-field dislocation creep model to P91 for calculation of creep curves and time-to-rupture diagrams. Materialia. 2022;**12**:100760

[6] Ghoniem N, Matthews J, Amodeo R. A dislocation model for creep in engineering materials. Research Mechanics. 1990;**29**:197-219

[7] Orowan E. Problems of plastic gliding. Proceedings of the Physical Society. 1940;**52**:8-22

[8] Yadav S, Sonderegger B, Stracey M, Poletti C. Modelling the creep behaviour of tempered martensitic steel based on a hybrid approach. Materials Science and Engineering A. 2016;**662**:330-341

[9] Kreyca F. State parameter based modelling of stress-strain curves in aluminium alloys [thesis]. Vienna: Vienna University of Technology; 2017

[10] Hirth J, Lothe J. Theory of Dislocations. 2nd ed. New York: Wiley; 1982

[11] Basirat M, Shrestha T, Potirniche G, Charit I, Rink K. A study of the creep behavior of modified 9Cr-1Mo steel using continuum-damage modeling. International Journal of Plasticity. 2012; **37**:95-107

[12] Askeland D, Fulay P, Bhattacharya D. Atomic and Ionic Arrangements: Essentials of Materials Science and Engineering. 2nd ed. Stamford: Edition Cengage Learning; 2009

[13] Cheng G, Xu W, Jian W, Yuan H, Tsai M, Zhu Y. Dislocations with edge components in nanocrystalline bcc Mo. Journal of Materials Research. 2013;**28**: 1820-1826

[14] Nitta H, Miura K, Iijima Y. Self-diffusion in iron-based Fe-Mo alloys. Acta Materialia. 2006;**54**:2833-2847

[15] Hasegawa Y. Grade 92 creep-strength-enhanced ferritic steel. In: Shibli A, editor. Coal Power Plant Materials and Life Assessment- Development and Applications. Sawston: Woodhead Publishing Series in Energy; 2014. p. 62

[16] CODATA. The NIST Reference on constants, units, and uncertainty. 2014. Available from: https://physics.nist.gov/cgi-bin/cuu/Value?k [Accessed: January 28, 2022]

[17] Pineau A, Antolovich S. High temperature fatigue: Behaviour of three typical classes of structural materials. Materials at High Temperatures. 2015;**32**: 298-317

[18] Schmid J. Modelling the microstructure of a creep resistant steel

[thesis]. Graz: Graz University of Technology; 2018

[19] Panait C, Zielińska-Lipiec A, Koziel T, Czyrska-Filemonowicz A, Gourgues-Lorenzon A, Bendick W. Evolution of dislocation density, size of subgrains and MX-type precipitates in a P91 steel during creep and during thermal ageing at 600°C for more than 100,000h. Materials Science and Engineering A. 2010;**527**:4062-4069

[20] Svoboda J, Fischer F, Fratzl P, Kozeschnik E. Modelling of kinetics in multi-component multi-phase systems with spherical precipitates: I: Theory. Materials Science and Engineering A. 2004;**385**:166-174

[21] Yadav S, Kalácska S, Dománkova M, Yubero D, Resel R, Groma I, et al. Evolution of the substructure of a novel 12 % Cr steel under creep conditions. Materials Characterization. 2016;**115**:23-31

[22] Jara D. 9-12% Cr heat resistant steels: alloy design, TEM characterisation of microstructure evolution and creep response at 650°C [thesis]. Bochum: Ruhr Universität Bochum; 2011

[23] Bendick W, Cipolla L, Gabrel J, Hald J. New ECCC assessment of creep rupture strength for steel grade X10CrMoVNb9-1(Grade91). International Journal of Pressure Vessels and Piping. 2010;**87**:304-309

[24] Swindeman R, Swindeman M, Roberts B, Thurgood B, Marriott D. Verification of allowable stresses in ASME section III, subsection for NH for for grade 91 steel, Technical Report. 2007. Available from: https://www.osti.gov/servlets/purl/974278 [Accessed: January 30, 2022]

[25] Kimura K, Sawada K, Kushima H, Toda Y. Influence of chemical composition and heat treatment on long-term creep strength of Grade 91 Steel. Procedia Engineering. 2013;**55**:2-9

[26] Greenwood JN, Miller DR, Suiter JW. Intergranular cavitation in stressed metals. Acta Metallurgica. 1954; **2**(2):250-258

[27] Kassner ME, Pérez-Prado MT. Fundamentals of creep in metals and alloys. Fundamentals of Creep in Metals and Alloys. 2015:1-272. Available from: https://linkinghub.elsevier.com/retrieve/pii/B9780080436371X50005

[28] Riedel H. Fracture at High Temperatures [Internet]. Berlin, Heidelberg, Springer Berlin Heidelberg; 1987. p. 430. Available from: http://link.springer.com/10.1007/978-3-642-82961-1

[29] Neves DLC, Seixas JR, Tinoco EB, Rocha AC, Abud I. Stress and integrity analysis of steam superheater tubes of a high pressure boiler. Materials Research. 2004;**7**(1):155-161

[30] Siefert JA, Parker JD. Evaluation of the creep cavitation behavior in Grade 91 steels. International Journal of Pressure Vessels Pipe. 2016;**138**:31-44. DOI: 10.1016/j.ijpvp.2016.02.018

[31] Kassner ME, Hayes TA. Creep cavitation in metals. International Journal of Plasticity. 2003;**19**(10):1715-1748

[32] Needham NG, Wheatley JE, Greenwood GW. The creep fracture of copper and magnesium. Acta Metallurgica. 1975;**23**(1):23-27

[33] Davanas K. Determination of creep cavity nucleation rates. Materials High Temperature. 2020;**37**(2):75-80

[34] He J, Sandström R. Formation of creep cavities in austenitic stainless steels.

Journal of Materials Science. 2016; **51**(14):6674-6685

[35] Volmer M, Weber A. Keimbildung in übersättigten Gebilden. Zeitschrift für Physikalische Chemie. 1926;**119**(1): 277-301

[36] Becker R, Döring W. Kinetische Behandlung der Keimbildung in übersättigten Dämpfen. Annals of Physics. 1935;**416**(8):719-752

[37] Frenkel J. A general theory of heterophase fluctuations and pretransition phenomena. The Journal of Chemical Physics. 1939;**7**(7):538-547

[38] Zeldovich YB, Sunyaev R. On the theory of new phase formation. cavitation. In: Barenblatt GI, Sunyaev RA, editors. Selected Works of Yakov Borisovich Zeldovich. Vol. I. Princeton University Press; 1992. pp. 120-137. Available from: https://www.degruyter.com/document/doi/10.1515/9781400862979.120/html

[39] Balluffi RW. The supersaturation and precipitation of vacancies during diffusion. In: Acta Metallurgica. Vol. 2. Mar 1954. pp. 194-202

[40] Raj R, Ashby M. Intergranular fracture at elevated temperature. Acta Metallurgica. 1975;**23**(6):653-666. Available from: https://linkinghub.elsevier.com/retrieve/pii/0001616075900474

[41] Hirth JP, Nix WD. Analysis of cavity nucleation in solids subjected to external and internal stresses. Acta Metallurgica. 1985;**33**(3):359-368

[42] Ahmadi MR, Sonderegger B, Yadav SD, Poletti C. Modelling and simulation of diffusion driven pore formation in martensitic steels during creep. Material Science Engineering A. 2018. 712: 466–477. Available from: https://doi.org/10.1016/j.msea.2017.12.010

[43] Farkas L. Keimbildungsgeschwin digkeit in übersättigten Dämpfen. Zeitschrift für Physikalische Chemie. 1926;**11**:236-242

[44] Russell KC. Nucleation in solids: The induction and steady state effects. Advances in Colloid and Interface Science. 1980;**13**

[45] Lee JK, Aaronson H. Influence of faceting upon the equilibrium shape of nuclei at grain boundaries—II. Three-dimensions. Acta Metall. 1975;**23**(7): 809-820. Available from: https://linkinghub.elsevier.com/retrieve/pii/0001616075901960

[46] Anderson PM, Rice JR. Constrained creep cavitation of grain boundary facets. Acta Metallurgica. 1985;**33**(3): 409-422

[47] Fernandez-Caballero A, Cocks ACF. Creep cavities: Theoretical formulation of cavity nucleation in the presence of dislocation defects. Euromat. 2021; **2021**

[48] Clemens BM, Nix WD, Gleixner RJ. Void nucleation on a contaminated patch. Journal of Materials Research. 1997;**12**(8):2038-2042

[49] Sonderegger B, Kozeschnik E. Generalized nearest-neighbor broken-bond analysis of randomly oriented coherent interfaces in multicomponent Fcc and Bcc structures. Metallurgical and Materials Transactions A: Physical Metallurgy and Materials Science. 2009; **40**(3):499-510

[50] Sonderegger B, Kozeschnik E. Size dependence of the interfacial energy in the generalized nearest-neighbor broken-bond approach. Scripta

Materialia. 2009;**60**(8):635-638. Available from: http://dx.doi.org/10.1016/j.scriptamat.2008.12.025

[51] Hull D, Rimmer DE. The growth of grain-boundary voids under stress. Philosophical Magazine. 1959;**4**(42):673-687

[52] Wagner R, Kampmann R, Voorhees PW. Homogeneous Second-Phase Precipitation. In: Kostorz G, editor. Phase Transformations in Metals. Weinheim, Germany: Wiley-VCH Verlag GmbH & Co. KGaA; 2001. pp. 309-407. Available from: http://doi.wiley.com/10.1002/9783527603978.mst0388

[53] Rajek H. Computer simulation of precipitation kinetics in solid metals and application to the complex power plant steel CB8. 2005. Available from: http://scholar.google.com/scholar?hl=en&btnG=Search&q=intitle:Computer+simulation+of+precipitation+kinetics+in+solid+metals+and+application+to+the+complex+power+plant+steel+CB8#0

[54] Prkoshkina D, Esin VA, Wilde G, Dviniski SV. Grain boundary width, energy and self-diffusion in nickel: Effect of material purity. Acta Materialia. 2013;**61**:5188-5197

[55] Janotti A, Krcmar M, Fu CL, Reed RC. Solute diffusion in metals: Larger atoms can move faster. Physical Review Letters. 2004;**92**(8):085901

[56] Kozeschnik E. Modeling Solid-state Precipitation. Momentum Press. 2004

[57] Arblaster J. Selected values of the crystallographic properties of the elements. ASM International. 2018

[58] Meixner F, Ahmadi MR, Sommitsch C. Modeling and simulation of pore formation in a bainitic steel during creep. Metallurgy Materials Transactions A. 2022;**53**:984-999

Chapter 4

A Review on the Dynamic Response of Liquid-Storage Tanks Associated with Fluid-Structure Interaction

Ayman Mohammad Mansour and Moustafa Moufid Kassem

Abstract

Water tanks are considered one of the most important facilities in firefighting systems and municipal water supply. These critical water storage and distribution facilities should remain operable even after a severe seismic event or sustain only damages that can be readily repaired. In recent years, the seismic design of storage tanks has been aimed at fulfilling safety requirements and the environmental impact on society. This paper provides a review of research work related to seismic response of liquid-filled tanks. Major contribution from previous research works related to dynamic behavior of liquid tanks are acknowledged in this review. This paper encompasses the phenomenon of fluid-structure interaction and reviews several equivalent mechanical models for liquid storage tanks that account for this phenomenon. The application of each modeling approach and its accuracy in accounting for the fluid-structure interaction are discussed based on available literature and applicable international standards. It was shown that different equivalent modeling approaches that consider the fluid-structure interaction effects can be used to reduce the computational cost and complexity of liquid-tank systems.

Keywords: liquid-storage tanks, infrastructure, fluid-structure interaction, liquid sloshing, dynamic response, equivalent mechanical models

1. Introduction

Earthquakes result from abrupt release of energy by the slippage of two tectonic plates. The sudden release of strain generates seismic waves that are transferred to the earth's surface and result in ground motions. These dynamic vibrations create lateral movement in structures, which affects their strength and behavior. The infrastructure system is very critical and should have extra immunity against possible disasters. And natural hazards due its essential function in remaining serviceable to satisfy the water demand for drinking and firefighting purposes [1].

The dynamic vibrations of liquid-containing structures create the phenomenon of Fluid-Structure Interaction (FSI), where the momentum of the oscillating fluid generate lateral pressure on the boundaries of the structure. The study of hydrodynamic pressure

IntechOpen

on structures can be traced back to the early 1930s. The research work by Westergaard on "Water Pressure on Dams during Earthquakes" is considered the earliest study on the behavior of FSI, where the impulsive pressure on vertical dams under the effect of earthquake excitations was evaluated [2]. Subsequently, the dynamic response of liquid contained tanks and the FSI phenomenon have been subjects for extensive experimental and numerical investigations by many researchers. Such studies emerged by the efforts of Jacobsen and Ayre [3], Housner [4], Veletsos [5] and later by other researchers such as Mansour and Nazri [6], Shakib and Alemzadeh [7], Elansary and El Damatty [8], and Moslemi et al. [9]. The aim of this study is to provide a comprehensive review for the equivalent mechanical models of liquid storage tanks that account for FSI, including the added-mass, single-lumped-mass, two-lumped-mass, spring-mass, three-mass, and other models that were featured in the literature. The theoretical background, application, application, and accuracy for each model were presented based on international standards and available literature.

2. Methods

To "review" has been defined as: "To view, inspect, or examine a second time or again" [10]. Review studies are amongst the most highly sought types of articles by researchers and are the ones that provide the most substantial contribution [10]. The body of a review study can be organized in a variety of ways depending on the type and the method of the review study. The literature review refers to the generic term of review which includes published resources that give an evaluation of recent or current literature. Can cover a wide range of topics at varying levels of comprehensiveness and completeness. It's possible to include study findings. Analysis of the literature review may have different structures, namely, chronological, conceptual, thematic, etc. In a thematic approach, recurring central themes exists, in which the literature review can be divided into subsections that address different aspects of the topic. For the current study, a literature review with a thematic approach was conducted, in which the study was divided into several section; a section for each equivalent mechanical model for liquid-storage tanks associated with fluid-structure interaction.

3. Equivalent mechanical models for the liquid-tank system

Analysis of hydrodynamic pressure in structures such as liquid-storage tanks is more complicated than that of other structures. In the 1960s, Housner [4] provided a practical idealization for evaluating the hydrodynamic pressure within rectangular and cylindrical tanks that are subjected to horizontal ground motion while assuming the tank walls to be rigid. The Chilean earthquake, that took place in 1960 and damaged several large water tanks, was the main plot behind the paper by Housner [4]. FSI can be simulated using different simplified modeling approaches (added-mass approach, two lumped-mass model, spring-mass model, etc.). Livaoğlu and Doğangün [11] presented a comparison and evaluation of some of these modeling methods.

3.1 The added-mass approach

The impulsive hydrodynamic pressure is usually accounted for by introducing added masses. The added mass concept is one of the simplest methods to account for the impulsive hydrodynamic component of liquid pressure. This method has been used for decades in the design of seismic resistant structures, such as gravity dams [2] and liquid tank containers. The Added-Mass Approach (AMA) relies on few main assumptions, which are water incompressibility and the rigidity of the boundary conditions of the structure. This approach neglects the stiffness effects in the fluid and in general leads to conservative results [12]. The use of AMA is proven to be a more appropriate technique for finite element modeling than other assumptions such as those characterized using the lumped mass models [13].

3.2 The single-lumped-mass model

The representation of a single-lumped-mass model for Elevated Water Tanks (EWTs) can be seen in **Figure 1**. This concept was introduced in the 1950s [14] and has two main assumptions. First, for a completely full liquid tank, the water sloshing behavior will not have any vertical movement thus allowing the system to behave as a system with a Single Degree-Of-Freedom (SDOF). Second, the supporting structure acts as a cantilever and is considered to have uniform rigidity along its elevation.

According to ACI-371R [15], this model should be used if the weight of water equals or exceeds 80% of the overall weight of the system. The lumped mass consists of the own-weight of the tank, two-thirds (at maximum) of the own-weight of the supporting structure, and the weight of the contained water.

Previous studies have shown that the use of single lumped-mass model representation of EWTs yields similar results relative to experimental testing and other mechanical models [13]. In addition, the convective mass may have negligible influence on the natural characteristics of EWTs depending on the geometrical shape of the tank.

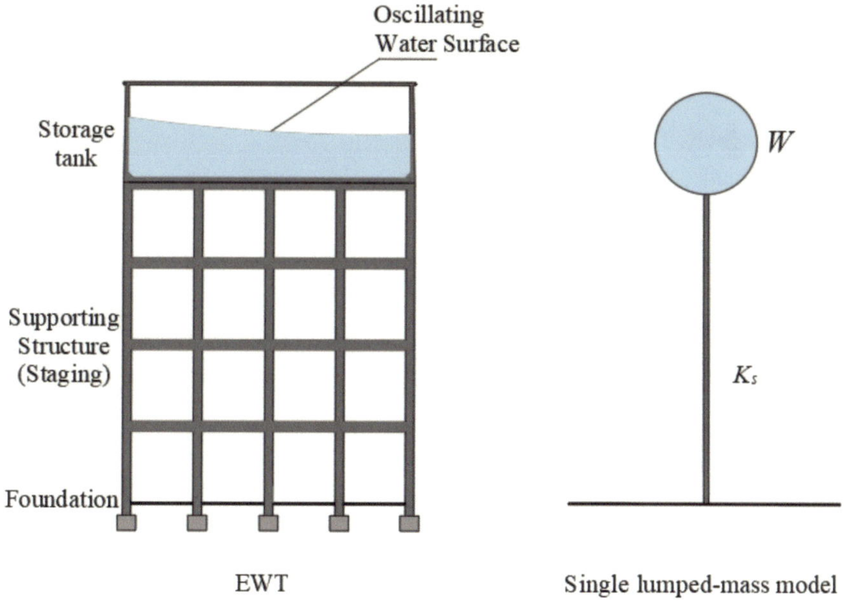

Figure 1.
An EWT and its single lumped-mass representation.

3.3 The two-lumped-mass model

The addition of a convective lumped mass relies on the assumption that was presented by Housner [16] on the relative motion between the storage tank, contained liquid, and the ground. According to his concept, analysis of EWTs can be performed by considering three conditions. First, if the water tank is empty then the sloshing water effect is absent, or if the tank is completely filled with water, then the sloshing effect in the tank is negligible. In this case the EWT will behave as a system with a SDOF, or a one-mass structure. However, the sloshing effect is not neglected where the tank is partially filled. Thereby, this gives the EWT an additional degree-of-freedom, making it a two-mass structure. Consequently, the dynamic analysis of equivalent models must include at least a two-mass system. More lumped convective masses may also be added for ground supported water tanks.

In the simplified analysis procedure for fixed-base EWTs ([17], ACI-371R), a two lumped masses model usually used to represent the fluid-tank system (**Figure 2**). Housner [4] assumed that the two masses to be uncoupled and the seismic forces on the support were evaluated by assuming two separate SDOF systems. The upper mass represents the convective mass of water, which characterizes the motion of the free-liquid-surface. The lower mass represents the impulsive mass of the fluid and the mass of the structure, which is derived by the own-weight of the storage tank plus a portion of the supporting structure's own-weight [11]. Lu et al. [13] showed in his study that the equivalent two-mass model can predict the natural characteristics of water sloshing effect with reasonable accuracy similar to that derived by the much more advanced fluid Finite-Element (FE) technique.

ACI 350.3 (ACI-350.32006) permits the idealization of EWTs as uncoupled single-lumped masses in order to estimate the natural characteristics of the convective and impulsive components of EWTs. Mansour and Nazri [6] evaluated the FSI effect on

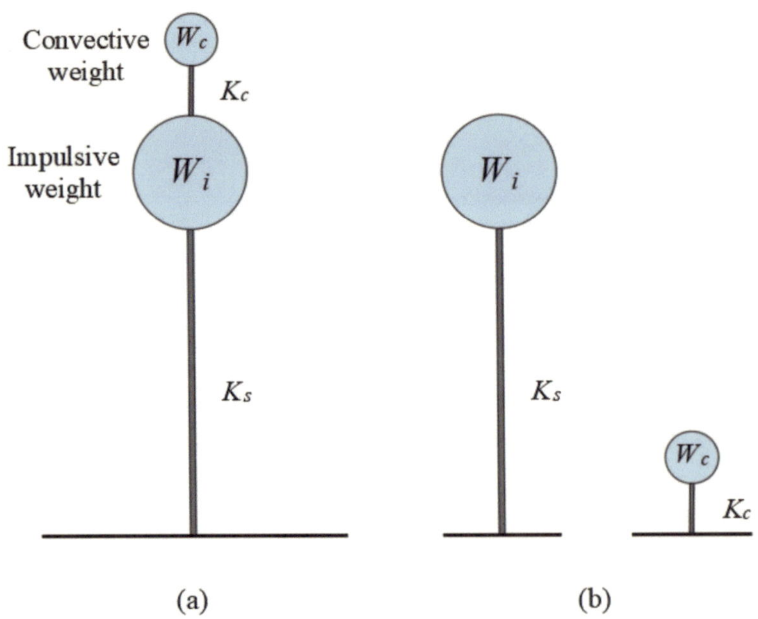

Figure 2.
(a) The two mass model for EWT proposed by Housner [4] and (b) the equivalent uncoupled system.

the dynamic response of EWTs. The periods of vibration corresponding to the impulsive and convective components of the EWTs were predicted using the equivalent two-mass model to validate the developed Three-Dimensional (3D) models. By considering the supporting structure of EWTs as a vertical cantilever, the periods of vibration the significant modes (i.e., T_i and T_c, which are m_i and m_s related, respectively) of EWT systems can be obtained using the equations presented in **Table 1**.

3.4 The spring-mass model

The FE model for a liquid-tank system can be represented by the spring-mass concept, which was originally proposed by Housner [4]. In the spring-mass model, also referred to as the equivalent mechanical model, the liquid is replaced by two lumped masses: the impulsive and convective masses. The impulsive mass is rigidly connected to the tank walls and the convective mass is connected using elastic springs. **Figure 3** illustrates the spring-mass model representation for EWTs based on the principle proposed by Housner [4]. This modeling technique has been used by researchers as a simplified approach for the assessment of the seismic vulnerability of liquid tanks as opposed to much more complex and computationally intensive approaches, such as the continuum liquid-medium models [4, 6, 18].

The parameters of the mechanical spring-mass model are calculated based on the aspect ratio of the liquid-filled tank [19]. According to American Concrete Institute (ACI) (ACI-350.32006), the parameters of this model models can be evaluated using the equations presented in **Table 2**.

The spring-mass model representation of liquid-tank system based on Housner's analogy is considered adequate for modeling of EWTs and is a widely used concept in many international guidelines for seismic design of tanks and buildings such as Eurocode, ACI, and NZS ([20, 21], ACI-350.32006, [15]). These design guidlines have tweaked Housner's method with a few changes due to the findings of the subsequent studies on the seismic design of liquid-containing tanks [22, 23]. The accuracy and efficiency of the two-mass representation of the EWTs was proved by Shepherd [24], who compared the theoretical results of a prestressed RC EWT to the experimental testing results. A detailed study conducted by Dutta, Dutta et al. [25] on RC EWTs integrating soil-structure-fluid interaction shows relatively small differences in in total structural response represented using Westergaard's AMA and the lumped-mass mechanical analogy.

The equivalent spring-mass model has been used by many researchers to simulate the dynamic behavior of EWTs using a simplified approach. Mansour et al. [1] investigated the non-linear seismic vulnerability of a set of EWTs with structural variables

Equations for the period of vibration		
The impulsive component	$T_i = 2\pi\sqrt{\frac{m_i+m_s}{K_s}}$	(1)
The convective component	$T_c = \frac{2\pi}{\sqrt{3.68\tanh\left(3.68\frac{H_L}{D}\right)}}\sqrt{\frac{D}{g}}$	(2)

**Where:* K$_s$ *is the horizontal translation stiffness of the EWT's supporting structure,* m$_s$ *is the lumped structural mass, whivelech includes mass of water tank and two-thirds of staging mass,* m$_i$ *is the impulsive mass,* g *is the acceleration due to gravity, equal to 9.81 m/s^2, and* HL *and* D *correspond to the tank's geometry, i.e., the height and the diameter.*

Table 1.
The vibration period of the significant modes of the EWT system.

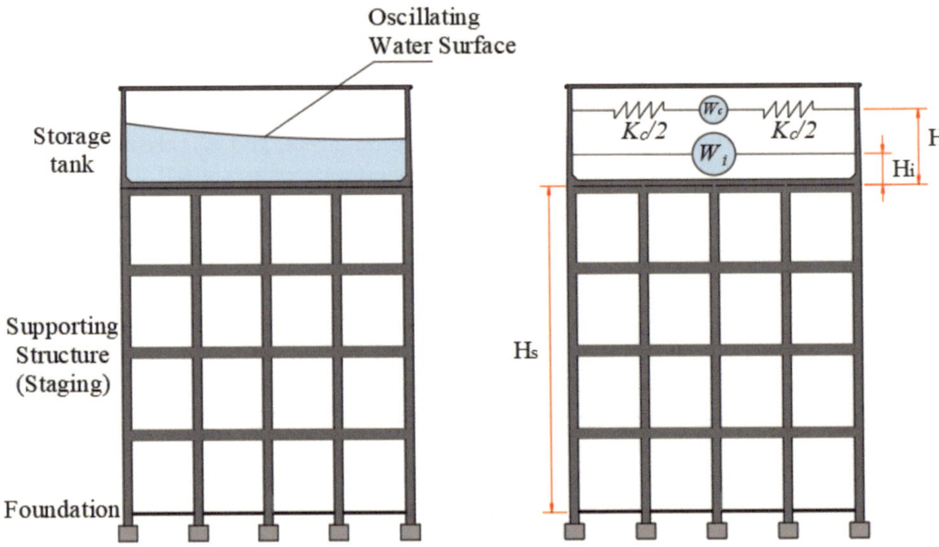

Figure 3.
The spring mass model representation of an EWT.

Equations for the parameters of the spring-model		
The impulsive and convective weights (W_i and W_c, respectively) can be obtained using Eqs. (3) and (4) where, D is the inside diameter of a circular tank, and H_L is the maximum water level.	$\frac{W_i}{W_l} = \frac{\tanh\left[0.866\left(\frac{D}{H_L}\right)\right]}{0.866\left(\frac{D}{H_L}\right)}$	(3)
	$\frac{W_c}{W_l} = 0.23\left(\frac{D}{H_L}\right)\tanh\left[3.68\left(\frac{H_L}{D}\right)\right]$	(4)
The heights of the impulsive and convective masses from the bottom of the tank wall can be determined from Eqs. (5)–(7) where, H_i is the height of the center of gravity of the impulsive mass measured from above the base of the tank wall, and H_c is the height of the center of gravity of the convective mass measured from above the base of the tank wall.	$\frac{H_i}{H_L} = 0.5 - 0.09375\left(\frac{D}{H_L}\right)$ for $\frac{D}{H_L} < 1.333$	(5)
	$\frac{H_i}{H_L} = 0.375$ for $\frac{D}{H_L} > 1.333$	(6)
	$\frac{H_c}{H_L} = 1 - \frac{\cosh\left[3.68\left(\frac{H_L}{D}\right)\right] - 1}{3.68\left(\frac{H_L}{D}\right)\sinh\left[3.68\left(\frac{H_L}{D}\right)\right]}$ for all tanks	(7)
The stiffness of convective mode can be obtained by Eqs. (8)–(10) where, λ is the circular frequency coefficient, g is the gravitational acceleration taken as 9.81 m/s^2, ω_c is the circular frequency of oscillation of the first sloshing mode (convective mode), and K_c is the spring stiffness of convective mode.	$\lambda = \sqrt{3.68g\tanh\left[3.68\left(\frac{H_L}{D}\right)\right]}$	(8)
	$\omega_c = \frac{\lambda}{\sqrt{D}}$	(9)
	$K_c = \frac{W_c}{g}{\omega_c}^2$	(10)
Alternatively, the stiffness of the convective mode can be obtained using the combined Eq. (11)	$K_c = 3.68\frac{W_c}{D}\tanh\left(3.68\frac{H_L}{D}\right)$	(11)

Table 2.
The equations and calculation procedure of the spring-mass model parameters.

(i.e., variable staging patterns and contained liquid fill levels) using the performance-based earthquake engineering methodology. The study utilized a developed seismic evaluation tool – the collapse margin indicator – to investigate the dynamic behavior of frame supported EWTs and considered the FSI effect by adding lumped masses that are connected to the tanks' walls either rigidly or elastically through oscillators as shown in **Figure 4**.

3.5 The three-mass model

Earlier studies considered rigid tank walls when evaluating the hydrodynamic pressure induced by ground motion records [4, 18, 26]. However, following a series of powerful earthquakes in Japan and the United States that caused severe damage to liquid storage tanks, it was realized that modeling tanks using the rigid-tank concept is insufficient since real tanks experienced significant deformation when subjected earthquake loads. Subsequently, multiple studies were conducted, and it was established that accounting for the tank flexibility and the interaction between the contained fluid and the vibration of the walls can significantly affect the hydrodynamic pressure and consequently the impulsive component of the structural response [27, 28]. Over time, assumptions concerning tank properties have been refined progressively to take account of the tank deformability and flexibility of the container and soil interaction effects [27].

Haroun and Housner [27] proposed the three-mass model representation for cylindrical tanks subjected to seismic loading as illustrated in **Figure 5**. The three masses in this equivalent mechanical model correspond to the impulsive mass, the convective mass, and the mass representing the tank wall's flexibility. In the following studies, Haroun and Ellaithy [23] implemented the three-mass model to evaluate the dynamic response of EWT and to assess the influence of the tank walls' flexibility on the dynamic behavior of EWTs. The effect of higher modes of convective masses on the pressure exerted on the vessel may be not significant, even when the fundamental frequency of the structure is close to the natural frequency of convective mode. A later study by Jaiswal et al. [29] show negligible differences in the parameters of the equivalent spring-mass mode obtained from rigid and flexible tank wall.

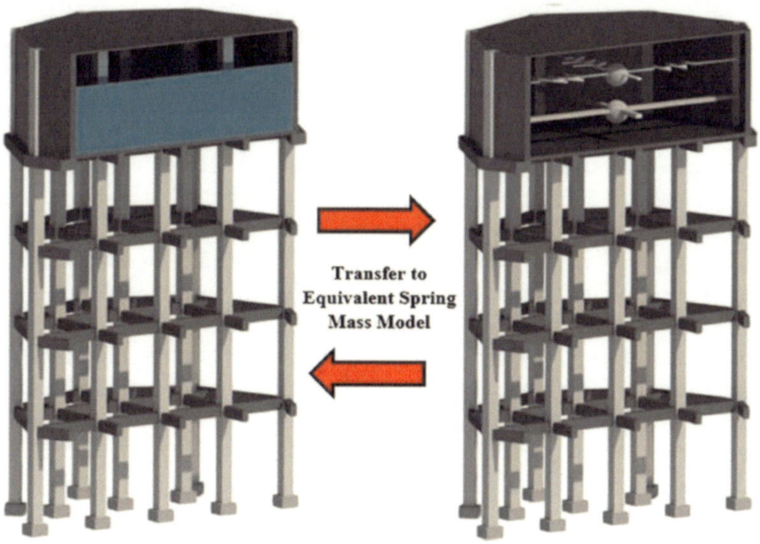

Figure 4.
The spring-mass model for an EWT [1].

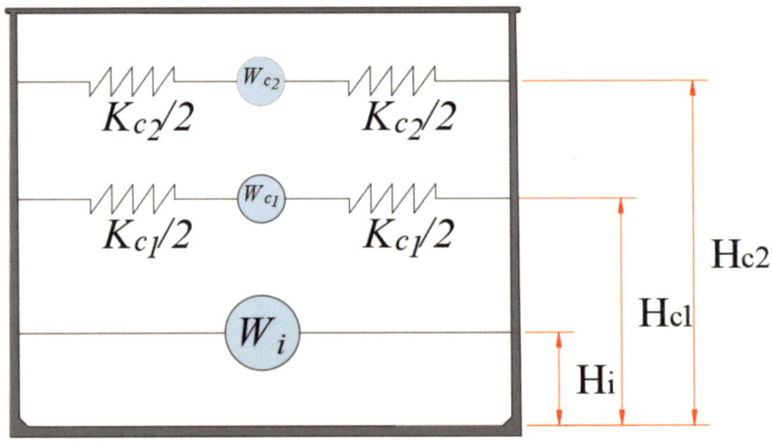

Figure 5.
Equivalent 2D system for liquid-filled storage tank with flexible tank wall behavior.

Some studies have identified the hydrodynamic pressure developing inside deformable cylindrical tanks. Haroun and Housner [27] analyzed the response of flexible liquid-containing tanks using modal superposition. The tank's walls were modeled as shell elements using the finite element method and the fluid domain was considered using a mathematical boundary solution technique. Previous studies showed that the flexibility of tank causes it to experience rocking wall and base translation, which result in longer impulsive periods and increased effective damping. However, due to convective mode having long period of oscillation, the convective mass can be computed without considering the tank wall and supporting soil flexibilities [30]. Using two-dimensional space FE modeling, Ghaemmaghami and Kianoush [31] examined the seismic behavior of two different tank configurations, tall and shallow, while taking the effects of FSI and wall flexibility into account. The results show that incorporating the fluid damping properties and the wall flexibility can drastically affect the dynamic response of the liquid tanks.

3.6 Other equivalent models

While simplified models, such as those developed by Housner [4] and Haroun and Housner [27], generate a dynamic response similar to that of a continuum liquid 3D-tank-model [32], it may not, however, take into account certain aspects that affect the accuracy of the analysis results. In a recent study, Papadrakakis and Fragiadakis [33] investigated the seismic performance of unanchored liquid-storage tanks having variable tank diameters and liquid-filling heights using two nonlinear FE computational methods; coupled Eulerian-Lagrangian and spring-mass analogy. Results show that the traditional equivalent masses-springs analogy does not consider the effect of uplifting history for ground unanchored liquid tanks and its influence on the tank's dynamic behavior.

Studies by Sweedan and El Damatty [34] and El Damatty et al. [35] verified the application of the previously established analytical and numerical models on combined conical tanks by experimentally identifying their dynamic characteristics. In order to further improve the seismic study of EWTs, Sweedan [36] suggested a mechanical model to duplicate forces produced in combined EWTs experiencing

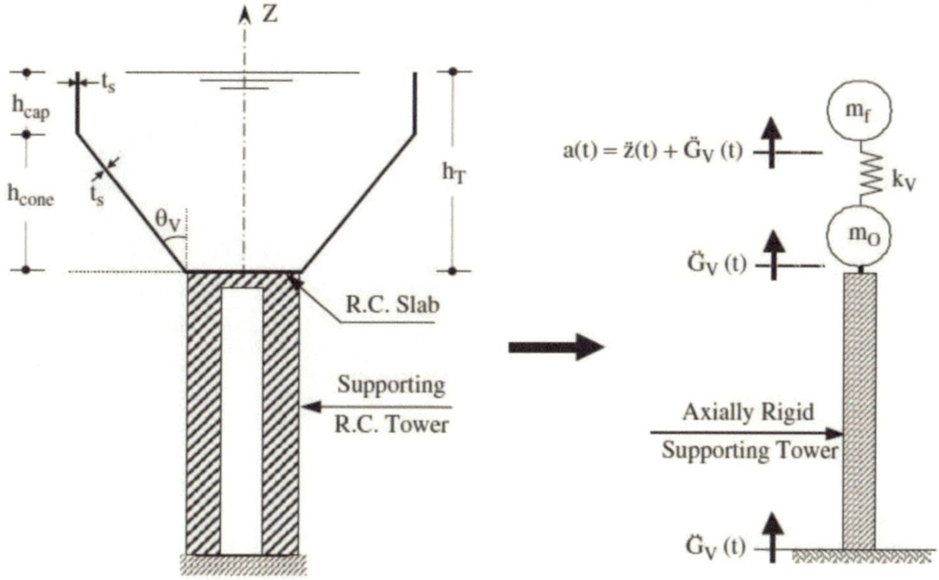

Figure 6.
Equivalent model for vertically excited combined tanks proposed by Sweedan [36].

vertical ground excitation. A schematic of the equivalent idealization is shown in **Figure 6**.

Vathi and Karamanos [37] studied the base uplifting behavior of ground cylindrical liquid tanks subjected to strong horizontal seismic excitations. A simplified liquid-tank model was developed using the spring-mass model improved by an appropriate rotational spring at its base to take into consideration the tank's rotation, or rocking, by the impulsive motion due to uplifting (**Figure 7**). The results from this study mark a significant influence of tank base uplifting on the dynamic response of unanchored tanks.

Algreane et al. [38] introduced an alternative impulsive masses configuration to the dynamic behavior of reinforced concrete EWTs. The proposed model suggests

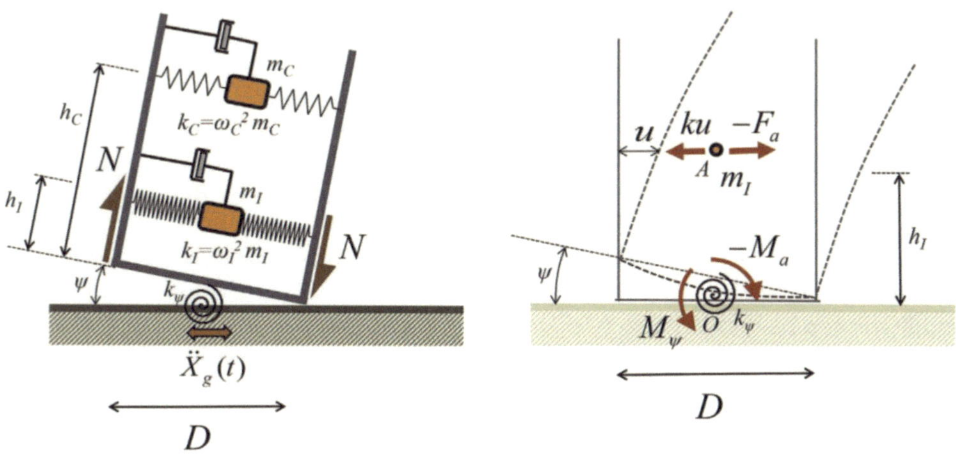

Figure 7.
A simplified model for an unanchored liquid storage tank accounting for base uplifting [37].

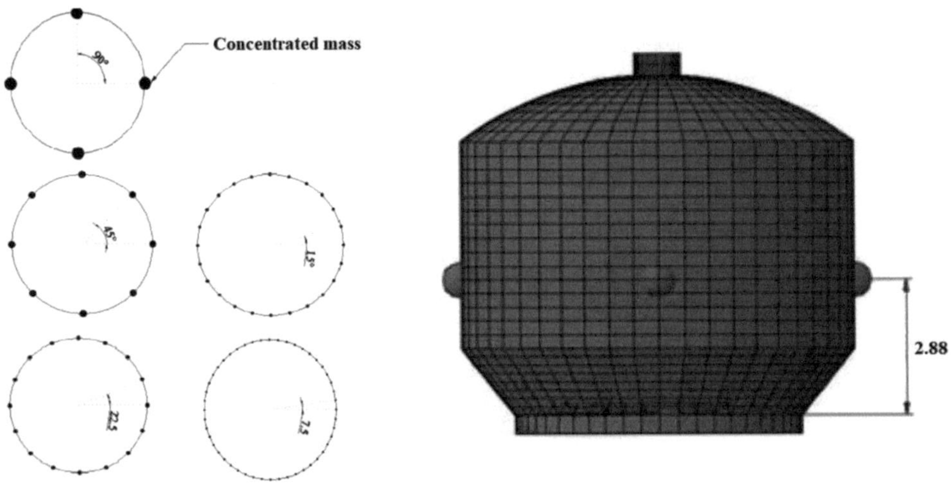

Figure 8.
Alternative masses distribution proposed by Algreane et al. [38].

the distribution of the impulsive mass by different alternative configurations in an effort to simplify Westergaard's AMA technique and reduce computational time. The impulsive mass is divided into 4, 8, 16, 24 and 48 masses, and distributed into wall panels of the tank at the center of gravity of an empty container as shown in **Figure 8**.

4. Conclusion

The functionality of liquid-storage tanks should be ensured during and after natural disasters, e.g., earthquakes, such that under intense ground motions the structural collapse is prevented. Therefore, it is crucial to quantify the safety margin against the structural collapse state for water tanks, which are considered as lifeline structures. Furthermore, the literature shows that the dynamic behavior of these structures is governed by many factors including the tank shape, fluid properties, structural flexibility. Soil characteristics, and the type of supporting structure.

Researchers tend to prefer simple and straightforward modeling techniques for the FE analysis of liquid-tank systems. Different simplified modeling approaches that consider the FSI effects can be used to reduce the complexity and computational cost of liquid-tank systems. The dynamic behavior of liquid-tank systems that is obtained using these modeling techniques can have high accuracy that can match that obtained from continuum liquid-tank systems. A simplified single-mass model could be used if the weight of water equals or exceeds 80% of the overall weight of the liquid-tank system. In the two-mass model representation of liquid-tank system, the hydrodynamic pressure developing within the liquid resulting from the dynamic motion of the liquid tank can be divided into two parts. The liquid mass in the top zone of the tank, called the convective mass, characterizes the motion of the free-liquid-surface. The liquid in the bottom zone of the tank, called the impulsive mass, represents the remaining mass of the fluid and the mass of the structure. A FE model can be represented by a spring-mass model based on Housner's analogy in which the liquid is replaced by two lumped masses: the impulsive and convective masses. The impulsive mass is connected to the tank walls using rigid links and the convective mass is connected through elastic springs. This modeling approach is effective in reducing the reduce the computational cost and complexity of liquid-tank systems while resulting in moments and forces that are comparable to that obtained from continuum liquid-medium models subjected to the same ground motion records.

Acknowledgements

The authors declare that they have no known competing financial interests that could have appeared to influence the work reported in this paper.

Author details

Ayman Mohammad Mansour and Moustafa Moufid Kassem*
School of Civil Engineering, Universiti Sains Malaysia, Penang, Malaysia

*Address all correspondence to: moustafa.mk@usm.my

References

[1] Mansour AM, Kassem MM, Nazri FM. Seismic vulnerability assessment of elevated water tanks with variable staging pattern incorporating the fluid-structure interaction. Structures. 2021; **34**:61-77

[2] Wilson EL. Three-Dimensional Static and Dynamic Analysis of Structures. Computers and Structures Inc.; 2002

[3] Jacobsen LS, Ayre RS. Hydrodynamic experiments with rigid cylindrical tanks subjected to transient motions. Bulletin of the Seismological Society of America. 1951;**41**(4):313-346

[4] Housner GW. The dynamic behavior of water tanks. Bulletin of the Seismological Society of America. 1963; **53**(2):381-387

[5] Veletsos A. Seismic response and design of liquid storage tanks. Guidelines for the seismic design of oil and gas pipeline systems. 1984:255-370

[6] Mansour AM, Nazri FM. On the influence of fluid–structure interaction and seismic design on frame-supported elevated water tanks. Structural Engineering International. 2021;**29**:1-15

[7] Shakib H, Alemzadeh H. The effect of earthquake site-source distance on dynamic response of concrete elevated water tanks. Procedia engineering. 2017; **199**:260-265

[8] Elansary AA, El Damatty AA. Seismic analysis of liquid storage composite conical tanks. Engineering Structures. 2018;**159**:128-140

[9] Moslemi M, Farzin A, Kianoush M. Nonlinear sloshing response of liquid-filled rectangular concrete tanks under seismic excitation. Engineering Structures. 2019;**188**:564-577

[10] Grant MJ, Booth A. A typology of reviews: An analysis of 14 review types and associated methodologies. Health Information & Libraries Journal. 2009; **26**(2):91-108

[11] Livaoğlu R, Doğangün A. Simplified seismic analysis procedures for elevated tanks considering fluid–structure–soil interaction. Journal of Fluids and Structures. 2006;**22**(3):421-439

[12] Wilson EL, Khalvati M. Finite elements for the dynamic analysis of fluid-solid systems. International Journal for Numerical Methods in Engineering. 1983;**19**(11):1657-1668

[13] Lu D, Liu Y, Zeng X. Experimental and numerical study of dynamic response of elevated water tank of AP1000 PCCWST considering FSI effect. Annals of Nuclear Energy. 2015; **81**:73-83

[14] Chandrasekaran A, Krishna J. Water towers in seismic zones. Proc. Third World Conf Earthquake Eng, New Zealand. IV. 1954

[15] ACI-371R. Guide for the analysis, design and construction of elevated concrete and composite steel-concrete water storage tanks. 2008

[16] Haroun M, Housner G. Earthquake response of deformable liquid storage tanks. 1981a

[17] ACI-350.3. Seismic Design of Liquid-Containing Concrete Structures (ACI 350.3-06) and Commentary (ACI 350.3R-06). Farmington Hills, MI: American Concrete Institute; 2006

[18] Housner GW. Dynamic pressures on accelerated fluid containers. Bulletin of

the Seismological Society of America. 1957;**47**(1):15-35

[19] Ibrahim RA. Liquid sloshing dynamics: Theory and applications. Cambridge University Press; 2005

[20] EC8. Design of structures for earthquake resistance—Part 1: General rules, seismic actions and rules for buildings (EN 1998-1: 2004). European Committee for Normalization, Brussels. 2004;**11**:17

[21] NZS. Code of practice for concrete structures for storage of liquids. New Zealand Standard. 1986;**3106**:1986

[22] Haroun MA. Stress analysis of rectangular walls under seismically induced hydrodynamic loads. Bulletin of the Seismological Society of America. 1984;**74**(3):1031-1041

[23] Haroun MA, Ellaithy HM. Seismically induced fluid forces on elevated tanks. Journal of Technical Topics in Civil Engineering. 1985; **111**(1):1-15

[24] Shepherd R. The two mass representation of a water tower structure. Journal of Sound and Vibration. 1972;**23**(3):391-396

[25] Dutta SC, Dutta S, Roy R. Dynamic behavior of R/C elevated tanks with soil–structure interaction. Engineering Structures. 2009; **31**(11):2617-2629

[26] Jacobsen LS. Impulsive hydrodynamics of fluid inside a cylindrical tank and of fluid surrounding a cylindrical pier. Bulletin of the Seismological Society of America. 1949; **39**(3):189-204

[27] Haroun MA, Housner GW. Seismic design of liquid storage tanks. Journal of the Technical Councils of ASCE. 1981b; **107**(1):191-207

[28] Haroun MA, Housner GW. Dynamic characteristics of liquid storage tanks. Journal of the Engineering Mechanics Division. 1982;**108**(5):783-800

[29] Jaiswal O, Rai DC, Jain SK. Review of code provisions on seismic analysis of liquid storage tanks. IITK-GSDMA Project on Building Codes, IITK-GSDMA-EQ04-V1.0. 2004

[30] Malhotra PK, Wenk T, Wieland M. Simple procedure for seismic analysis of liquid-storage tanks. Structural Engineering International. 2000;**10**(3): 197-201

[31] Ghaemmaghami A, Kianoush M. Effect of wall flexibility on dynamic response of concrete rectangular liquid storage tanks under horizontal and vertical ground motions. Journal of Structural Engineering. 2010;**136**(4): 441-451

[32] Sezen H, Hur J, Smith C, Aldemir T, Denning R. A computational risk assessment approach to the integration of seismic and flooding hazards with internal hazards. Nuclear Engineering and Design. 2019;**355**:110341

[33] Erkmen B. Evaluation of code provisions for seismic performance of unanchored liquid storage tanks. In: Proceedings of the 6th ECCOMAS Thematic Conference on Computational Methods in Structural Dynamics and Earthquake Engineering (COMPDYN 2017), Rhodes Island. 2017

[34] Sweedan A, El Damatty A. Experimental identification of the vibration modes of liquid-filled conical tanks and validation of a numerical model. Earthquake engineering &

structural dynamics. 2003;**32**(9):1407-1430

[35] El Damatty A, Saafan M, Sweedan A. Experimental study conducted on a liquid-filled combined conical tank model. Thin-Walled Structures. 2005;**43** (9):1398-1417

[36] Sweedan AM. Equivalent mechanical model for seismic forces in combined tanks subjected to vertical earthquake excitation. Thin-Walled Structures. 2009;**47**(8–9):942-952

[37] Vathi M, Karamanos SA. A simple and efficient model for seismic response and low-cycle fatigue assessment of uplifting liquid storage tanks. Journal of Loss Prevention in the Process Industries. 2018;**53**:29-44

[38] Algreane GA, Osman SA, Karim OA, Kasa A. Dynamic behaviour of elevated concrete water tank with alternate impulsive mass configurations. WSEAS International Conference. Proceedings Mathematics and Computers in Science and Engineering, WSEAS. 2009

Chapter 5

Perspective Chapter: Defining and Applying the FMEA Process Method in the Field of Industrial Engineering

Cristina-Ileana Pascu, Raluca Malciu and Ilie Dumitru

Abstract

The analysis of failure modes and effects (FMEA) is a method of analyzing the potential failure of a product or process and developing an action plan aimed at their prevention and increased quality of products, processes, and job production environments. As a method of critical analysis, FMEA has very clear objectives: determination of the weaknesses of a technical system; initiating causes of failure-seeking components; analysis of the environmental impacts, safety of operation, the product value; provision of corrective actions to remove the causes of the occurrence of defects; provision of a plan to improve product quality and maintenance; determining the needs of technology and modernization of production; increasing the level of communication between departments of working people at hierarchical levels. FMEA should be used before taking the product. Subsequently, there is no point, only because the customer demands it, to achieve FMEA. Therefore, FMEA must be within organizational conduct. This chapter describes the FMEA method and presents studies about the improvement of the quality process for some products from industrial engineering by using FMEA, such as: axis for packaging, assembly "stator Housing," composite parts used in the railway field. The potential causes of the defects were studied, and improvement measures were proposed.

Keywords: FMEA, process, quality, occurrence, improving, RPN

1. Introduction

1.1 Method definition

Worldwide, due to customer expectations coupled with the continuing increase in the complexity of products, declining design and launch periods have necessitated systematic quality planning [1]. In the last decade, quality control and improvement have become a priority in the development strategy of companies in all fields: industry, distribution, transport companies, health, government agencies, financial organizations, etc. Achieving and maintaining a high level of quality of products or services offer a

IntechOpen

competitive advantage that allows a company to dominate its competitors in the field in which it operates. Thus, a company can dominate its competitors by continuously improving processes and applying quality control [2]. The application of quality control at all stages of product process fabrication is considered a zero priority in the automotive field [3]. The premise of traditional quality assurance, based on the detection and elimination of defective products, is no longer relevant, the argument being extremely simple and intuitive: defects that can be avoided before launching the product do not need to be corrected later [4]. Modern methods of systemic quality design are the answer to new quality requirements. They must allow the analysis and elimination of potential defects from the design and implementation stage [5], so that more and more often the notion of "quality design" is encountered [6]. One of these methods that has become increasingly used in recent decades is the failure modes, effects, and analysis (FMEA) [7, 8].

One of the most widely used methods in the field of quality engineering is failure modes, effects, and analysis (FMEA), with applicability from the manufacturing design stage to the prototyping and zero series production stage [9].

The FMEA is a systematic method of determining and preventing errors, defects and risks that may occur, applicable to a process, product or equipment used in the process. This method consists of detecting possible defects, inventorying the causes that could produce these defects, the effects of defects on users, in order to plan the necessary measures to prevent their occurrence [10]. The English name of the method has a correspondent both in Romanian and in French, where it is called L'Analise des Modes de Defaillance, de leur Effet et de leur Criticite (AMDEC), or in German DAMUK [11]. This method consists of detecting possible defects, inventorying the causes that could cause these failures, demerits effects on users, in order to plan the necessary measures to prevent their occurrence.

The method was originally developed by the US military, as evidenced by the 1949 MIL-P-1829 military procedure entitled "Procedures for Failure, Effects and Critical Analysis" applicable to projects aimed at ensuring the maximum availability of strategic military equipment [12, 13].

The first notable applications of AMDEC techniques are related to NASA (1960s), and later, in the 1990s, by the top three US automakers: GM, Ford, and Chrysler by including them in the prescriptions of the QS 9000 quality standard [14], **Figure 1**.

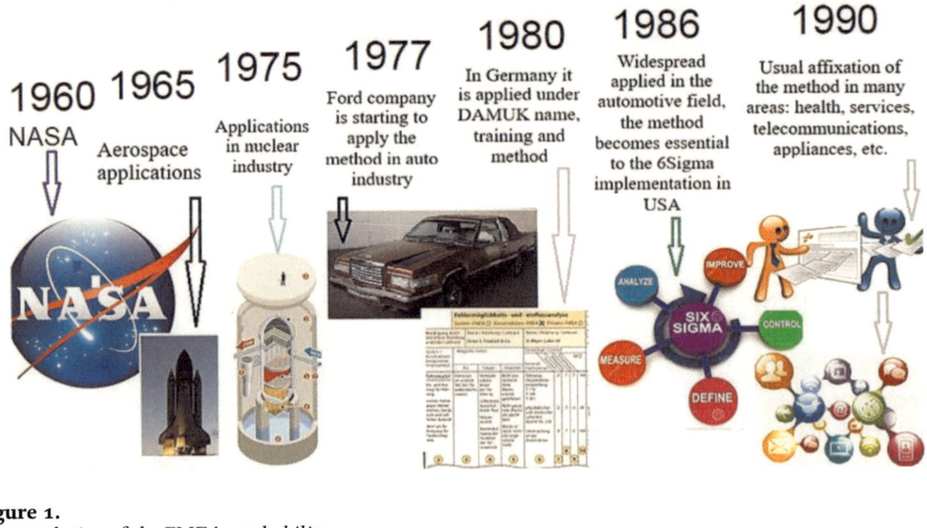

Figure 1.
Time evolution of the FMEA workability.

From **Figure 1** it can be seen that since the 1970s, the FMEA method has been used in the aerospace and nuclear industry. Since 1977, the American company Ford has been widely applying the FMEA method for solving car projects, and since 1986, with the launch of 6Sigma (6σ) technique, the method has become one of the basic tools used in the US automotive industry [14].

According to ref. [15], the French claim the discovery of this method in 1995, by applying a previous AMDEC method to shorten the production time of the prototypes for the mirage fight planes.

In 1980, Renault adopted the AMDEC method to make the Clio car, which became the first feasible project obtained by applying the method to both the process and the assembly, and since 1984, the French car manufacturer has adapted and improved the method using various other names: AMDEC CONNECTIQUE, DELTA 2, etc.

The correct use of the method allows the analysis and minimization of potential risks. By applying it, all types of defects and/or potential defects in terms of causation and effects are predicted.

Once the appropriate and correct actions have been implemented, the causes of the defects can be hampered or even avoided. The FMEA is considered to be an effective tool for quality assurance prevention.

1.2 The purpose and objectives of FMEA

The purpose of the FMEA is to ensure the development of high-quality products from the design and prototyping phases, before the transition to series production.

The FMEA provides the necessary preconditions for early detection of quality problems and the prevention of their occurrence through appropriate measures.

This makes it possible to meet product quality requirements and, at the same time, reduce the costs of errors and the consequences of errors. The objectives of the FMEA are shown in **Figure 2**.

Fault analysis and effects analysis (FMEA) is a systemic analysis of potential failures of a product, process, or machine used during the process, with the goal of developing an action plan to prevent their occurrence and to improve the quality of the products, work processes, and production environments.

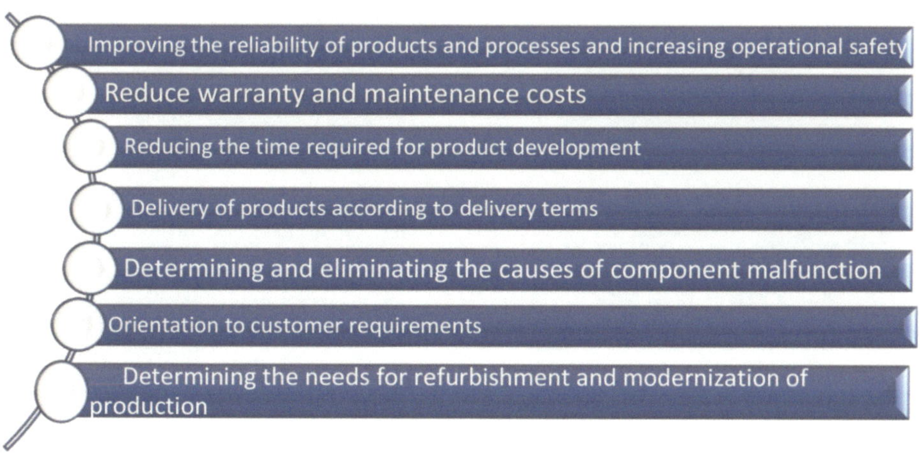

Figure 2.
FMEA objectives.

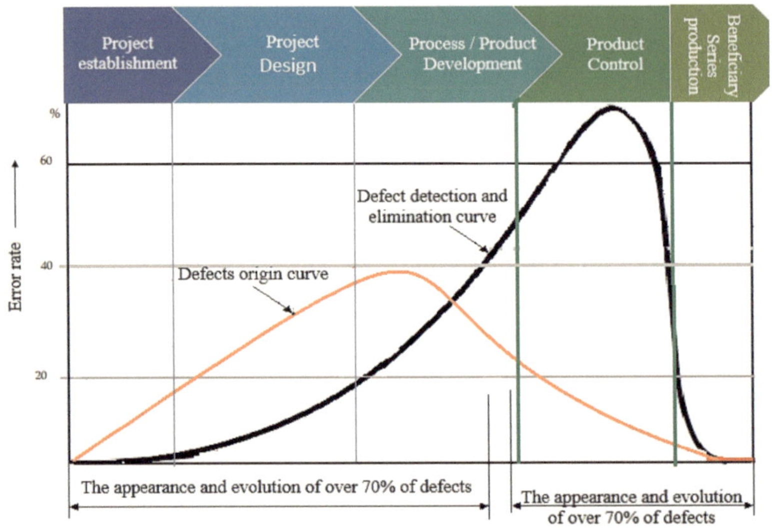

Figure 3.
The relationship between the origin of defects and their detection during the manufacturing cycle of a product.

We start from the elements to determine the triplet Cause – Mode – Effect. **Figure 3** shows the relationship between the origin of the defects occurring and their detection during the manufacturing process of a product [16].

It is mentioned that the FMEA method highlights possible risks, but does not solve the problem. However, the correct approach of the FMEA can result in "zero defects" but not "zero errors."

1.3 FMEA types

There are two main types of FMEA [17]: product design, DFMEA, and process development, PFMEA, **Figure 4**.

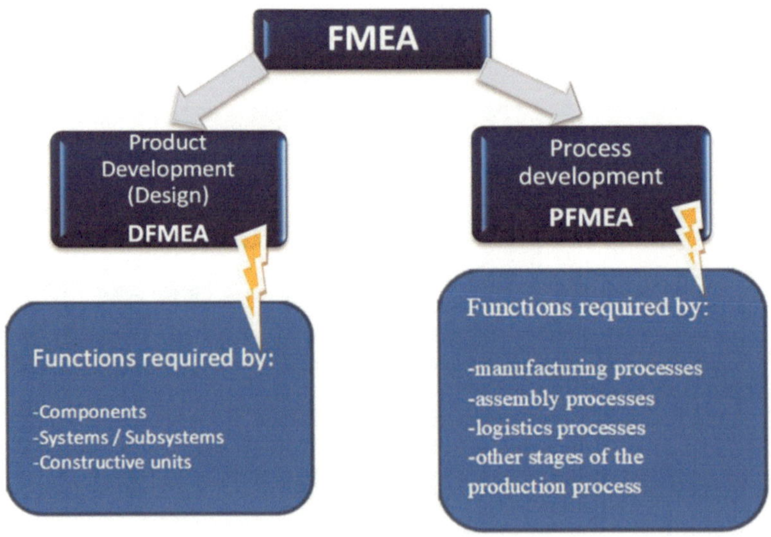

Figure 4.
Main types of the FMEA method.

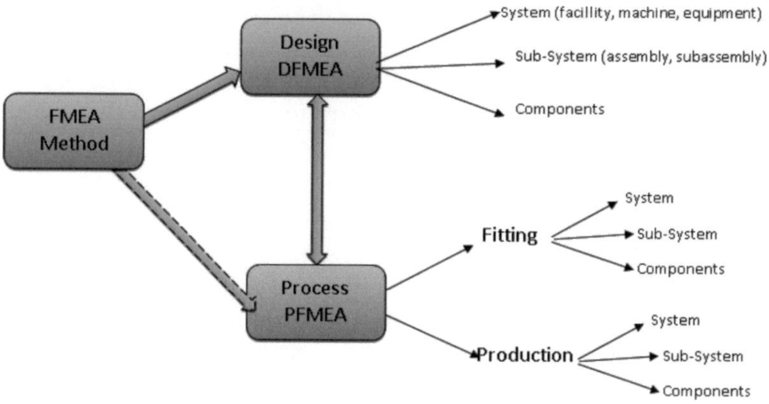

Figure 5.
DFMEA and PFMEA.

Accordingly to the AIAG, VDA manuals [18, 19], in addition to the two types of FMEA listed above (DFMEA and PFMEA), explained in **Figure 5**, the following are also known and used:

- FMEA system focused on the study of the functions of the components of the subsystem, machine, machinery, or technological equipment;
- FMEA service that is used to analyze the functions of the service;
- FMEA software focused on studying the functions of software and computer components.

However, the FMEA method must be performed whenever errors or malfunctions may occur that could cause potential harm to the user (customer) for whom the product is intended.

The scope of the FMEA method is wide and diverse. This method can be implemented not only in industry but also in the field of services.

The FMEA is aimed at [13, 18]:

- product-project;
- product-process;
- machine/work equipment or system.

The product-design FMEA is applied immediately after the elaboration of the constructive design documentation (design) in order to follow and analyze the products from the conception and design stage.

The product-process FMEA is performed after the preparation of the technological documentation of the process necessary to make the product. It also allows the validation of the technological process of making a product in accordance with the desired quality and efficiency expectations.

FMEA machine/work equipment is executed after the system of machines, machines, and technological equipment is established. Based on that, the technological process consists of the analysis of the means of production aiming to reduce the number of scrap, failure rate and increase availability and reliability of the product. The development of the FMEA lies in the inventory of the way of detecting errors, component problems, analyzing the causes of occurrence, and evaluating their effects on the set of functions of the system.

1.4 Fields of application of the FMEA method

In the current context, the need to apply the FMEA method derives from the quality requirements, with the emphasis shifting from detecting noncompliant products to preventing errors and defects before prototyping the final product or switching to series production.

As can be seen in **Figure 6**, FMEA is a method of preventive quality recommended in the IATF16949: 2016 standard, AIAG, VDA, norms and regulations manuals, being a basic tool for 6Sigma.

The FMEA method is applicable to, **Figure 7**:

- products, parts, with quality problems;
- processes with non-conformities;
- modernization of an existing technology or implementation of a new technology;
- products that require a high level of security;
- launching a new type of product or process;
- evaluation of the probability of occurrence of errors and/or failures, at important components from the point of view of the safety of the whole;

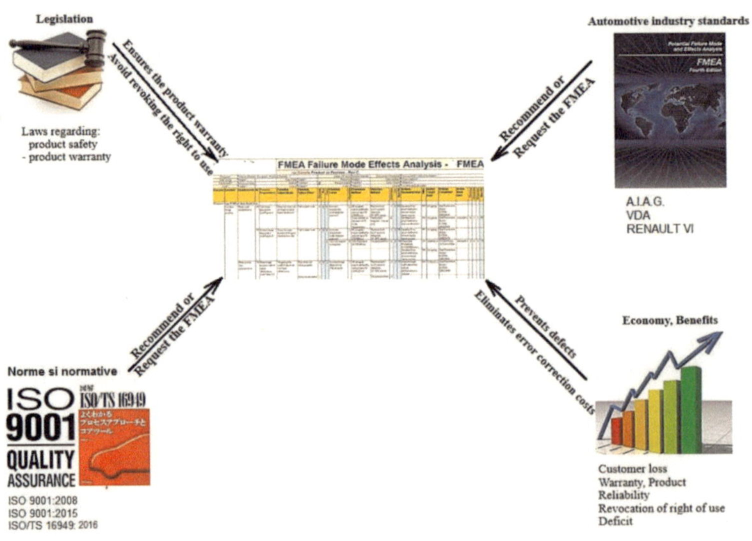

Figure 6.
The requirement to apply the FMEA method.

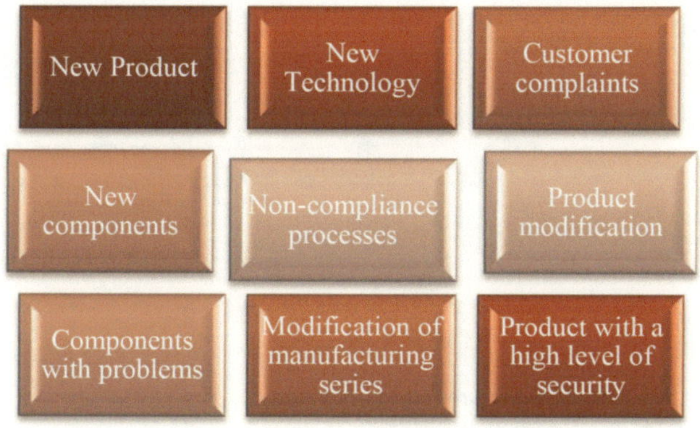

Figure 7.
The area of FMEA applicability.

- adapting products to new operating conditions;
- modification of manufacturing series.

By applying the FMEA method, the risk of non-conformities in the design and manufacture of products is reduced. The implementation of the method allows to reduce costs in all stages of the quality spiral, **Figure 8**: in design, by better reflecting customer expectations in design quality; in supply, by avoiding difficulties due to improper selection of suppliers; in manufacturing, by preventing the occurrence errors and defects and avoidance of critical points in the field of service, by reducing customer grievances and complaints, etc.

1.5 Methodology for FMEA analysis implementation

FMEA must be performed before the product prototype is made, **Figure 9**. No sense in applying FMEA afterward, upon the request of the client.

Figure 8.
The spiral of quality.

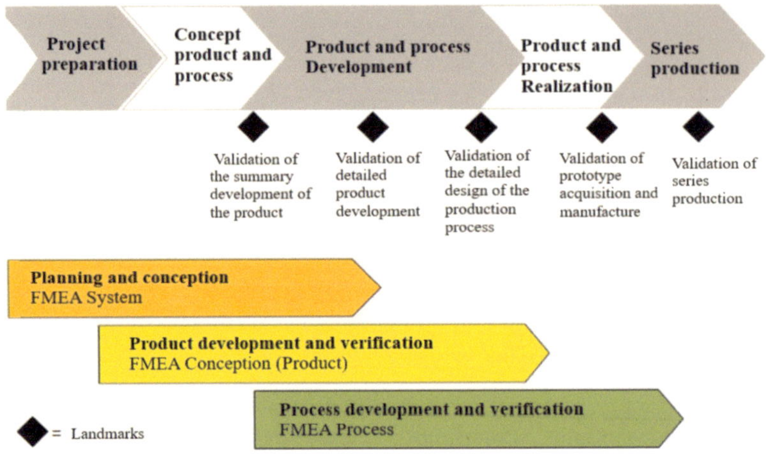

Figure 9.
Validation of FMEA analysis.

The FMEA must be performed before the product is prototyped as it is a living, dynamic organism that must be updated throughout the life of the product or process.

This is the reason why the approach and implementation of the FMEA must be part of the organizational development of the product manufacturing, starting before prototyping and ending before the launch of series production.

In carrying out the FMEA method, the main "5 steps" must be completed.

These important "5 steps" in the development and implementation of the FMEA are, **Figure 10**:

Step 1—The elaboration of the structure of the system consists in the description of the system, with the definition of its component elements and the structural determination of the whole.

Step 2—The introduction of functions in the system structure involves the functional description, the structural determination of the system with functions, the correlation of functions, which will lead to obtaining a network of functions, respectively, the formation of a functional structure.

Step 3—Defects and faults analysis consists of the introduction of the "failure" functions in the system structure and correlating these functions, obtaining a "network of failures," respectively, the functional structure of the faults.

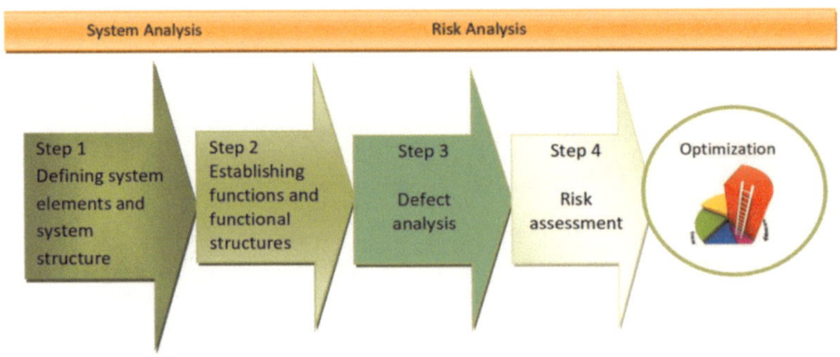

Figure 10.
Stages of FMEA application.

Step 4—The risk assessment lies in the FMEA form through avoidance and detection measures, with the assessment form being completed.
Step 5—Optimization is the stage in which changes are processed in the FMEA form, the risks of errors and defects are reduced by new measures, the persons responsible and the deadlines in which the recommended measures must be organized and carried out are designated, and the optimization form is made.

1.6 The FMEA team

The FMEA research is carried out in interdisciplinary groups in which the departments involved in the realization of the product participate under the leadership of a moderator.

The FMEA moderator is usually a person from outside the company or from the Quality Assurance department.

Tasks of the FMEA moderator:

- Participate in FMEA planning.
- FMEA preparation/organization.
- Moderation of the work team.
- Documentation of the analysis.
- Evaluation and presentation of FMEA results.
- Ensuring methodological correctness.
- Participate in improving the efficiency of the FMEA.
- Exchange of FMEA experience.

Employees of the design-development, manufacturing process planning, manufacturing, control, customer service, and quality assurance departments generally participate in the implementation of the FMEA, and their number will not exceed 6–8 people.

This ensures that all departments involved in the manufacturing of their products bring their experience in the analysis. The success of the FMEA largely depends on the creativity of the team.

The product is systematically broken down by a "top-down" process into components or functions and subsequently researched to meet the constructive requirements and to maintain these requirements during manufacture. The systematic analysis procedure is supported by using an appropriate form.

In order to perform an FMEA analysis, the operation of the analyzed system must be very well known, or the appropriate means must be available to obtain the necessary information from the owners.

The FMEA team consists of, **Figure 11**:

- The project coordinator (decision-maker) is the person with responsibility in the company who has the power to exercise a final choice. They will make the final decisions regarding cost, quality, and deadlines:

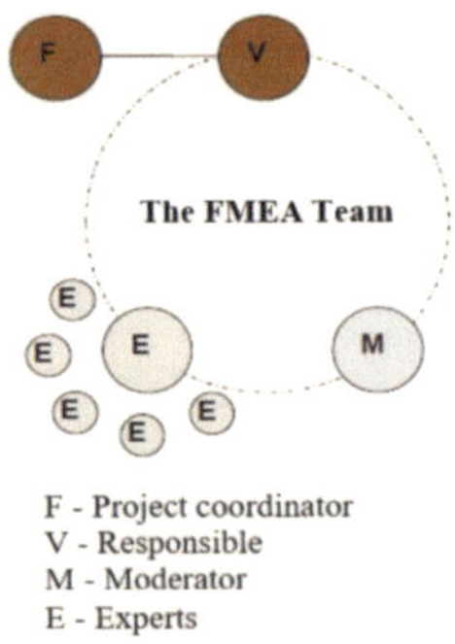

Figure 11.
The FMEA team.

- Enforcement decisions
- Information collection support
- FMEA approval

Team members should, as required, provide the following:

- The person in charge (Initiator) with the FMEA analysis is the person or service that has the initiative to start the study and to choose the subject of the analysis. In general, it has the following tasks:
- Co-participation in FMEA training
- Contributing information based on experience gained from already known processes
- Co-participation in the selection of measures

The moderator (animator) is the guarantor of the method, the organizer of the group activity. He specifies the agenda of the meetings, conducts the meetings, provides the secretariat, and monitors the progress of the study.

Often, it is a person outside the company, or at least outside the department, to animate the members of the group.

These first three people generally do not have precise technical skills.

Experts (analysis group) in number of 2–5 people will be people responsible and competent, with good knowledge of the studied system and who could bring the necessary information for the analysis (one can only discuss what is well known).

Depending on the study, the following will be:

- maintenance service staff;
- staff of the quality assurance department;
- production operators;

- members of the design offices;
- experts in the field studied.

Their tasks are:

- Presentation within the team of the project/development stage
- Contributing with information based on experience gained from already known processes
- Introduction of optimization measures established in the project stage

The FMEA designated design team consists of 5–8 people.

Figure 12 shows the phases necessary for the preparation and planning stage of the FMEA analysis session.

1.7 Stages of FMEA application

The necessary steps for the development and implementation of the FMEA method are presented in **Figure 13** [13].

Identification of the functions of the product/process subject to FMEA analysis

When applying the FMEA to the product, identify the functions of the product, part, or component considered for the study.

In relation to these functions, potential failures are determined, assessing their severity (criticality).

The causes of errors and malfunctions are then determined, and measures are taken to prevent them from occurring. The Ishikawa Diagram or the "5 Why" technique is usually used to identify the causes of errors or malfunctions.

The application of the FMEA process involves, in a first stage, the description of the process functions. Starting from these functions, the potential faults are identified, and the critical stages of the process are highlighted. The necessary corrective measures shall be taken to prevent damage.

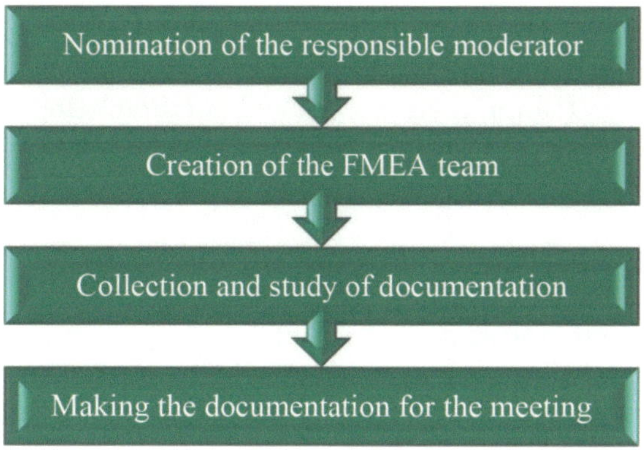

Figure 12.
The necessary steps for the preparation and planning phase of the FMEA analysis session.

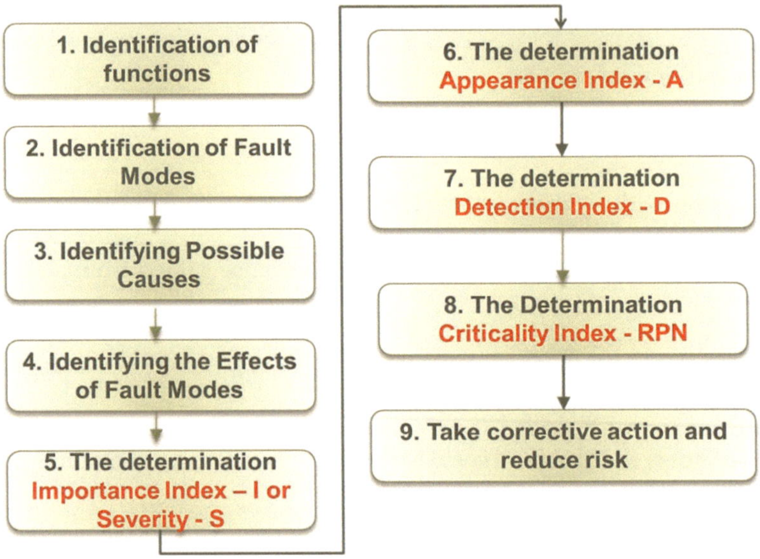

Figure 13.
Steps required to develop and implement the FMEA method.

Identifying fault modes consists of inventorying all possible faults of the product, part, component (DFMEA), or process (PFMEA) and establishing fault modes. This is usually done by specialists, but in some cases it is possible to use working groups, capitalizing on the experience gained in the field by the group members (company workers, workers serving the equipment used during the technological process). The modes of failure can be multiple: wear, deformation, corrosion, rupture, buckling, crushing, etc.

1.8 Assessing the effects and importance (criticality) of failures

Defects are usually assessed depending on two criteria: probability of occurrence (A) and probability of detection (D), which are expressed using the same notation scale.

The quantification of these probabilities depends on the type of product or process analyzed.

Severity (importance) means the value that characterizes how serious the established effect is for the failure mode and how it affects the customer, in terms of product/process failure/failure, their effects, and notation being presented in **Table 1**.

In assessing the significance (severity) of the faults, the following general rules must be observed:

- the importance of a failure is that for all potential causes of failure;
- defects that generate the same effects will be of equal importance;
- for different causes of a failure, the probabilities A and D may be different;
- the defect that has the highest probability of being identified by the customer will be marked with the maximum score (10 points)

The assessment of the significance of the defects is done using the notation scale. Based on probabilities A and D, and importance I, the RPN risk coefficient is determined using the relationship:

$$RPN = A \cdot D \cdot I \tag{1}$$

This coefficient has values between 0 and 1000. It is generally considered that measures are needed to prevent potential damage when the RPN risk coefficient is greater than 100.

In **Table 1**, the evaluation of the significance (severity, seriousness) of the defects ("I") when FMEA of the product or process is applied is presented.

The appearance index (A) estimates the probability of occurrence of the defect as a product of the probability of occurrence of the cause and the probability that this cause will provoke the considered defect [13].

Table 2 shows how to score the occurrence index according to its probability of occurrence.

The detection index (D) estimates the probability of detecting noncompliance before it reaches the customer.

Table 3 shows the scoring values of the detection index.

After evaluating and scoring each of the three indices, in the next stage, indications are given regarding the need for improvement measures, depending on the values of indices A, D, and I.

The assessment of the need for improvement measures (general guidance) is presented in **Table 4**.

The results of the analysis are written in tabular form, similar to that described in **Figure 14**

Advantages of applying FMEA:

- Improving the quality, reliability, and safety of a product/process
- Early identification and avoidance of possible defects in the various phases of product planning and manufacturing, as well as at the structural level of the process.
- Reducing the costs of development times;

I (S)	Criteria for Assessing the Importance of Failure Mode
1	Minimal deficiency, the client does not notice it
2–3	Minor deficiency, which the customer can detect and which can cause him a slight dissatisfaction, but without any degradation of product performance.
4–5	Defect of medium severity that bothers or disturbs the customer
6–7	Serious defect, which causes a significant degradation of product performance, causing customer dissatisfaction
8	Serious defect that causes great dissatisfaction to the customer and requires high repair costs
9–10	Particularly serious defects that involve customer security issues and affect product or process safety
10	Deficiency involving safety issues, possible accident

Table 1.
Ways to assess the importance index (I).

A	Probability of occurrence		
1	From 0	to 3/100000	The defect is unlikely to occur
2	3/100000	1/10000	
3	1/10000	3/10000	
4	3/10000	1/1000	
5	1/1000	3 /1000	
6	3/1000	1/100	
7	1/100	3/100	
8	3/100	10/100	
9	10/100	30/100	
10	30/100	100/100	The defect is inevitable

Table 2.
Evaluation of the Occurrence Index (A).

D	Probability of error detection		The probability of the defect reaching the customer	
1	100%	Errors are detected	From 0%	to 2%
2	Very high	The probability of error detection is very high	2%	12%
3	High	The probability of error detection is very high. Checks are safe.	12%	22%
4			22%	32%
5	Moderate	The probability of error detection is medium. Checks are relatively secure.	32%	42%
6			42%	52%
7	Low	The probability of error detection is low. Checks are inaccurate.	52%	62%
8			62%	72%
9	Improbable	The probability of error detection is almost impossible. No checks are possible.	72%	82%
10			82%	100%

Table 3.
Scoring the detection index.

- Collecting information in order to reduce future failures and defects;
- Increasing problem prevention;
- Minimizing late changes that could be made to the product/process, as well as the related costs, etc.;
- Quickly carrying out of the necessary changes and avoiding the unnecessary ones, thus reducing manufacturing times, respectively reducing quality costs in all areas

A	D	I	Description	Measures
1	1	1	The ideal case (goal)	Not required
1	1	10	The situation is certainly under control	Not required
1	10	1	The defect does not affect the customer	Not required
1	10	10	The defect may affect the customer	Required
10	1	1	The defect is more common; it will definitely be detected by the customer	Required
10	1	10	The defect is more common; it may affect the customer	Required
10	10	1	More frequent defect, of great importance	Required
10	10	10	Totally inappropriate situation	Required

Table 4.
The need for improvement.

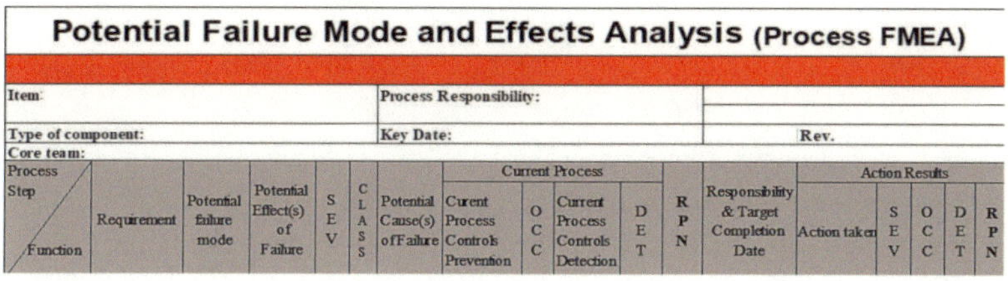

Figure 14.
FMEA spreadsheet example.

- Extremely simple use and neutral application in all industries, both for technical and organizational issues and for services;
- Successful completion of new, verified work techniques, such as Quality Function Deployment value analysis;
- Correct use of existing expert knowledge;
- Improving the company's image and competitiveness;
- Improving communication, cooperation, and collaboration between customers, suppliers, and various internal departments of an organization;
- Increasing consumer satisfaction.

The main advantages of applying the FMEA are shown in **Figure 15**.

2. Studies about the implementation of the FMEA Process

2.1 Study about implementing the FMEA Process for a steel structures components assembly

In **Figure 16**, the technical details of the components of the "Stator Housing" assembly are shown, which were considered for the FMEA Process research,

Figure 15.
The main advantages of applying FMEA for a company.

respectively the welds used to make the housing are shown [20]. It is specified that the welds used to make the housing were considered very important.

The components of the "Stator housing" assembly shown in **Figure 16** are: 1 – support plate, 2 – flat bar, 3 – stiffening plate, 4 – flat bar, 5 – housing base, 6 – ring segment, 7 – front wall, 8.9 – ribs, 10 – shell, 11 – shell segment, 12 – intermediate wall, 13 – front wall, 14 – cable guide, 15 – reinforcement, 16 – pipe, and 17 – supporting bush.

Figure 17 shows details of the elements of the assembly that will be considered in the FMEA study, respectively the welds used to make the housing.

Figure 18 shows the positioning of the welded elements along the stator housing.

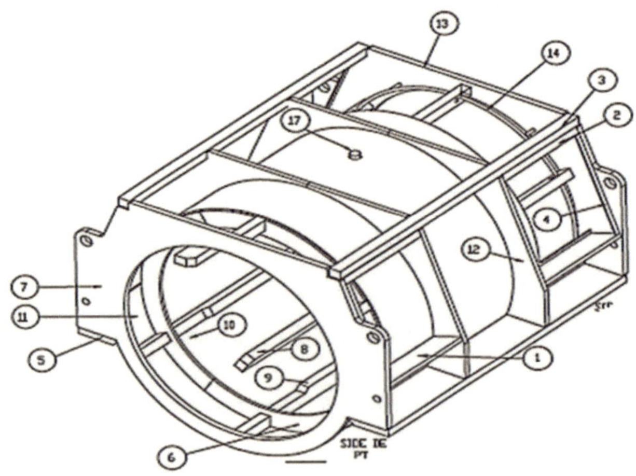

Figure 16.
Elements of assembly "Stator Housing" considered for the FMEA study.

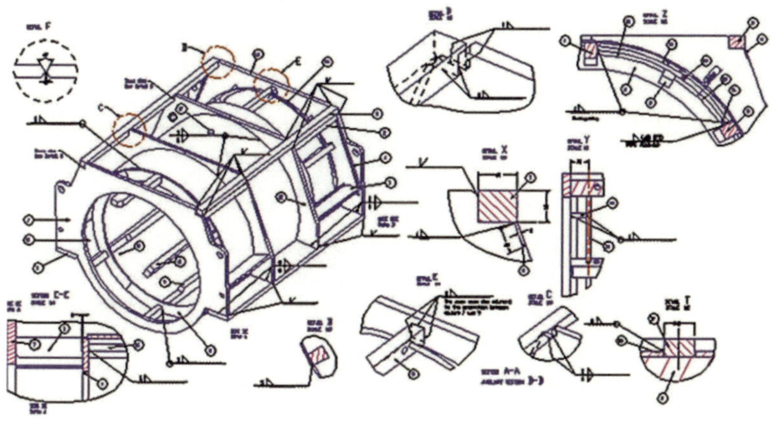

Figure 17.
Elements of the ensemble considered for the FMEA study.

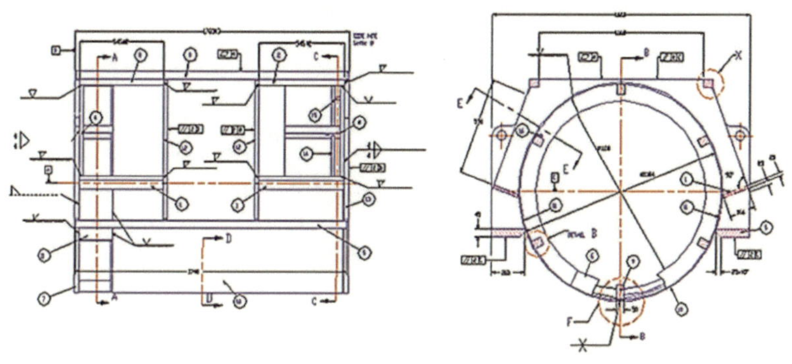

Figure 18.
Positioning the welded elements on the stator housing assembly.

Regardless of the product/manufacturing process to which the FMEA method is applied, the steps required to perform an FMEA Process analysis in order to obtain a "zero defect" production are those shown in **Figure 19**.

FMEA-Process analysis includes several stages: process/product analysis, process diagram determination, FMEA preparation, control plan development, statistical data analysis, document package update, FMEA team information, use of documentation made by FMEA application.

In general, the steps required to prepare an FMEA-Process analysis are as follows: planning and preparation, risk analysis, assessment and, subsequently, risk minimization. These steps are shown in **Figure 20**.

Figure 21 shows the stages of the technological process of manufacturing the "Stator housing" assembly. The necessary operations and stages in the chronological order of their development are: qualitative inspection—material reception, CNC cutting, saw cutting, adjustment, semiautomatic cutting, saw cutting, parting off on guillotine, components adjustment, shell rolling and spot welding, rings rolling and spot welding, assembly and sharpening, ribs assembling and welding, welding and assembling, manual marking out and parting-off, final welding, final assembling, first sandblast cleaning, final adjustment, final sandblast cleaning, priming.

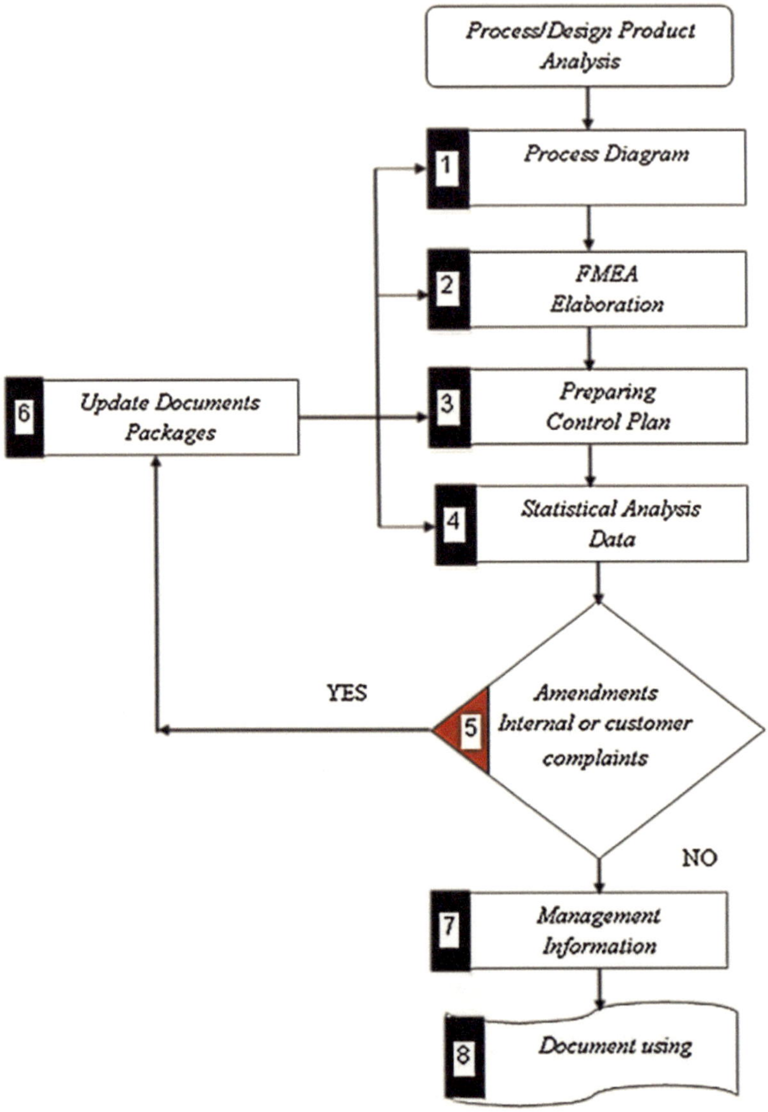

Figure 19.
FMEA roadmap [19].

From **Figure 21** it can be observed on the left side of the figure the presence of the 6M: Man, Machine, Method, Measure, Mother Nature (Environment), Material.

The influence of each factor of the 6M on the quality of the product, as a result of a process (output) is shown in **Figure 22** [13].

The influence of the equipment on which the manufacturing process takes place is minimal, because the initial settings remain unchanged for a long time.

The influence of the human factor is remarkable, being on a higher class. Employees can be trained and motivated to perform, manage, and verify manufacturing processes and products; however, the mental factor related to their health or family problems cannot be quantified, producing inattention, lack of interest, etc.

Figure 20.
The stages of preparing the FMEA analysis.

The most significant and uncontrollable influence has the environment (M Nature) in which the manufacturing processes take place, being included here is both the internal working environment of the company and the external business environment, the domestic and international economic market.

Figure 23 shows the technological flow used to build the "Stator Housing" assembly.

The FMEA-process analysis was performed in order to improve the quality of the welded elements of the "Stator housing" assembly, this being partially shown in **Table 5**.

As it can be seen in **Table 5**, there are several operations that have a cumulative score for Severity, Occurrence, and Detection, respectively RPN major, of over 60 points. But the operations that could cause serious disruptions to the technological process are those with a high RPN of over 100 points (cells colored red as in **Table 5**), these being the following:

- when chamfering small surfaces, due to the low weld mechanical strength over time (Severity—point 8, Appearance—point 3, Detection—point 5), RPN is 120; it is recommended to purchase a chamfering machine;

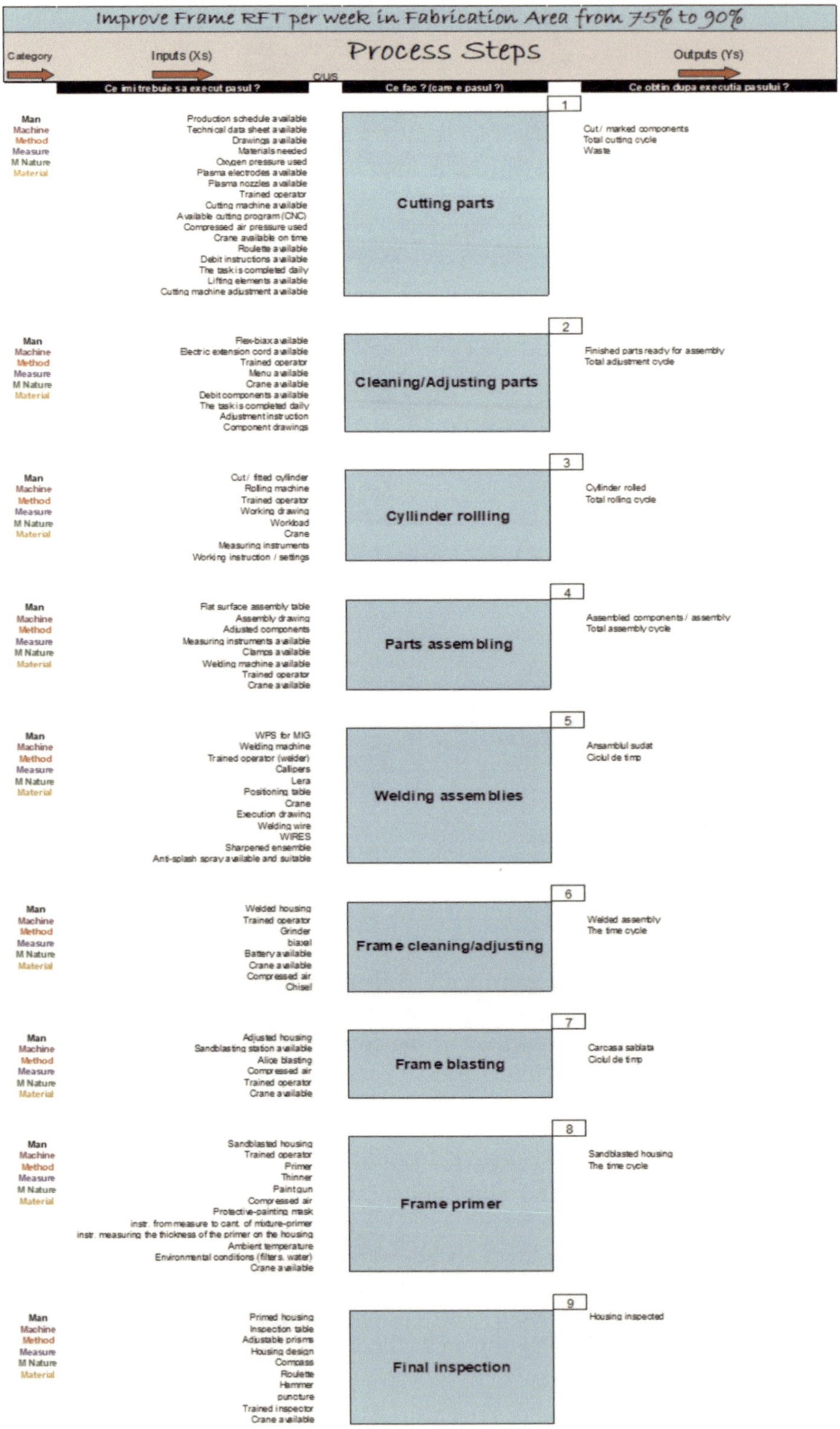

Figure 21.
The stages of the technological process of making the ensemble.

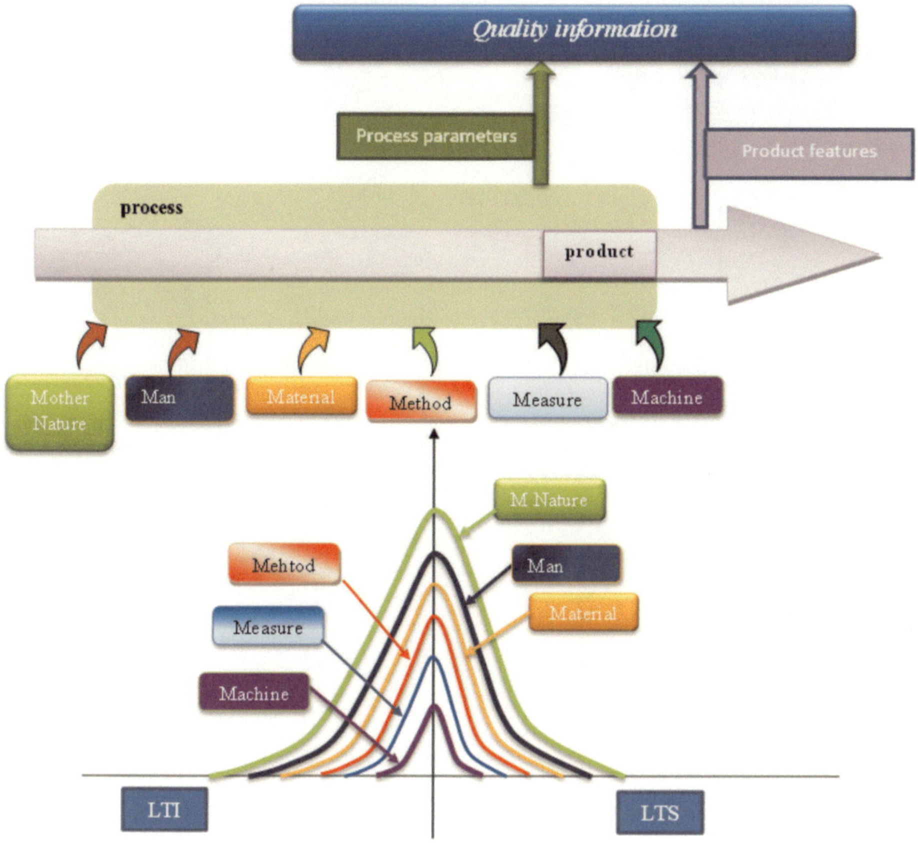

Figure 22.
Process model and influencing factors.

- when chamfering along the entire length of the part, due to the irregular shape of the chamfer, having the effect of low mechanical resistance of welding over time (Severity – score 8, Appearance – score 4, Detection – score 5), RPN is 160, or due to the quality of the welding bead, (Severity – score 5, Appearance – score 4, Detection – score 5), RPN is 100; when chamfering smaller surfaces, due to the small size of the chamfer, the effect is the low mechanical resistance of the weld over time (Severity – score 8, Appearance – score 3, Detection – score 5), RPN is 120; for all this it is recommended to buy a chamfering machine;

- in the case of the operation of assembling reinforcements/stiffeners and frameworks/terminal box console, due to geometric deviations (flatness, parallelism, perpendicularity) there is an angle of less than 90^0 having as result the impossibility of mounting blinds/grids/other parts or subassemblies (Severity – score 8, Appearance – score 5, Detection – score 3), RPN is 120, EMM/SDV replacement is recommended.

It was found that RPN decreased to zero and therefore achieved finished products with "zero defects," after taking the necessary steps to implement the recommendations for each operation with a high RPN.

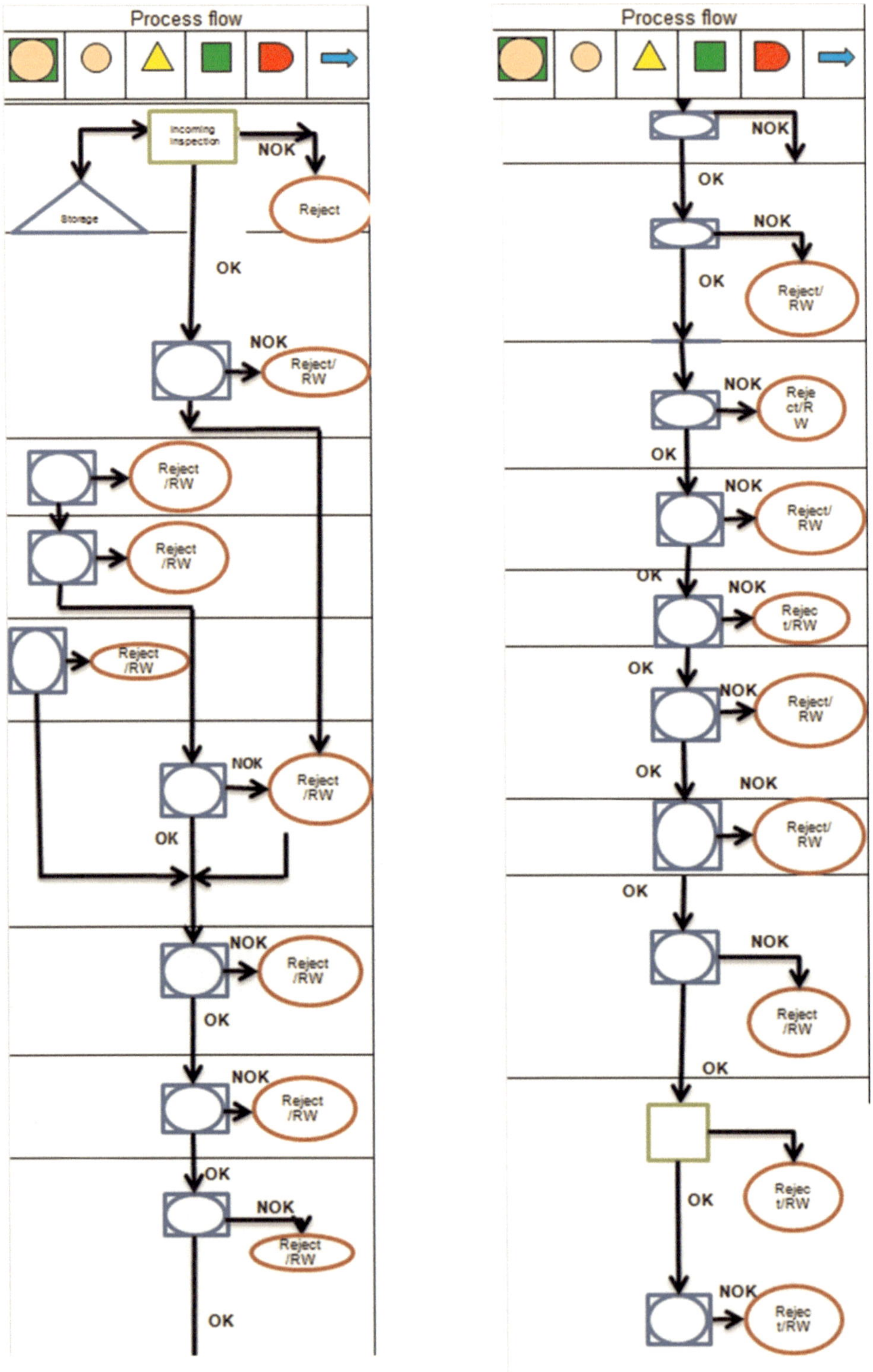

Figure 23.
Process flow for the realization of the "Stator Housing."

Potential failure mode and effects analysis (Process FMEA)								
Process step / Function	**Potential failure mode**	**Potential Effect(s) of Failure**	**Severity "S"**	**Potential Cause(s)/ Mechanism(s) of Failure**	**Occurrence "P"**	**Detection "D"**	**R.P.N.**	**Recommended Action(s)**
Cutting parts on saw	sloping surfaces	additional consumption of labor (additional technological operations)	4	incorrect positioning of the semi-finished product on the machine table	5	3	60	purchase of a specialized table for the placement and taking over of the semi-finished product
Chamfering	larger chamfered surface dimensions (inclination angles, chamfering dimensions)	weld bead appearance and quality	5	skill of the operator	3	5	75	purchase of chamfering machine
	smaller chamfered surface dimensions (inclination angles, chamfer dimensions)	appearance of welding defects (non-penetration at the weld root)	6	skill of the operator	3	5	90	purchase of chamfering machine
	smaller chamfered surface dimensions (inclination angles, chamfer dimensions)	low mechanical strength of the weld over time	8	skill of the operator	3	5	120	purchase of chamfering machine
	irregular shape of the chamfer	weld bead appearance and quality	5	skill of the operator	4	5	100	purchase of chamfering machine
	irregular shape of the chamfer	low mechanical strength of the weld over time	8	skill of the operator	4	5	160	purchase of chamfering machine
	large unevenness on chamfered surfaces	additional labor (for removal by polishing)	3	skill of the operator	4	5	60	purchase of chamfering machine
Ribs assembling	small number of provisional welds	breaking of temporary welding and displacement of the rib	3	incomplete work instruction	10	2	60	

Potential failure mode and effects analysis (Process FMEA)								
Process step / Function	**Potential failure mode**	**Potential Effect(s) of Failure**	**Severity "S"**	**Potential Cause(s)/ Mechanism(s) of Failure**	**Occurrence "P"**	**Detection "D"**	**R.P.N.**	**Recommended Action(s)**
Ribs welding	excesive convexity (cod 503/ISO6520-1:2007)	low mechanical strength of the assembling over time	8	Too small current	2	4	64	automatic system welding
	excesive convexity (cod 503/ISO6520-1:2007)	low mechanical strength of the assembling over time	8	Too small voltage	2	4	64	
Bases assembling	the bases are not in the same plane	no machining allowance	6	front wall assembly error	3	3	54	grinding and checking the assembly table
	the bases are not in the same plane	no machining allowance	6	cutting parts error	6	2	72	Purchase of material cutting software (Fast Cam)
	the bases are not in the same plane	too large machining allowance	6	cutting parts error	6	2	72	Purchase of material cutting software (Fast Cam)
	smaller distance from the horizontal axis to the base surface	the overall dimension is not achieved	8	front walls cutting error	3	3	72	
Assembling of intermediate walls / gussets	the intermediate walls are not in the same plane	deviation from linearity between the four walls	8	cutting parts error	3	3	72	
	the intermediate walls are not in the same plane	no machining allowance for the top milling	8	cutting parts error	3	3	72	
	the intermediate walls are not in the same plane	too large machining allowance for the top processing	8	cutting parts error	3	3	72	
	the intermediate walls are not in the same plane	unequal dimensions of the walls after machining	8	cutting parts error	3	3	72	
	the intermediate walls are not in the same plane	large gap when assembling the intermediate walls with the bases	8	cutting parts error	3	3	72	

Potential failure mode and effects analysis (Process FMEA)								
Process step / Function	**Potential failure mode**	**Potential Effect(s) of Failure**	**Severity "S"**	**Potential Cause(s)/ Mechanism(s) of Failure**	**Occurrence "P"**	**Detection "D"**	**R.P.N.**	**Recommended Action(s)**
Assembling reinforcements / stiffeners and frameworks/ terminal box console	geometric deviations (flatness, perpendicularity, parallelism): angle less than 90 degrees	impossibility of mounting blinds / protection grilles / other parts or subassemblies	8	defective / worn EMM	5	3	120	EMM replacement – action in progress
	no machining allowance to the upper part of the housing	additional remediation operations	5	cutting parts error	3	4	60	Purchase of material cutting software (Fast Cam)
Sandblasting before final welding	incorrectly sandblasted areas	possible welding faults	5	skill of the operator	3	4	60	
Final welding	too thin welding beads	low mechanical strength of the weld over time	8	non-compliance with WPS	3	3	72	specific WPS develop-ment
Final welding	shape of the weld: excessive convexity (cod 503/ ISO6520-1:2007)	low mechanical strength of the assembling over time	8	Too small current	2	4	64	
			8	Too small voltage	2	4	64	
Adjustment	surfaces with welding spatter	unsightly product	6	hard to reach areas	3	3	54	purchase of special chisels for cleaning welding spatter
Final Innspection	the appearance of the weld beads outside the documentation tolerances	correction	8	skill of the operator	4	2	64	Operators monitoring

Table 5.
Aspects of FMEA-process analysis for "Stator Housing".

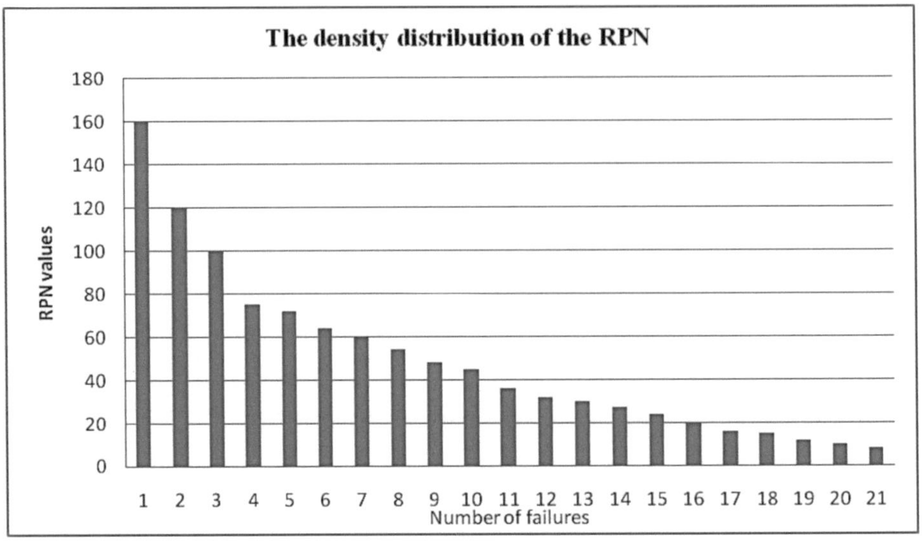

Figure 24.
Diagram of RPN distribution.

The RPN values in descending order for the assembly subjected to the FMEA analysis are represented in the diagram shown in **Figure 24**.

The risk prioritization figure (RPN) is obtained by multiplying the assessment factors established for Severity, Probability of Occurrence and Detection.

The maximum RPN value for the product in question was:

$$RPN = 8x4x5 = 160 \qquad (2)$$

This maximal value for RPN was obtained when chamfering on the entire length of the workpiece due to irregular shape of chamfering.

The solution that has been proposed for the correction of this defect was purchasing a chamfering machine, which significantly reduced the RPN value.

From the diagram shown in **Figure 24**, it may be noticed that measures are required for the defects that have RPN values greater than 120. Thus, for the "Stator housing" assembly, for the chamfering operation along the entire length of the part, due to the irregular shape of the chamfer, a low mechanical resistance of welding over time results (Severity – score 8, Appearance – score 4, Detection – score 5), RPN is 160 or, due to the quality of the weld bead, (Severity – score 5, Appearance – score 4, Detection – score 5), RPN is 100. For smaller dimensions and surfaces chamfering, the effect is low weld mechanical strength over time due to the small size of the chamfer (Severity – score 8, Appearance – score 3, Detection – score 5), RPN is 120.

The recommended improvement measure is the purchase of a chamfering machine, with the setting of the responsible department and the deadline for application. Following the application of the last measure, a risk assessment is carried out again. The main potential defects analyzed are: low mechanical strength of the weld; deformations; welding lines; large unevenness on chamfered surfaces; excessive weld convexity; high values of geometric deviations in excess of the prescribed tolerances; assembling errors; non-uniformity of the color of the part surface; cracks; air gaps; welding spatter, scratches.

The potential causes of the defects have been studied and improvement measures have been proposed. These include:

- implementation and monitoring of preventive maintenance program;
- preparation and compliance with specific welding rules;
- purchase of a specialized table for placing and taking over the semi-finished product;
- purchase of a chamfering machine;
- changing the welding installation with an automatic welding installation;
- purchase of a device for measuring humidity and its use;
- grinding and checking the assembly table;
- purchase of special chisels for cleaning welding spatter;
- monitoring operators through actions to monitor operator non-compliance and their regular training.

Figure 25 shows a comparison between the actual process RFT (Right First Time) index for the number of landmarks inspected weekly, after applying the FMEA analysis, and the waiting level.

Figure 25 shows that, after applying the FMEA Process for the Stator Housing assembly, at the weekly check, for Monday, from January to August 2014, the RFT index was always over 80%, and starting from week 21, it was located at 100%, far exceeding the proposed level of expectation of 50%.

In conclusion, by proper use of the analysis of FMEA Process expensive amendments of the assembly "Stator Housing" technological process could be avoided by

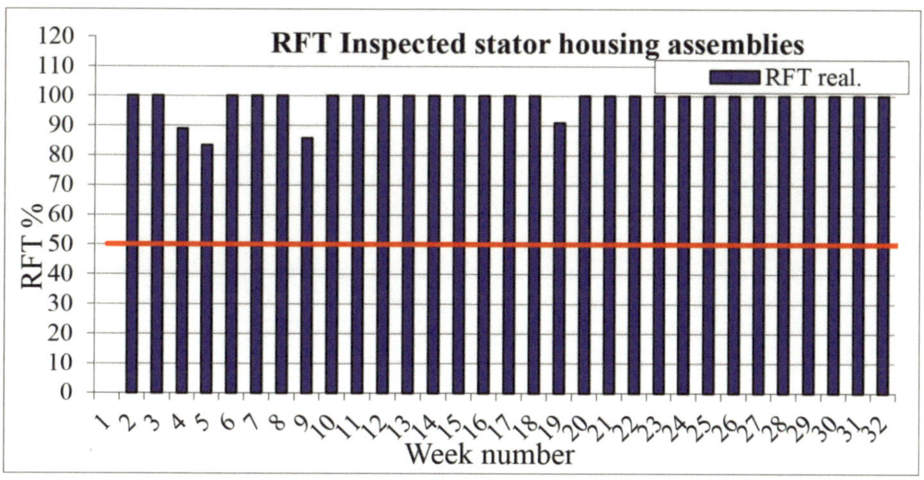

Figure 25.
Actual process RFT index for the number of assemblies inspected.

identifying potential defects, to avoid them, and also, by assessing risks and potential consequences of failures/defects [19].

The main potential defects that were analyzed were: low mechanical strength of the weld; deformations; weld line; large irregularities on the tapered surface; excessive convexity of the weld; high values of the geometric deviations exceeding the tolerances specified values; assembly errors; irregularity of the part surface nuance; cracks; bubbles; weld spatter, scratches.

Potential causes of defects were studied and, after that, improvements were proposed. These include: implementation and tracking preventive maintenance program; preparation and specific compliance for welding; purchase of specialized setting and reception table for workpieces; buying a chamfering machine; changing welding system with automatic welding system purchasing a humidity measuring device and its uses rectification and verification of table assembly; purchase of special chisels for cleaning weld splashes; operators monitoring by monitoring nonconformities operators and their regular training

2.2 Research about implementing FMEA Process for a Packing Shaft

A study on the application of FMEA Process analysis was performed for the packing axis shown in **Figure 26** [21].

Figure 27 shows the execution drawing of the packing shaft, where there also appear the sections with risk factors prone to.

Figure 28 presents the process flow of the shaft, with the operations and stages necessary to achieve the benchmark, in chronological order of their development.

In **Figure 29**, the method of estimating the effects of each process and the influences on the final product is presented [20].

The FMEA Process analysis for the packing shaft was performed by a team of experts from the company.

Table 6 shows the beginning of the table completed after the accomplishment of the analysis.

It was found that there are several operations that have a cumulative score for Severity, Occurrence and Detection, respectively the major RPN risk coefficient of over 90 points.

There are operations that lead to serious disruptions of the technological process, with a high RPN, of more than 100 points (cells colored red, as in **Table 6**), these being the following:

Figure 26.
Packing shaft.

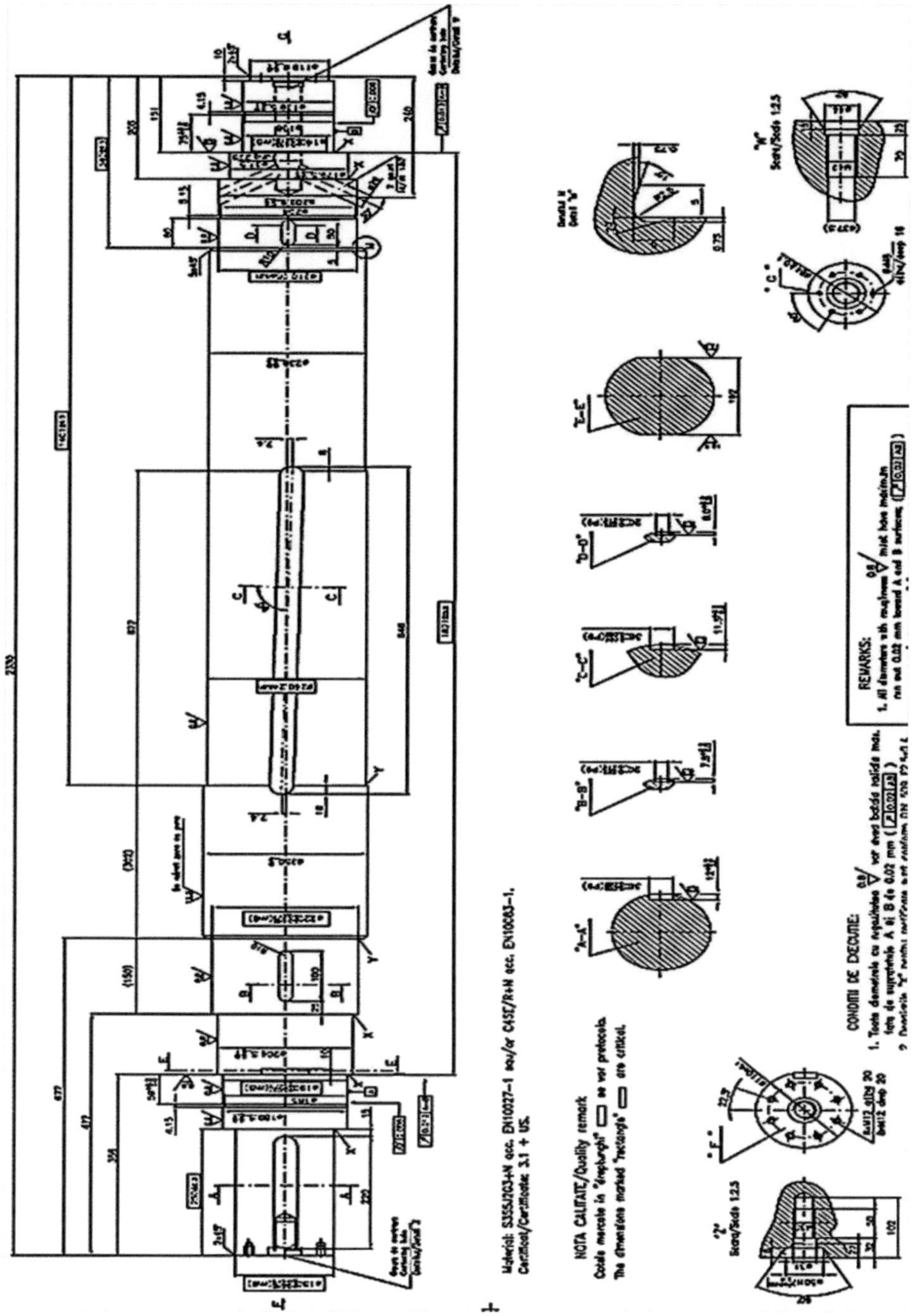

Figure 27.
Execution drawing of the packing shaft.

- for cutting-off operation, when the semi-finished product shows arrow curvature due to deviation from coaxiality because of large successive temperature variations (Severity – score 5, Appearance – score 3, Detection – score 10), so RPN is 150, it is recommended to have material certification from the supplier.

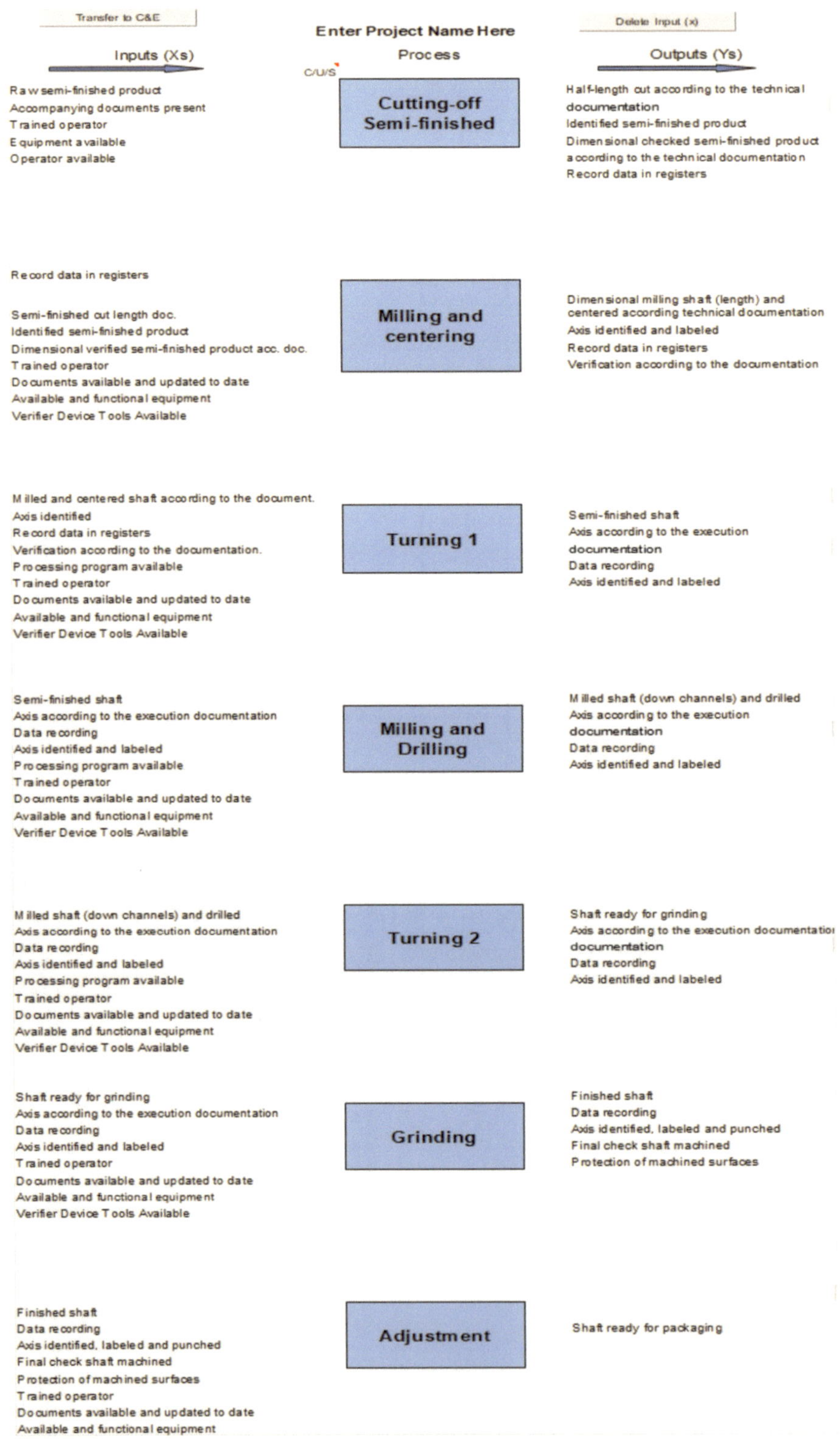

Figure 28.
Process flow of packing shaft subjected to FMEA analysis.

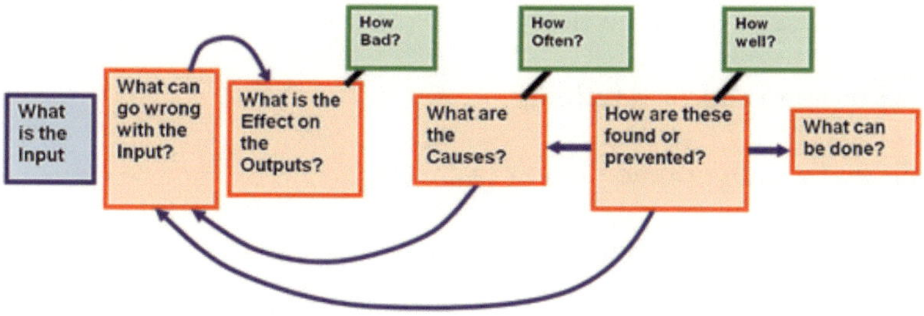

Figure 29.
Method for establishing the influences that occur during technological process [20].

- for cutting-off operation, when the documents are incorrectly completed or incomplete, having the effect of mounting the shaft made of other material than the one mentioned in the specification because of the incorrect filling of the form by the material suppliers (Severity – score 9, Appearance – score 4, Detection – score 3), so RPN is 108, it is recommended to have material certification and control from the supplier;

- for cutting-off operation, when the cut semi-finished product is longer than necessary, which may affect other equipment due to non-compliance with the working instructions (Severity – score 10, Appearance – score 4, Detection – score 3), so RPN is 120, training, regular retraining and self-monitoring on the flow are recommended.

- for turning operation no 1, when the milled and drilled shaft is not in accordance with the documentation and may cause the shaft to be rejected or reclassified due to human error, respectively the operator is trained but does not comply with the manufacturing process (Severity – score 10, Appearance – score 5, Detection – point 4), so RPN is 200 and standard working instructions describing the process steps are recommended.

- for turning operation no 1, when the overhaul is outdated and the shaft may be reshaped, reclassified or rejected due to delayed launch by the design department, respectively the operator is trained but does not comply with the manufacturing process (Severity – score 10, Appearance – score 3, Detection – score 4), so RPN is 120, it is recommended to modify the file later and audit.

It is specified that the abbreviation OTD, which appears in **Table 6** means "on time delivery."

The highest density of risks of occurrence of defects with major result on the product are signaled in the parting-off phases of the semi-finished product, but the highest value of RPN = 250 is recorded in the turning operation, when the milled and drilled shaft does not comply with the documentation, the shaft may be rejected or reclassified due to human error, respectively the operator is trained but does not comply with the manufacturing process.

When recommending measures to improve the process, in order to completely reduce the RPN, the following were considered:

- requesting the material certificate from the supplier;

- self-control and/or check on flow;
- periodic training and/or retraining;
- periodic audit and periodic randomly checks;
- modernization of the milling machine or processing of the shaft by milling at another company;
- purchase of a new milling machine or the modernization of the existing one or processing by third parties;
- change plan for machining tools, work holding devices, and measuring instruments;
- internal orders registration form;
- implementation and supply plan for machining tools, work holding, and tool holding devices and measuring instruments;
- purchasing a new numerically controlled lathe, modernizing the existing one or processing to third parties;
- update for the software of the processing program—introduced in the program, with the processing and revision of the drawings;
- purchase of a program for simulating the turning operation, periodic testing of the program in production;
- periodic testing (once a month) by introducing higher quotas – the first shaft in the series is processed in phase-by-phase mode for error detection;
- update for the software of the processing program—introduced in the program, with the processing and revision of drawings, elaboration data, periodic retraining, clear text verification—self-control;
- periodic trainings and annual audits;
- standard working instructions that describe the process steps.

The RPN values in order of their appearance are presented in the diagram in the **Figure 30** [21].

It can be seen from diagram 30 that the operations with major risk, respectively those with an RPN> 100 appear in the first stages of the shaft processing process, respectively at cutting and turning, a fact also found in **Table 6**.

In **Figure 31**, the RPN values in descending order of parts subjected to FMEA analysis are represented.

After taking the necessary measures to implement the recommendations for each operation with a high RPN, it decreased to zero, which led to the obtaining of finished products with "zero defects."

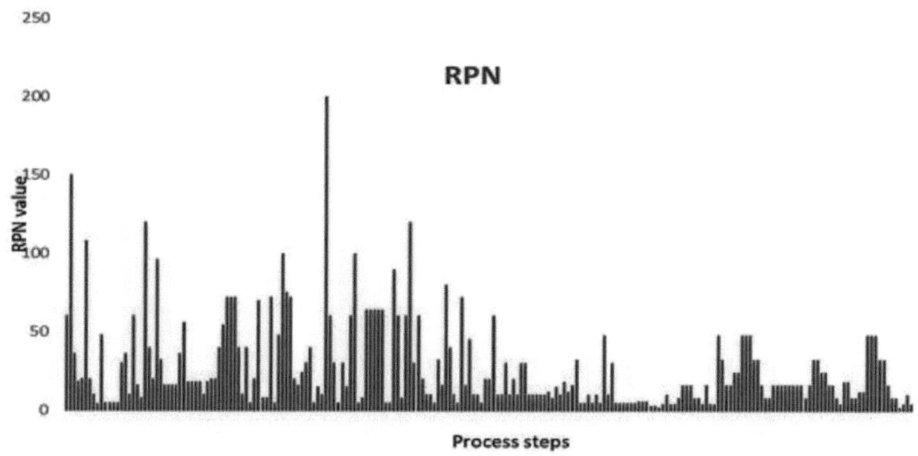

Figure 30.
The RPN values in order of their appearance [20].

The main potential defects analyzed were: deviation from coaxiality due to large successive temperature variations, mounting of the shaft made of a material other than the one mentioned by the specification, milled and drilled shaft is not in accordance with the documentation, shaft reshaping, reclassifying or rejecting, part dimensions damage.

The potential causes of the defects have been determined and improvement measures have been proposed, including:

- implementation and follow-up of preventive maintenance program;
- modernization of the existing milling machine;
- modernization of the numerically controlled lathe on which the turning operations are performed;
- update for the software of the processing program—introduced in the program, with the processing and revision of the drawings;
- purchase of a program for simulating the turning operation, periodic testing of the program in production;
- standard working instructions describing the process steps;
- the operators supervising through noncompliance monitoring actions and operators regular training.

After the application of these measures, a new risk assessment was carried out, finding that the target of having a production with "zero defects" was reached.

Timely use of FMEA-Process analysis can avoid costly changes to the technological process of making the product "packaging tree" by identifying potential defects, avoiding them, and assessing the risks and potential consequences of defects and obtaining "zero defects" as target products [21].

Potential failure mode and effects analysis (process FMEA)								
Process step / Function	**Potential failure mode**	**Potential Effect(s) of Failure**	**Severity "S"**	**Potential Cause(s)/ Mechanism(s) of Failure**	**Occurrence "P"**	**Detection "D"**	**R.P.N.**	**Recommended Action(s)**
Cutting-off the semi-finished product	Missing identifi-cation	Incorrect material Cutting-off	6	Failure to follow the identification procedure	2	5	60	Self-control Check on flow
	Semi-finished material with deflection	Coaxiality deviation	5	Successive large temperature variations	3	10	150	Certificate of material from the supplier
	Wrongly completed or incomplete documents	Mounting of shaft made of a material other than that specified in the documen-tation	9	Errors in document completion made by the suppliers of the semi-finished material	4	3	108	Certifica-tion and control of material from the supplier
	Unreadable, incom-plete data registra-tion	It affects OTD (On Time Delivery)	5	Non-compliance with the document control procedure	4	3	60	Self-control on flow, training
	Longer cut semi-finished product	It affects OTD	10	Failure to follow work instruction	4	3	120	Training, periodic retraining, self-control on flow
	Longer cut semi-finished product	Increasing internal material losses – reclassification	8	Failure to follow work instruction	4	3	96	Training, periodic retraining
	Wrongly identified semi-finished product	It affects OTD	4	Direct promotion order on flow	7	2	56	
Milling and centering	Semi-finished product cut to size with shorter length	It affects OTD	10	Failure to follow working instructions	4	3	120	Periodic training, versatility
	Semi-finished product cut to size with shorter length	Increasing internal material losses	10	Failure to follow working instructions	4	3	120	Periodic testing (once a week) by entering higher dimensions
	Wrong identified semi-finished product	It affects OTD	5	Direct promotion order on flow	5	8	200	Periodic training, versatility

Potential failure mode and effects analysis (process FMEA)								
Process step / Function	**Potential failure mode**	**Potential Effect(s) of Failure**	**Severity "S"**	**Potential Cause(s)/ Mechanism(s) of Failure**	**Occurrence "P"**	**Detection "D"**	**R.P.N.**	**Recommended Action(s)**
	Outdated revisions	Material rejection/ reclassifica-tion	9	Wrong technical data sheets	4	2	72	Preventive mainte-nance. Periodic training
	Outdated revisions	Material rejection / reclassifica-tion	9	The archive did not change the drawings	2	4	72	
	Outdated revisions	It affects OTD	9	Material rejection/re-classification	4	2	72	
	Defective machine	It affects OTD	5	Machine wear	7	2	70	Modernization of the milling machine. Periodic retraining
	Defective machine	Deterioration / damage of part dimensions	8	Machine wear	3	3	72	Modernization of the milling machine. Periodic retraining
	Unavai-lable machining tools, work-holding devices and measuring instruments	It affects OTD	5	Poor management of the area	5	4	100	Purchase of a modern milling machine. Calibration of measuring instruments
	Unavailable machining tools, work-holding devices and measuring instruments	It affects OTD	5	They were not supplied on time	5	3	75	Modernization of the milling machine. Periodic retraining
	Damaged machining tools, work-holding devices and measuring instruments	Rejection / reclassifica-tion	8	Advanced wear	3	3	72	

Potential failure mode and effects analysis (process FMEA)								
Process step / Function	**Potential failure mode**	**Potential Effect(s) of Failure**	**Severity "S"**	**Potential Cause(s)/ Mechanism(s) of Failure**	**Occurrence "P"**	**Detection "D"**	**R.P.N.**	**Recommended Action(s)**
Turning 1	The shaft is in accordance with the documenta-tion	It affects OTD	5	Outdated revision of document-tation	4	5	100	Control and up-date of technical documenta-tion
	Unidentified shaft	It affects OTD	5	Non-compliance with the process	5	4	100	In the long run, acquisition of the operation simulation program, periodic testing of the program in production
	Lack of working documenta-tion	Remanu-facture/ reclassify-cation/ waste	10	Work according to another specification	3	3	90	
	Outdated overhaul		10	Launch delayed by Design Department	3	4	120	
	Proces-sing program available, but incorrect		10	Program-ming error	2	4	80	
	Trained operator, but he does not comply with the manufactu-ring process		10	Insertion of wrong corrections in the machine control	5	5	250	Periodic testing (once a week) by entering higher dimensions
	Trained operator, but he does not comply with the manufactu-ring process	It affects OTD	5	Insertion of wrong corrections in the machine control	5	5	125	

Table 6.
Aspects of the FMEA-process analysis performed for the packing axis.

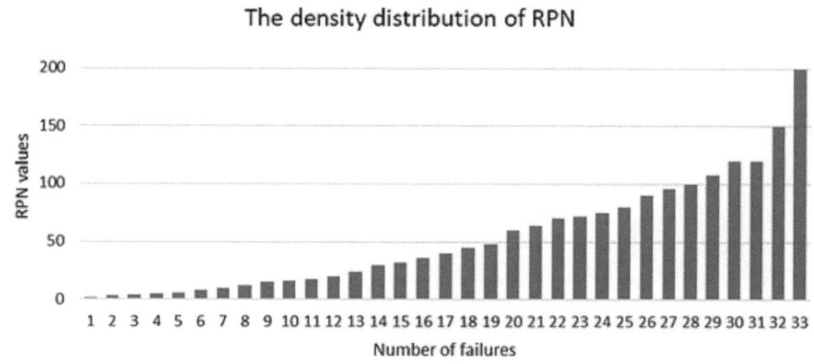

Figure 31.
Diagram of RPN distribution [20].

2.3 Research about the FMEA-Process implementation at First Article Inspection (FAI) for materials composites pieces

Another study focused on some aspects of quality control at the First Inspection Article (FAI) for composite parts used in the railway sector.

An FAI is performed in a planned manner for each first piece or first set of parts made in a project. In many industrial areas, including the railway sector, in the prototyping phase of composite products, this requires strict quality control. The last step in the reproducibility of the validation process is to check that the parts made comply with the technical drawings and specifications of the requirements [22]. Quality control for a prototype, FAI is performed in a planned manner for each first part or first set of parts made in a project. Process performance must be measured by a performance indicator [23].

The FAI will include all details and subassemblies that make up the final product manufactured or ordered. All requirements must be verified by documentation and also all characteristics will be inspected and verified [24]. Approval in the first article is done to demonstrate that the manufacturing process and quality control methods of the products used are appropriate to make a product that meets customer requirements [25].

One of the most important steps in the application of product design FAI is the implementation of the method of efficient process error analysis (PFMEA) for product and process validation [26]. The method of organization and performance is the responsibility of the production department [27]. **Figure 32** shows a detail of the technical drawing of the composite part under study for FAI.

The materials used to obtain the product were [28]:

- low-viscosity polyester resins Giralithe® Ditra GL 2109-10 XP manufactured by Mäder AG Composites, Germany;
- white polyester gelcoat system, based on a NUVOPOL® Gelcoat 37-03 TGP unsaturated polyester unsaturated polyester resin produced by Mäder AG Composites, Germany;
- 3M Scotch-WeldTM 7260 B/A two-component epoxy adhesive manufactured by 3M Deutschland GmbH Industrie-Klebebänder.

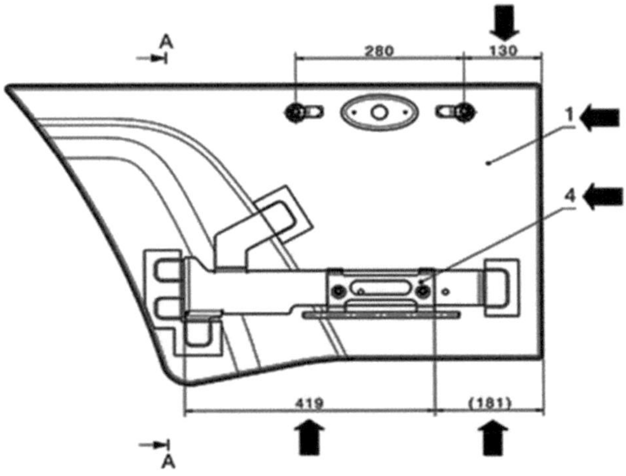

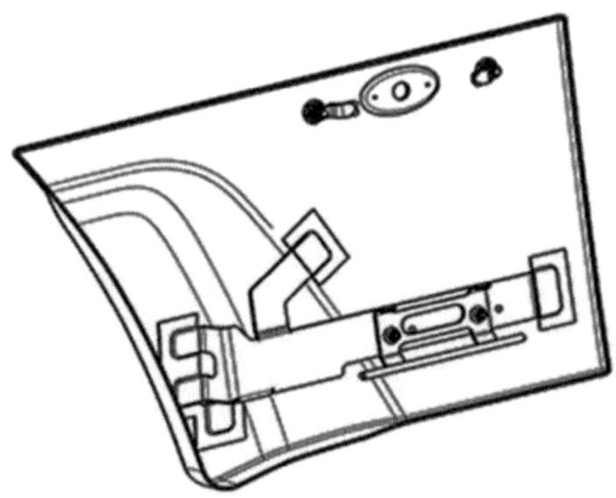

Figure 32.
Details from the technical drawing of the composite piece subjected to FAI study.

Giralithe® Ditra GL 2109-10 XP is a self-extinguishing, self-extinguishing resin made of low-viscosity, low-viscosity polyester resins. It has been specially obtained for applications in public transport engines and vehicles.

NUVOPOL® Gelcoat 37-03 TGP is a specially designed, intumescent, white gelcoat polyester system, based on an isophthalic unsaturated polyester resin.

3M Scotch-WeldTM 7260 B/A Series is a two-component adhesive product that is used at room temperature. These adhesives are designed for bonding hard durable surfaces, elastic hard steel materials, aluminum or composite fibers, and other materials.

Figure 33a and **b** show the appearance of the final composite part used in the outer front area of the locomotive.

The analysis of product requirements in order to implement FAI for the final composite product is described in the diagram in **Figure 34** [28].

Next, **Figure 35** shows the process diagram for making the composite product.

In order to eliminate the risks of errors and defects during the development of the technological process for the obtaining of the composite part for the railway industry, it was considered absolutely necessary to apply the PFMEA-Process method.

The steps required to develop the PFMEA method were performed according to [29].

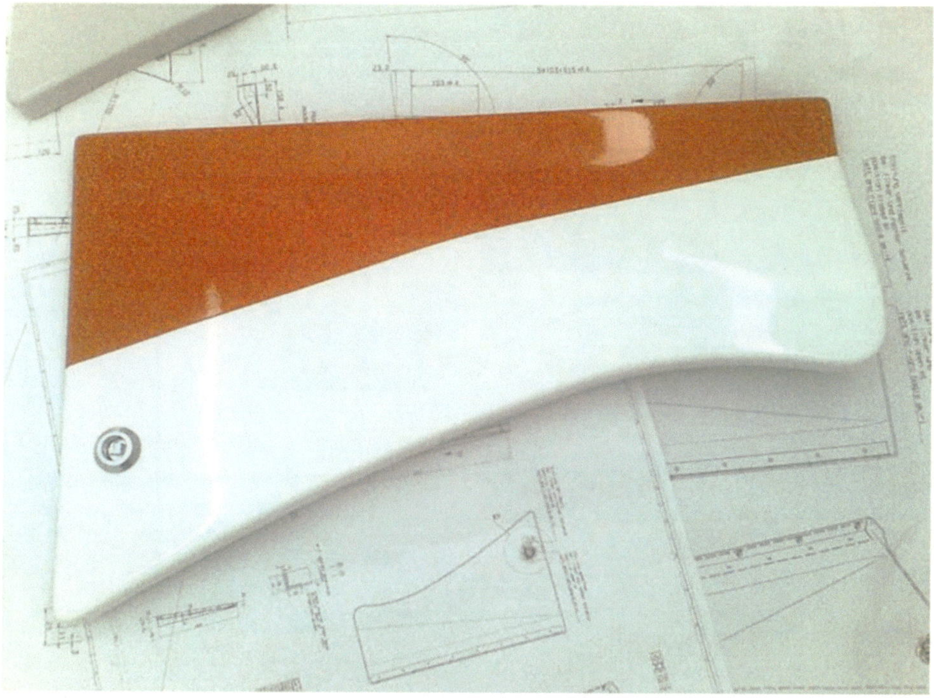

(a)

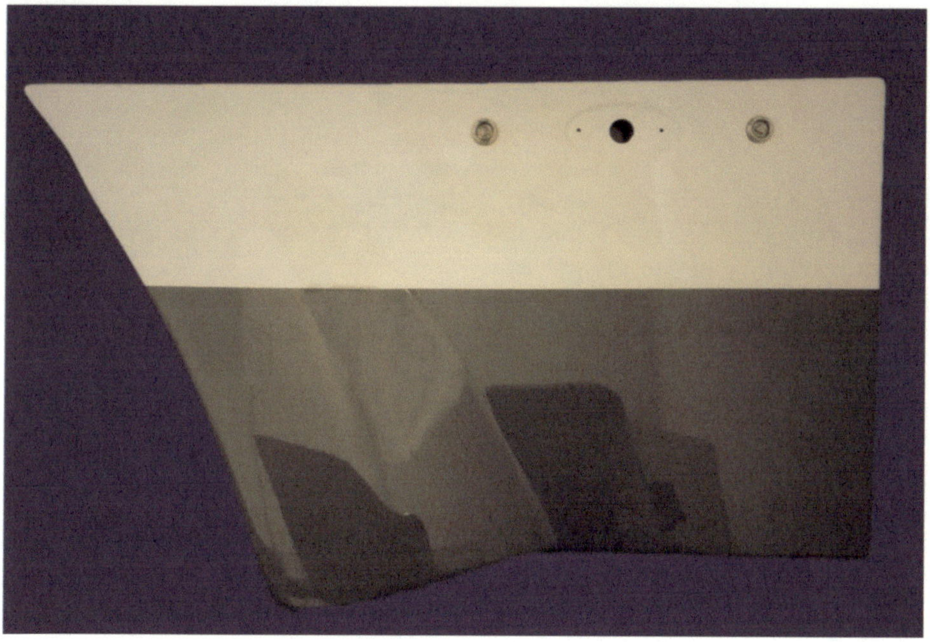

(b)

Figure 33.
The final piece obtained by composite materials.

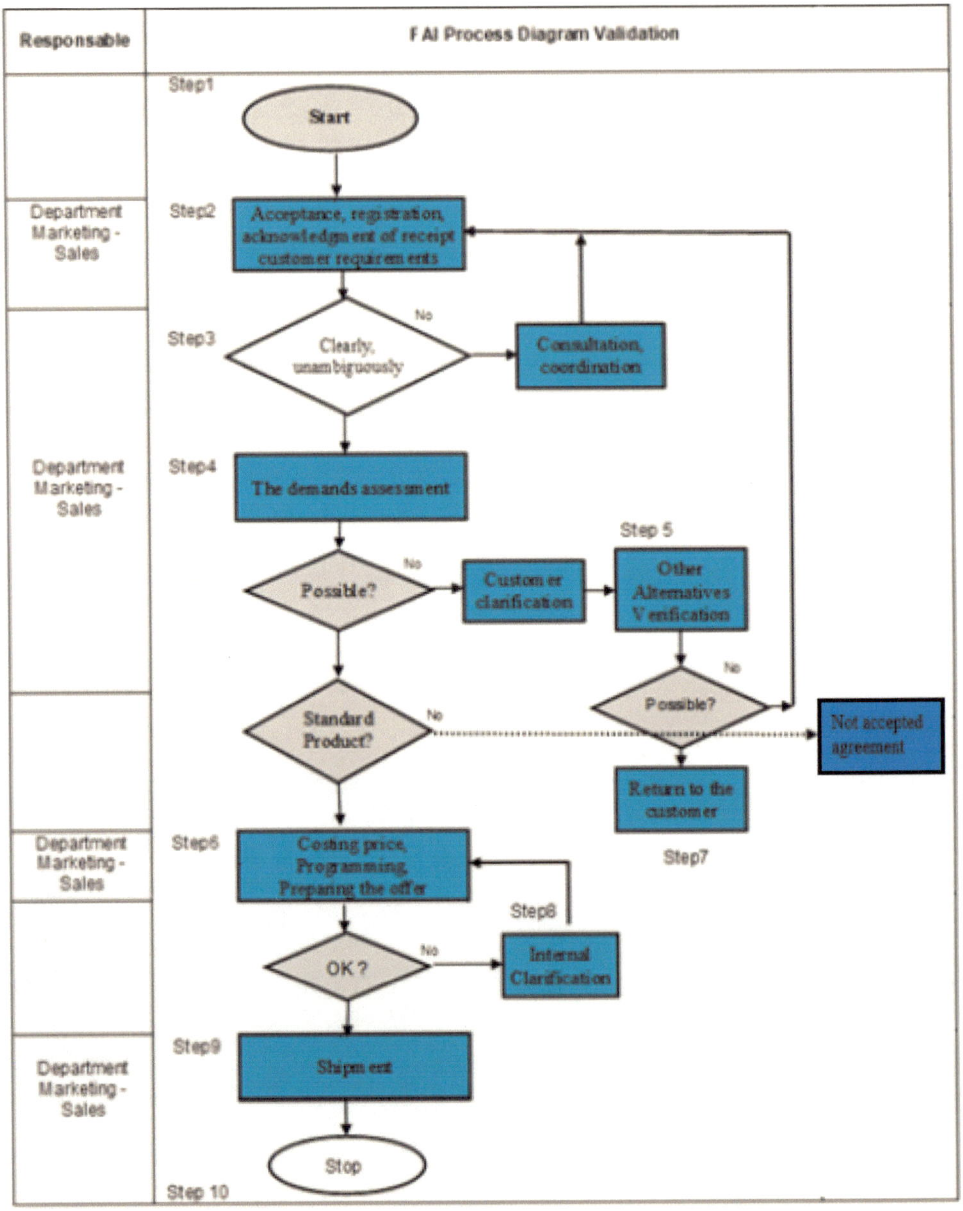

Figure 34.
Logical diagram of the analysis of product requirements [27].

Table 7 shows the important aspects of the PFMEA analysis when making the part, respectively the operations in which the RPN value was higher than 70.

Table 7 shows that, for each operation in which the value of the RPN index was higher than 70 (yellow and red areas), measures were proposed to improve the value of the RPN, finding that, after their application, the value of the criticality index has been greatly diminished.

In **Figure 36**, the distribution of the RPN index for each phase of the technological process is presented [28].

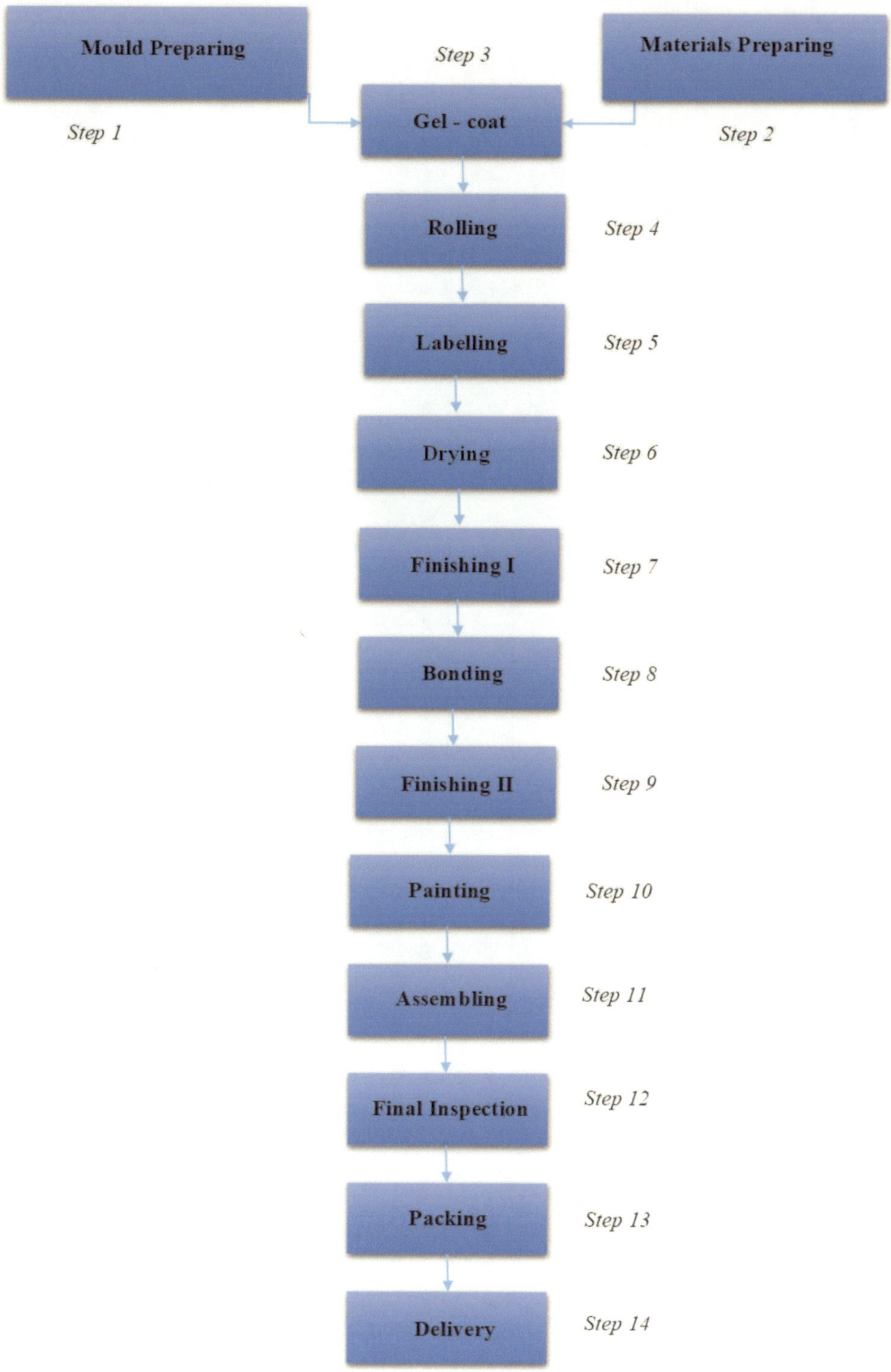

Figure 35.
Diagram of the technological process for the obtaining of the composite product.

Potential failure mode and effects analysis (process FMEA)								
Process step / Function	**Potential failure mode**	**Potential Effect(s) of Failure**	**Severity "S"**	**Potential Cause(s)/ Mechanism(s) of Failure**	**Occurrence "P"**	**Detection "D"**	**R.P.N.**	**Recommended Action(s)**
Material storage	Materials damaged by storage conditions	The part may have manufacturing defects	7	Improper storage conditions	4	4	112	Periodic check of storage conditions
Mold heat treatment	Inadequate treatment of the active surface	The part remains attached to the surface of the mold	8	Operator error	5	4	160	Check in the register of mold records; the last treatment
Materials preparation	The materials are not in validity term	Part does not meet the standards	8	Improper stock turnover	4	4	128	Operator training for stock monitoring and stock monitoring
	The materials are not valid	Part does not meet the standards	8		3	3	72	Operator training and storage conditions check
Gelcoat application	Gelcoat application without catalyst	Part does not meet the standards (not hardening)	7	Broken installation	4	4	112	Application of preventive mainte-nance to the installation
Rolling	Properly unhomoge-nized resin	Part does not meet the standards	8	Broken mixing installation	4	4	128	Traceability sheet
Rolling	Improper manufactu-ring recipe	Part with variable thickness (part does not meet the standards)	8	Operator error	3	3	72	Operator training and checking, tracking
Finishing	Part does not meet the standards	Affecting dimensions when deburring	5	Operator error	4	4	80	Operator training and tracking
Final check	Part does not meet the standards	Shape and dimensional deviations	8	Operator training and tracking	4	4	128	Application of acceptance and verification criteria

Table 7.
Aspects of the FMEA-process analysis performed for the FAI.

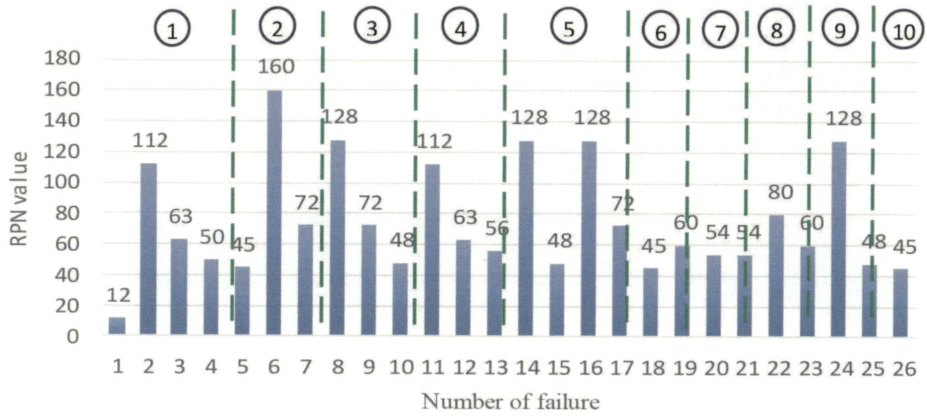

Figure 36.
The RPN index distribution for each phase of the echnological process [28].

In **Figure 36**, the numbers in the circles represent the steps of the technological process: 1 is for the material preparation; 2 is for mold preparation; 3 is the mold treatment; 4 is materials preparation; 5 is the gel-coat applying; 6 is for rolling; 7 is labeling; 8 is for drying; 9 is the finishing; and 10 is the final inspection.

After applying the PFMEA method, the value of the RPN index was reduced below 20 due to the application of the recommended measures. Some of them are presented in **Table** 7.

In conclusion, by applying the FAI, both the product quality and the process were monitored according to customer requirements. Also, during the implementation of the FAI procedures, the risks of errors and defects occurrence were analyzed by applying the PFMEA method and technical and control procedures were developed: Process Diagram, Control Plan, Execution and Control Technology.

The quality of the product is certified by a protocol resulting from the First Article Inspection (FAI) signed between the parties

3. Conclusions

Modern methods of systemic quality design are answers to new quality requirements, they must allow the analysis and elimination of potential defects from the design and implementation stage [7], so that more and more often the notion of "quality design" is encountered.

One of these methods that has become increasingly used in recent decades is the FMEA. Failure modes, effects and analysis (FMEA) is a systematic method of determining and preventing errors, defects, and risks that may occur, applicable to a process, product, or machine used in the process [30].

This method consists of detecting possible defects, inventorying the causes that could cause these failures, demerits effects on users in order to plan the necessary measures to prevent their occurrence. FMEA method is regarded as a basic instrument in project management, the maintenance, and the total quality and could be considered mainly a qualitative analysis.

FMEA is a systematic technique that identifies and prevents product process and system problems (defects) before they occur. It is a method that focuses on preventing problems, therefore increasing safety for improving customer satisfaction.

FMEA can be used in the following situations [31]:

- development of products or processes;
- amendments to existing products processes or system;
- assess the probability of failures, in case of important components in terms of overall safety;
- adapt products to new conditions;
- reduce costs by identifying system, product, process and system improvements early in the development cycle.
- evaluate the design and processes from a new advantageous point.

By applying this method, the risk of failures in the design and manufacturing of products is reduced

On this basis, it ensures reduced costs in all stages of quality spiral: the design through a better reflection of customer requirements in quality design, supply, to avoid problems caused by improper selection of suppliers, production by preventing critical points in the service door, reducing customer complaints, etc.

The method is applied in two main variants [32]: in the product design or process development stage product and process.

FMEA design product to the design of products (components) or their redesign to prevent errors in design and implementation of future product failures is applied. Responsibility for FMEA design product implementation is the responsibility of design department.

After presenting the methodology of FMEA, the application and implementation of some proper research in this area were presented.

The studies developed on the implementation of quality methods in manufacturing processes in automotive engineering focused on the following directions:

- application of the FMEA-Process method;
- quality control on inspection of the first article (FAI);

For the FMEA-Process analysis (PFMEA), it was considered of interest to approach several studies on the application of the method for:

- a packing shaft
- a "stator housing" assembly
- parts made of composite materials used in the railway sector.

After applying the FMEA method, it was found that for both elements studied, after applying the necessary measures, considered after the analysis, in order to execute the recommendations for each operation with a high RPN, it decreased to zero, which led to some finished products with "zero defects."

Another research carried out was the study of aspects on quality control at the inspection of the first article (FAI) for parts obtained from composite materials used in the railway sector. By applying FAI, both product quality and process were monitored in accordance with customer requirements.

Also, during the implementation of the FAI procedures, the risks of errors and defects were analyzed by applying the FMEA process method, and technical and control techniques were developed: Process diagram, control plan, execution, and control technology.

After applying the PFMEA method, the value of the RPN index was reduced below 20 due to the application of the recommended measures, and the product quality was confirmed by the elaboration and signing by the parties (manufacturer and beneficiary) of the protocol resulting from the inspection of the first article.

Acknowledgements

The authors are grateful to the University of Craiova, Faculty of Mechanics, for advice and support. This research was funded by University of Craiova, Faculty of Mechanics, 13 A.I. Cuza Street, RO-200585, Craiova.

Author details

Cristina-Ileana Pascu*, Raluca Malciu and Ilie Dumitru
Faculty of Mechanics, University of Craiova, Craiova, Romania

*Address all correspondence to: ileana.pascu@edu.ucv.ro

References

[1] Swamidass PM. Encyclopedia of Production and Manufacturing Management. Boston: Springer; 2000

[2] Montgomery D. Statistical Quality Control. 7th ed. New York: John Wiley& Sons; 2013. p. 768

[3] Lixandru CG. Supplier quality management for component introduction in the automotive industry. Procedia –\ Social and Behavioral Sciences. 2016;**221**:423-432

[4] Pekuri A, Haapasalo H, Herrala M. Productivity and performance management – Managerial practices in the construction industry. International Journal of Performance Measurement. 2011;**1**:39-58

[5] Ott ER, Schilling EG, Neubauer DV. Process Quality Control: Troubleshooting and Interpretation of Data. 4th ed. Milwaukee: ASQ Quality Press; 2005. p. 628

[6] Ashley RR. Total Quality Management (TQM). Bloomington: AuthorHouse; 2008. p. 324

[7] Mhetre RS, Dhake RJ. Using failure mode effect analysis in a precision sheet metal parts manufacturing company. International Journal of Applied Sciences and Engineering Research. 2012;**1**(2): 302-311

[8] Segismundo P, Cauchick M. Failure mode and effects analysis (FMEA) in the context of risk management in new product development: A case study in an automotive company. International Journal of Quality & Reliability Management. 2008;**25**:899-912

[9] Carlson CS. Understanding and applying the fundamentals of FMEAs. In: Annual Reliability and Maintainability Symposium (RAMS); 27-30 January 2014. USA: Colorado Springs; 2014. p. 32

[10] Marvin RM, Hoylan A. System Reliability Theory: Models, Statistical Methods, and Applications. 2nd ed. New York: Wiley Series; 2004. p. 672

[11] VDA 4-Chapter FMEA: Product and Process FMEA, 2nd ed. 2019

[12] Bouti A, Aitkadi D. A state-of-the-art review of FMEA/FMECA. International Journal of Reliability, Quality and Safety Engineering. 1994;**1**(4):515-529

[13] Pugna AP, Potra S. Controlul și Asigurarea Calității. Ghid de redactare a Documentelor Calității. Timişoara: Editura Solness; 2015. p. 165

[14] Ford Design Institute. Failure Mode 60 and Effects Analysis. FMEA Handbook Version 4.1. Copyright © 2004. Dearborn, MI: Ford Motor Company. p. 290

[15] Guide-Methodologique-Audit-Processus-Renault. 1st ed. 2002

[16] Oprean C, Kifor CV. Quality Management. Germany: Callidus Publishing House; 2008

[17] Zheng LY, Liu Q, McMahon CA. Integration of process FMEA with product and process design based on key characteristics. In: Proceedings of the 6th CIRP-Sponsored International Conference on Digital Enterprise Technology, Advances in Intelligent and Soft Computing (DET 2009); 14-16 December 2009. Hong Kong. pp. 1673-1686

[18] AIAG. Statistical Process Control (SPC), Reference Manual. 2nd ed. 2005.

DaimlerChrysler Corporation, Ford Motor Company, and General Motors Corporation. p. 232

[19] AES Standards. Available from: http://quality.aes-standards.com/fmea.htm [Accessed: April 28, 2022]

[20] Pascu CI, Paraschiv D. Study about Improving the Quality Process Performance for a Steel Structures Components Assembly using FMEA Method. In: Proceedings of International Congress Automotive, Safety and Environment (SMAT), 23-25 October 2014. Craiova, Romania. pp. 103-109

[21] Pascu CI, Paraschiv D. Research about using the Failure Mode and Effects Analysis method for the quality process performance. IOP Conference Series: Material Science and Engineering. 2020; **898**:235-240

[22] Guidelines for the Development of Process Specifications, Instructions, and Controls for the Fabrication of Fiber-Reinforced Polymer Composites. 2003. Available from: http://www.tc.faa.gov/its/worldpac/techrpt/ar02-110.pdf. [Accessed: April 28, 2022]

[23] Morris R, Enhance R. First Article Inspection. New York: E-Publishing Inc; 2007

[24] Pennella CR. Managing Contract Quality Requirements. Wisconsin, USA: Quality Press Publishing; 2006. p. 233

[25] Chrysler Corporation, Ford Motor Company, General Motors Corporation. Advanced Product Quality Planning (APQP) and Control Planning. Reference Manual. 2nd ed. New York: Chrysler LLC, Ford and General Motors; 2008. p. 117

[26] Stamatis DH. FMEA: A General Overview, Failure Mode and Effect Analysis: FMEA from Theory to Execution. 2nd ed. Milwaukee, WI, USA: ASQ Quality Press; 2003. p. 488

[27] Nannikar AA, Raut DN, Chanmanwar RM, Kamble SB. FMEA for manufacturing and assembly process. In: Proceedings of International Conference on Technology and Business Management (ICTBM-13), March 18-20. Dubai; 2012. pp. 501-509

[28] Pascu CI, Gheorghe S, Popa D. Aspects about the quality control at first article inspection (FAI) for parts obtained by materials composites used in the railway sector. In: Advanced Engineering Forum. Switzerland: TransTech Publication; 2018. pp. 262-267

[29] Chrysler Corporation, Ford Motor Company, General Motors Corporation. Potential Failure Mode and Effects Analysis (FMEA). Reference Manual. 2nd ed. Detroit, Michigan, USA: Chrysler LLC, Ford and General Motors; 1995. p. 64

[30] Paciarotti C, Mazzuto G, D'Ettorre D. A revised FMEA application to the quality control management. International Journal of Quality & Reliability Management. 2014;**31**(7): 788-810

[31] Jain K. Use of failure mode effect analysis (FMEA) to improve medication management process. International Journal of Health Care Quality Assurance. 2017;**30**(2):175-186

[32] Lakhmi CJ, Lim CP. Handbook on Decision Making. Vol 1: Techniques and Applications. Springer-Verlag Berlin Heidelberg; 2010. p. 545

Chapter 6

Dimensioning of Fractal Fracture on a Concrete Slab

Francisco Casanova-del-Angel

Abstract

What is presented here is a methodology that allows to study fractures in any material, especially in concrete elements. The importance lies in the moment in which the fracture that occurs in a structural element (slab in our case study) causes a negative effect on its structural behavior. Traditionally, its study is developed from the perspective of fracture mechanics, with which the energy at the tip of the crack or fracture is calculated, and it cannot go further. But the combination of the theories of fracture mechanics and fractal geometry allows us to obtain the patterns of its behavior in the future. Thus, the research was based on tests made on structural elements of concrete, from a slab led to structural failure and choosing a fracture. This was vectorized in order to obtain a fractal axis, which was called state 0, taking three parts or sections of that state and repeating them on a smaller scale. Subsequently, the research was based on five of the methods used for the study and calculation of the fractal dimension: box dimension, perimeter-area dimension, information dimension, mass dimension, and ruler dimension. Most of the fractal dimensions obtained under these methods were similar to each other.

Keywords: curved fractal, mechanical fracture, dimension of fracture, concrete, fractal geometry

1. Introduction

I think it is necessary to start this article with some comments that have to do with the icon that represents a fracture, fraction, and the concept of fractal. For those seasoned in fractures and fractals, they can be childish, but not for architects and civil engineers. Why this brief comment? Well, the research covers two gigantic areas of knowledge: fractals and fracture mechanics. The molecular composition in humans, that is, small structures in large quantities, is the same in a single being. Geometric shapes with the same structure, which if split into equal parts, result in the same smaller part that remains the same or many of them to create a larger one with the same features.

A fractal is a geometric object with a basic structure, whether fragmented or irregular, repeated at various scales. Mathematically, a fractal may be expressed as follows: Let $I = [0, 1]$ and let $A_1 = I - \left(\frac{1}{3}, \frac{2}{3}\right)$ be that subset of I obtained by removing those points which lie in the open middle third of I; that is, $A_1 = \left[0, \frac{1}{3}\right] \cup \left[\frac{2}{3}, 1\right]$. Let A_2

IntechOpen

be that subset of A_1 obtained by removing the open middle third of $\left[0, \frac{1}{3}\right]$ and of $\left[\frac{2}{3}, 1\right]$. Continue this process and define $A_{3,}$ $A_{4,}$ … . The set $C = \cap_{n=1}^{\infty} A_n$ is called the Cantor set. Some characteristics are (i) C is a compact set having Jordan content zero, (ii) C is uncountable, and (iii) the characteristic function of C is Riemann integrable on [0, 1], see [1].

We should bear in mind that the fractal dimension may be calculated in various ways. One is through the Hurst exponent: many structures in nature have the feature to be originated in two dimensions and end in a fractional dimension between two and three [2, 3]. A fractal object should have at least one of the following features: (i) it should show similarity among details on small and large scales; (ii) it should not be able to be represented through classical geometry; (iii) it should have a fractional dimension, that is, a non-whole dimension; and (iv) it should be able to be recursively defined; see [4].

The classical theory of fracture mechanics allows to predict the quick propagation of a macro crack through a homogeneous, isotropic, and elastic material. In this theory, the stress intensity factor, K_I, which is a function of the geometry of the crack and stress, is applied. Failure occurs when K_I reaches a critical value K_{IC}, known as the critical stress intensity factor in plane strain conditions. Therefore, K_{IC} is a measure of the fracture hardness of the material. To properly measure K_{IC} in a material, the sample to be examined must be of a size that allows the ensuring of maximum restriction (plane strain) on the tip of the crack. To apply linear elastic fracture mechanics (LEFM), the value of K_{IC} must be a constant of the material, regardless of the geometry of the specimen (as are other constants of materials, such as yield strength).

Regarding the material used in the laboratory, it must be said that it has the adhesion and cohesion properties required to bond inert aggregates and make a solid mass with proper resistance and durability, as well as that for the fabrication of structural concrete the so-called hydraulic cements are exclusively used. In the current case, Portland cement has been used since it is a grayish material finely powdered, mainly formed by calcium aluminum silicates, see [5, 6]. When the cement is mixed with water to create a soft paste, it gradually becomes rigid until becoming a solid mass. This process is known as casting and hardening.

2. Laboratory tests

The test included applying a load concentrated on the gravity center of a 70 cm wide and 500 cm long portion of a prefabricated slab, 20 cm thick with two open web joists on the ends and a small vault covering the span, **Figure 1**. The test required the construction of a slab portion, 70 cm wide and 500 cm long with two joists on the ends and small vaults covering the spaces between them, with a 6*6–10/10 electro-welded mesh and a 3 cm thick concrete layer in the compression zone. The slab was supported by two 70 cm long and 170 cm high concrete blocks, low walls with reinforcing bar grids, and closure slabs armored with *armex* 15*20–4, while to confine such low wall headframes were placed on the ends with *armex* 15*20–4. The resistance to the compression of the concrete was 200 kg/cm^2, see [7].

The steel required for the area was calculated for the construction of the 5 m long joist, with a 17 cm slope. Once the steel areas were obtained, preassembling, armoring, and casting of the joints were carried out together with the curing process, waiting for

Figure 1.
View of the slab before the first load.

the concrete to reach its maximum strength. The joints and the small vaults were taken to the laboratory where the load frames were placed to place the prefabricated slab. First, the low walls to support the slab were constructed with a concrete block, founded on a reinforcing bar grid with a 15 cm base and 20 cm slope, and armored with *armex* 15*20–4. To confine the low wall, headframes were placed on the ends, 15 cm by 20 cm, armored with *armex* 15*20–4, and the same was used for the closure slab. After that, the required formwork was placed, the joists were fitted into the closure, and the small vaults were put in place. In the center of the span, a temperature nerve was placed and for the compression layer a 6*6–10/10 electro-welded mesh was used and a 3 cm thick layer was placed at the closure, allowing the concrete to harden, *ibid*.

After 7 days, the formwork was removed from the slab, the concrete having been cured for days previously. After this and after the concrete reached its maximum strength, which occurred 28 days after casting the slab, the test started. To be precise in obtaining the results, the load cell was first calibrated in the computer. To do this, a wooden pole was placed on which the load was applied. In the computer, it was observed if the load and strain were registered, which were correct. Then, the load was taken away and all data involved in the test were marked as zero. A smooth steel rod was placed on the central axis of the slab and with such a steel a square profile was placed to reach the height of the frame load piston, which in turn supported the 200-ton load cell. At last, we have the piston applying the load on the frame. Under the slab and on the sides of the low walls supporting the slab, micrometers were placed to measure warping and lateral strain. To easily observe the cracks, the slab's lower part was painted white.

To start, a 10-ton load was applied. When the first cracks appeared, they were colored with a green marker. Then, a second 10-ton load was applied. New cracks appeared and the existing ones were enlarged. Those were colored with a red marker. A 5-ton load was then applied. More cracks appeared and the existing ones were

Figure 2.
Evolution of cracking as the load was applied.

enlarged, which were then colored blue. Finally, a final 5-ton load was applied. No new cracks appeared, only the existing ones were enlarged. Those were colored brown. Days after that, the test was continued applying load until the stab collapsed, **Figure 2**.

3. Fractal dimension

In fractal geometry, the fractal dimension *D* is a real number that generalizes the concept of ordinary dimension for geometric objects that do not admit tangential space and is an exponent, which accounts for the wholeness, that seems to fill a fractal in space as it moves toward finer and finer scales. The problem here is that there exists a series for its calculation which frequently but not always is equivalent. Among the ones mostly used are: Hausdorff-Besicovitch dimension, packing dimension, homothetic dimension, and Rényi dimensions. None of these is universally used, and one of the reasons for this is that the discrepancy among them is related to differences in the structure of the fractal. Although for a great number of classical fractals the values of the various definitions of fractal dimension and all those dimensions coincide, they are not equivalent. The positive aspect of this is that some definitions of fractal dimension are easy to calculate and are probably more used. However, they do not always have the most desirable mathematical properties.

3.1 Dimensioning of fractal fracture on a concrete slab

In the study of fractures, from a fractal point of view, the previously specified slab was used, **Figure 1**, on which loads were applied. Propagation of the crack depends on important factors, such as stress and intensity. It also depends on the curing of concrete and the resistance of such structural elements. To carry out such dimensioning of the slab on which loads were applied, a crack was chosen among all cracks

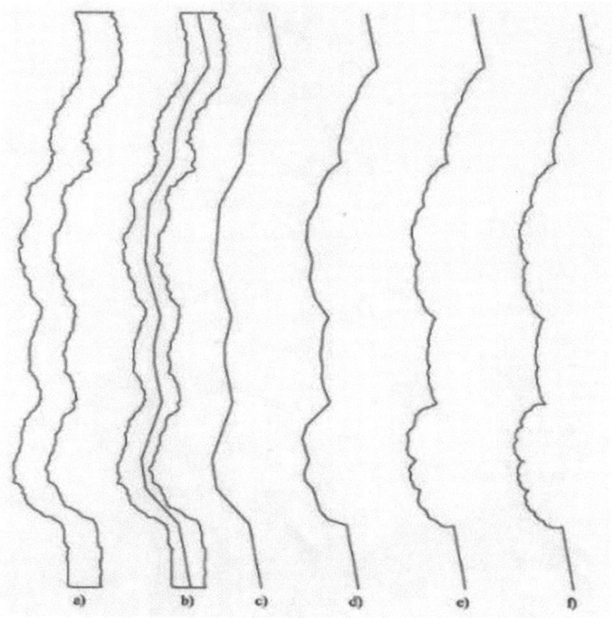

Figure 3.
(a) Vectorization of the (b) image with fractal axis and 4 fractal radius increases statuses of the fracture; (c), (d), (e), and (f) (Statuses 0, 1, 2, and 3).

created and vectorization of the outline of the chosen fracture was carried out, **Figure 3a**, with CorelDraw 4X to obtain the axis of the fracture.

From here, a set of parts of that axis was sectioned and chosen to create the generator, which will be known as the starting point of the fractal. In this case, three strokes were chosen, which are constantly repeated throughout the fracture, **Figure 3a** and **b**. These strokes are repeated at a lower scale than the original one, within each separate section, called status. The main axis of the fracture is *status 0*, following scale or lower repetition, *status 1*, and so on, repeatedly, up to *status n*. In this study, we only used up to *status 3* of the fracture, since the scale with higher status becomes imperceptible. These repetitions at a lower scale were made with AutoCAD software.

Upon obtaining *status 0*, certain symmetry and proportionality were observed in the fracture. It showed a 70° tilt degree and a proportional radius increment in circles as the length of the fracture increased, **Figure 4**. Upon meshing with the vertical and horizontal axes, regarding the tilt of the radius increasing angle, it was observed that *Status 0* matched with the beginning and the end of the vertical axis and that the increments in radius upon writing X against Y showed almost the same proportion, **Figure 4**. After this process, the fractal dimension for each status was calculated using five of the well-known theoretical methods: (i) box dimension, (ii) perimeter-area dimension, (iii) information dimension, (iv) mass dimension, and (v) ruler dimension. The fundamental explanation for why all these methods were used for the calculation of the fractal dimension of the fractured slab is that each method provides different results for the fractal dimension of the fracture, and it has been wanted to manifest that the values obtained are very similar but slightly different. To estimate the fractal dimension, a fractal dimension calculator should be used, and this is carried out through the algorithm *box-counting* or *box-counting dimension*.

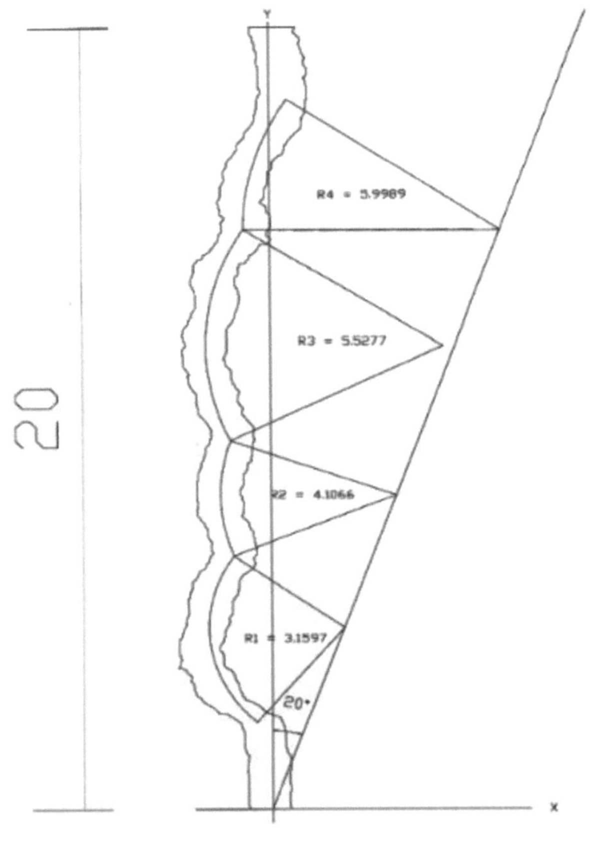

Figure 4.
Fracture tilt angle.

3.2 Fractal dimension through the box-counting method (box-dimension method)

This is an application created in the beginning of the 1990s. It establishes that if a line is split in half, we will have two parts to recreate the original line. If it is split into four parts, we will use that to cover the original line. We may deduce that if we split a line's time, we will have such a number of segments covering the original line. Therefore:

It is also true that, if we take a square and split each side in half, we obtain four smaller parts of the original square. If we split each side into ¼, we will have 16 parts of the original. Therefore, if we repeat this for: $D = 2 : N_s = \left(\frac{1}{s}\right)^2$ and for $D = 3 : N_s = \left(\frac{1}{s}\right)^3$.

1, 2, and 3 exponents in the above examples are fundamental for the concept of dimension. This may be generalized by Eq. (1):

$$N_s = \left(\frac{1}{s}\right)^D \tag{1}$$

Fractal dimension	Standard deviation	R-square	Equation of straight line	Estimation method
1.95322	0.0109513	0.9964	$y = 1.81E{+}06{*}x^{-1.95}$	Box dimension
1.03266	0.1647710	0.9855	$y = 0.105{*}x^{1.94}$	Perimeter-area dimension
1.96417	0.0010445	0.9921	$y = 1.78E{+}06{*}x^{-1.96}$	Information dimension
1.95776	0.0093884	0.9898	$y = 3.9{*}x^{1.96}$	Mass dimension
1.01606	0.0009957	0.9823	$y = 6.14E{+}03{*}x^{-1.02}$	Ruler dimension

Table 1.
Values of calculation of fractal dimension for each method used.

where D is the dimension (whole or unit) as in the beginning, but if we take logarithms on both sides, we have Eq. (4):

$$\log\left(\mathrm{N_s}\right) = D\log\left(\frac{1}{s}\right) \tag{2}$$

In other words, we may estimate the dimension graphing $\log(N(s))$ to $D\log(1/s)$. Its slope is the dimension. If the said dimension is not a whole, then it is a fractioned dimension.

The box-dimension method is the most popular one and counts boxes containing part of the fractal as a calculation procedure. Given an A fractal structure contained in a d-dimensional volume, the counting method implies distributing the space of the structure on a fixed d-dimensional mesh with square boxes of the same size r. Several algorithms [8–10] based on this method have been widely developed and used to estimate Df, and may be applied to systems with or without self-similarity. The fractal dimension obtained was 1.95322, **Table 1**.

3.3 Fractal dimension through the perimeter-area dimension method

The perimeter-area fractal dimension is attractive because it reflects the complexity of a shape through a range of spatial scales (size of fragments). However, as it's homologous to the fragment level (FRAC), the perimeter-area fractal dimension is only significant if the log-log ratio between perimeter and area is linear on the full range of fragment sizes. If it is not (and this should be determined separately), the fractal dimension is calculated separately for the range of patch sizes or fragments on which it is constant. Note that as this index uses regression analysis, it is subject to false results when there are small sample sizes. In mosaics with only some patches or fragments, it is common to obtain values above the theoretical limits of this index. PAFD (perimeter-area fractal dimension) is equal to 2, divided by the slope of the regression line obtained through regression of the mosaic piece mosaic area logarithm (in square meters) times the piece mosaic perimeter logarithm (in meters). That is, 2 divided by the coefficient b_1 derived from a regression of minimal squares for the equation $\ln(\text{area}) = b_0 + b_1 * \ln(\text{perimeter})$. Note that PAFD excludes any background mosaic. It has no units and its range is $1 \leq \text{PAFD} \leq 2$.

The formula for the calculation of the fractal dimension through this method is Eq. (5).

$$PAFD = \frac{2}{\frac{\left[n_i \sum_{j=1}^{n} \left(\ln\left(p_{ij}\right) * \ln\left(a_{ij}\right)\right)\right] - \left[\left(\sum_{j=1}^{n} \ln\left(a_{ij}\right)\right)\left(\sum_{j=1}^{n} \ln\left(a_{ij}\right)\right)\right]}{\left(n_i \sum_{j=1}^{n} \ln\left(p_{j=1}^{2}\right)\right) - \left(\sum_{j=1}^{n} \ln\left(p_{ij}\right)\right)^2}} \tag{3}$$

a_{ij} is the area (in square meters) of the fragment or piece ij, p_{ij} is the perimeter (in meters) of the fragment or piece ij, and n_i is the number of pieces in the mosaic patch or fragment i (class). The fractal dimension thus obtained is 0.164771, **Table 1**.

3.4 Information-dimension method

The concept of fractal dimension may be seen as a basic vision of complex construction. Let us define the information function as Eq. (6), see [11]:

$$I = -\sum_{i=1}^{N} P_i(\epsilon)\ ln\left[P_i(\epsilon)\right] \tag{4}$$

where $P_i(\epsilon)$ is the natural measure or the probability that the element i is the index of the normalized population such that: $\sum_{i=1}^{N} P_i(\epsilon) = 1$. The dimension of information d_{inf} is Eq. (7):

$$d_{inf} \equiv -\lim_{\epsilon \to 0^+} \frac{I}{\ln(\epsilon)} = \lim_{\epsilon \to 0^+} \sum_{i=1}^{N} \frac{P_i(\epsilon)\, ln\left[P_i(\epsilon)\right]}{\ln(\epsilon)} \tag{5}$$

The fractal dimension, thus obtained is 0.0010445, **Table 1**.

3.5 Mass-dimension method

Unifractal shows something similar to a random Cantor dust. This is the version of the two dimensions of the Cantor set, attributed to the mathematician George Cantor [12]. The method was supported when Berger and Mandelbrot found data showing strong hierarchical grouping in a high number of superimposed levels. The model published is a discretized Lévy dust, but the authors here consider the first a model based on Cantor dust, see [13].

Entries N and b are not present in more realistic models, much less in reality, and should not be measured separately only through the similar dimension Eq. (8).

$$D = \frac{\log N}{\log b} \tag{6}$$

In the case of Lévy dust, no counter-physical grid and cascade are involved; hence, N and b are not needed. Here, the sole parameter D is an easily measurable concrete quantity. It is not a similarity dimension, but a mass dimension. That is, an interval of length R centered on the dust contains a mass of the order of R^D. The fractal dimension obtained with this method is 0.0093884, **Table 1**.

3.6 Ruler-dimension method

A standard method used to estimate the length of a curve in a plane is counting how many lengths of a ruler are used to move from the start to the end of the curve. Since the ruler has a finite length, the details of the curve are smaller than the ruler with which we obtain the jumps, and therefore, the length we measure is generally lower than the real length. We intuitively believe our rulers are shorter, but our estimation of the actual length is more exact and increases.

The estimation of the length on the right is closer to reality since successive rulers are closer. The estimation improves each time the length of the ruler decreases.

The ruler or compass dimension of a curve is calculated between the ratio of a ruler with length L and the number of rulers required to measure the length of a curve, given as $N(L) = c\,L^{-D}$, where c is a constant. Taking the logarithms of both sides, we have: $\log(N(L)) = \log(c) - D\log(L)$. Therefore, if we graph $\log(N(L))$ *vs* $\log(L)$ within a range valued as L, the slope will be an estimate of the fractal dimension D, see [14, 15]. It must also be said that after making the corresponding scaling of *Status 0* of the fracture chosen up to *Status 3*, the necessary calculations were made through the methods mentioned above. The fractal dimension, thus obtained, is 1.01606, **Table 1**.

Table 1 shows that there is a great similarity among three of the methods regarding the results: box dimension, information dimension, and mass dimension; because the fracture is proportional as it expands along the slab. We may observe that the fractal dimension is between 1 and 2. In addition to this, from the values in **Figure 6**, it was

x	Slope equation $y = a + bx$
1.68	4.71675451
2.91	8.16668121
4.02	11.2800297
5.37	15.0665346

Table 2.
Results of calculation of the slope of fracture radius dot increase.

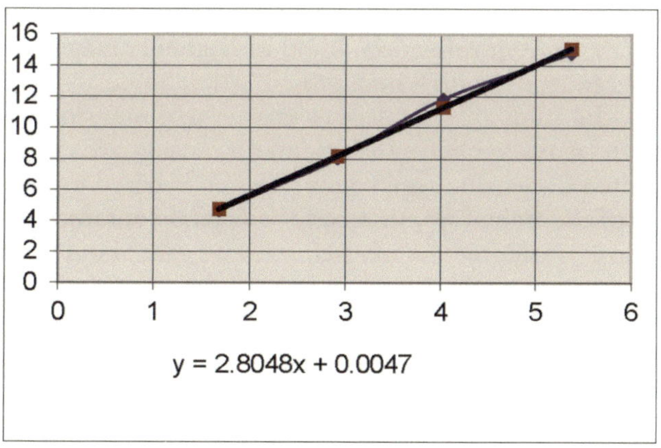

Figure 5.
Increase in the slope of straight line and equation of the slope.

(a) (b)

Figure 6.
(a) Fracture chosen and (b) its vectorization.

possible to obtain the equation of the pending line of the increments that the fracture had, **Table 2**. The values obtained are represented in **Figure 5**.

4. Intersection of fractal curves

Let us now focus on a phenomenon commonly present in this type of tests or cases: the intersection of fractal curves, as shown in **Figure 2**. Since the intersection of fractals F and G is often a fractal, it is natural to relate the dimension of such an intersection with one of the original sets, but, in general, it may not be said that A is true. This is because if F is limited, there may exist a copy F_1 of F such that dim $(F \cap F_1)$ – dim-F exists, but in the copy, if the sets are disjoint, dim $(F \cap F_1) = \Phi$. In addition, if F and F_1 are collinear, then $F \cap F_1$ is a line segment. However, if they intersect at a given angle (as in **Figure 2**), such an intersection is a singular or sole point (also shown in **Figure 2**). The latter case is the most frequent. If the transformation σ of the plane is congruent; that is, a rigid movement, it transforms any set F into a congruent $\sigma(F)$ without reflection. Rigid movements may be parameterized through coordinates (x, y, θ), with θ being the rotation angle.

The objective of the study of fractals is to know the dimension of fractal curves intercepted. Since the ways to obtain the fractal dimension are varied, we will use the Hausdorff-Besicovitch dimension (see below), which is based on considering an open cover of n-spheres of the fractal set, that is, for a fractal contained in the Euclidean plane open circles are considered (in mathematics we talk about balls, but everything depends on the dimension in which such exist). Of the possible coatings, the minimum formed by balls with a smaller diameter equal to a given size ε, Eq. (9) is considered. Once such a minimum is calculated, its limit is considered when $\varepsilon \to 0$, therefore formally defining Hausdorff content as:

$$\mathcal{H}_\delta^s(F) = min\left\{\sum_{i=1}^{\infty} |U_i|^s\right\} \ni \mathcal{H}^s(F) = \lim_{\delta \to 0} \mathcal{H}_\delta^s(F) \tag{7}$$

where $|U_i|$ = diam $(U_i) < \delta$, complying with the fact that Hausdorff content defines a function of the set power $\mathfrak{R}^n$ in non-negative real numbers, Eq. (10).

$$\mathcal{H}^s : P(\mathfrak{R}^n) \to \{0\} \cup \mathfrak{R}^+ \cap \{\infty\} \tag{8}$$

For any set, the function defined in Eq. (10) has the property to be null for any $s > s_0$ and infinite for $s < s_0$, and for $s = s_0$ the Hausdorff-Besicovitch dimension is obtained, shown below dim_{HB}.

Given F and G are two fractal curves to which the dimension mentioned above has been applied, then, in general, it is true that Eq. (11):

$$dim\ (G \cap \sigma(F)) \leq max\ \{0, dimG + dimF - n\} \tag{9}$$

And it is also true that Eq. (12):

$$dim(G \cap \sigma(F)) \geq \mathrm{dim}G + dim\ F - n \tag{10}$$

Therefore, it is possible to obtain higher limits for $\mathrm{dim}(G \cap \sigma(F))$ with a given translation or similarity group. Such limits are kept for large groups of translations and similarities.

Based on **Figure 7**, and as was mentioned in the first paragraph of this section, parameterization allows a natural measure in the space where the rigid movement is made with the measure of a rigid movement set given in space, for instance, tridimensional, and if we assign the origin to a certain point in the rectangle [1, 2]*[0, 2] we obtain measure $1^*2^*2\pi$.

Algebraically, in $\mathfrak{R}^n$ where two smooth variables G and F are completely intersected, in general, they intersect in a sub-variety of dimension $max\ \{0, dim - G + dim - F - n\}$. If $\mathrm{dim}G + dim\ F - n > 0$, therefore Eq. (13):

$$dim(G \cap \sigma(F)) = \mathrm{dim} - G + dim - F - n \tag{11}$$

For a rigid movement set with positive measure, but it is 0 for everything else.

The study of fractal intersections moving between them is known as integral fractal geometry. In the optics of classical geometry, see [16], they may be

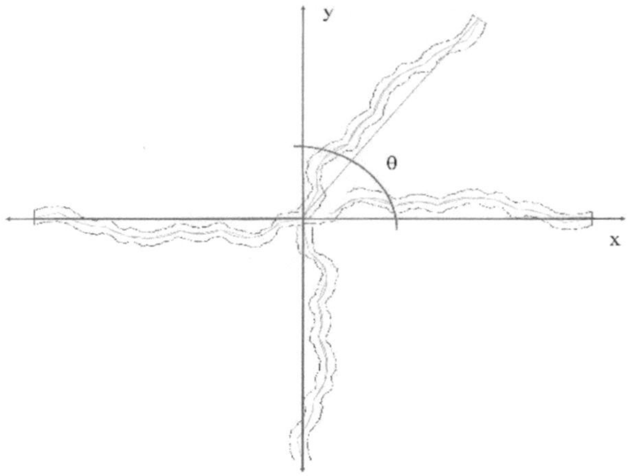

Figure 7.
Intersection of fractals.

consulted, and the ones who have contributed the most to the fractal intersection are seen [17–20].

Although the vectorization of the fractal image was chosen here, there is an article that proposes a method of rapid fractal coding based on the fractal dimension. The fractal dimension in fractal theory can be used to describe the texture of the image. The larger the fractal dimension, the rougher the surface of the corresponding graph, and vice versa. The method is based on the differential box count that is chosen specifically for texture analysis. Since the search space is reduced and the sorting operation is simple and computationally efficient, improving the encoding speed and preserving the quality of the decoded image; see [21].

5. Conclusions and impact of the research

From the civil engineering point of view and regarding the laboratory tests, it must be established that the low walls supporting the slab bent within the permissible limits since there were already headframes confining them. On the other hand, there was no reinforcing steel in the temperature nerve, so it did not function as such and that was a noticeable aspect. In addition, the surface of the slab on which the smooth steel rod was placed was not smooth enough, that is, the surface of the slab was rugose, which caused the load again not to be completely vertical, but rather, that the load might have had a certain unknown tilt angle but is supposed to be lower than the angle generated in the first test, so that applied load may be diminished in a certain percentage lower than the previous one. Taking into account the deformations obtained in this test, they were also higher than the permissible ones, above all, in the second and third loads. In the first test, it was almost the permissible one: 2.61, 4.73, and 5.98 cm, respectively.

The element worked satisfactorily, and while it is true that its deformations were not permissible, this was because the loads applied are not the ones usually received by a slab, and in cases where it is necessary to support a load of such magnitude, it has to be replaced with an element with higher load capacity, such as a slab or if necessary a column, **Figure 8**.

Regarding the fractal part, the methods used in this work are related to the fractal line or curve taken as a unit and the number of segments into which it is divided and the fractal dimension is always between one and two, but not higher than two. Logarithmic units are used, since smaller scales of the unit are progressively used, which are proportionally mirrored, that is why the logarithm is used, due to the smaller proportionate scales that appear. In the equations of straight lines shown in **Table 1**, we may see that the exponents to which the variable x is raised are almost identical, except for the one obtained through the ruler-dimension method.

Whether using the box-dimension method, the ruler-dimension method, or the mass-dimension method, these methods are related to the logarithm of N (which is the unit or the whole of a segment) between the numbers of s (which is the number of parts into which N is divided).

The increase of regression lines has also been observed, but they do not change since they follow the same increase direction. This means that the fracture never gets to failure, which will increase infinitely at higher scales, and which, to the eye, one might think is a curve, but from a closer point of view, one would realize that there are the same fractal lines but very small. These are imperceptible to the human eye and are only visible through other instruments or programs that may increase the scale of

Figure 8.
Large deformation and cracking a short time before the collapse.

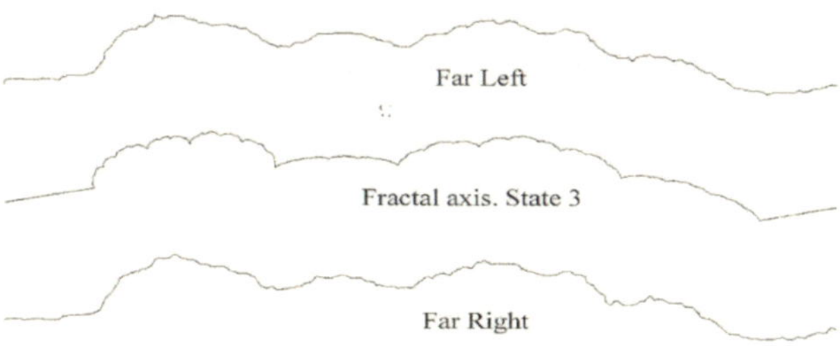

Figure 9.
Comparison of the ends with the fractal axis in its status 3.

the drawing. In addition, it has been observed that *Status 3*, which is still the fractal axis but at a higher scale, shows us that it is very similar to the borders of both ends, which at a higher scale would be more similar to the fractal axis, with a higher *status n* on the borders of the fracture. This shows that no matter the distance of the opening of the fracture, it will always be the same and/or similar to the fractal axis on both ends, **Figure 9**.

Regarding the theoretical part related to the fractal intersection, it has been proven that if there are two collinear fractal curves, their intersection is a segment of a line. If they are cut with a certain angle, **Figure 2**, such an intersection is a singular point and the rigid movements may be parameterized through coordinates (x, y, θ); with θ being the rotation angle. In addition, the fractal geometry obtained with the fractal in **Figure 3** is impressive upon considering the large intersections between them.

Acknowledgements

I want to thank Velasco Aguilar, C. S., and Casanova-del-Angel, F., who have allowed me to use one of the different tests carried out on various reinforced concrete structures (columns, beams, slabs, and frameworks), published in [10]; Santos Vázquez, R. E., who has made some of the calculations to dimension the fracture under study [22]; and the IPN-SIP research projects 20181253 and 20196119.

Conflict of interest

The author declares no conflict of interest.

Author details

Francisco Casanova-del-Angel
ESIA, Unit ALM of the Instituto Politécnico Nacional. Mexico

*Address all correspondence to: fcasanova49@prodigy.net.mx

References

[1] Apostol Tom M. Mathematical Analysis: A Modern Approach to Advanced Calculus. Department of Mathematics. California Institute of Technology. Addison-Wesley Publishing Company Inc; 1965

[2] Hurst HE. Long-term storage capacity of reservoirs. Proceedings of the Institution of Civil Engineers. 1951;**116**: 770-808

[3] Hurst HE, Black RP, Simaika YM. Long-term Storage: An Experimental Study. London: Constable; 1965

[4] Barnsley M. Fractals Everywhere. Orlando, FL: Academic Press Inc; 1988

[5] ASTM C150/C150M-12. Standard Specification for Portland Cement. West Conshohocken, PA: American Society for Testing and Materials International; 2012

[6] Wang L, Jin M, Zhou S, Tang S, Xiao L. Investigation of microstructure of C-S-H and micro-mechanics of cement pastes under NH_4NO_3 dissolution by ^{29}Si MAS NMR and microhardness. Measurement. 2021;**185**:110019

[7] Velasco Aguilar CS, Casanova-del-Angel F. Estudio de las fracturas con fractales en elementos de concreto reforzado: pruebas de laboratorio. Journal El Portulano de la Ciencia. 2004;**12**:421-434

[8] Li J, Du Q, Caixin S. An improved box-counting method for image fractal dimension estimation. Pattern Recognition. 2009;**11**:2460-2469

[9] Nirupam S, Chaudhuri BB. An efficient differential box-counting approach to compute fractal dimension of image. IEEE Transactions on Systems, Man, and Cybernetics. 1994;**24**(1): 115-120

[10] Pentland AP. Fractal-based description of natural scenes. IEEE Transactions on Pattern Analysis and Machine Intelligence. 1984;**6**: 661-674

[11] Baker GL, Gollub JB. Chaotic Dynamics: An Introduction. 2nd ed. Cambridge, England: Cambridge University Press; 1996

[12] Cantor G. On the power of perfect sets of points (De la puissance des ensembles parfait de points). Acta Mathematica. 1884;**4**:381-392

[13] Berger JM, Mandelbrot BB. A new model for error clustering in telephone circuits. IBM Journal of Research and Development. 1963;**7**(3):224-236

[14] Otto PH, Richter PH. The Beauty of Fractals. Germany: Springer-Verlag; 1986

[15] Peitgen HO, Jürgens H, Saupe D. Fractals for the Classroom: Part One Introduction to Fractals and Chaos (Corrected Edition). Springer 1991. ISBN-10: 038797041X. ISBN-13: 978-0387970417

[16] Santaló LA. Integral Geometry and Geometric Probability. Reading, MA: Addison-Wesley; 1976

[17] Kahane JP. Aspects of Mathematics and Its Applications. Amsterdam: North-Holland; 1986. pp. 419-430

[18] Mattila P. Hausdorff dimension and capacities of intersections of sets in n-space. Acta Mathematica. 1984;**152**: 77-105

[19] Mattila P. On the Hausdorff dimension and capacities on intersections. Mathematika. 1985;**32**: 213-217

[20] Mattila P. Geometry of Sets and Measures in Euclidean Spaces. Cambridge: Cambridge University Press; 1995

[21] Wang X-Y, Lang Y. A fast fractal encoding method based on fractal dimension. Fractals. 2009;**17**(04): 459-465

[22] Santos Vázquez RE. Dimensionamiento de la fractura fractal en el concreto [Tesis de licenciatura]. Mexico: Escuela Superior de Ingeniería y Arquitectura, Profesional Unit Adolfo López Mateos of the Instituto Politécnico Nacional; 2016

Chapter 7

Perspective Chapter: Probabilistic Modeling of Failure – Nonlinear Approximation

Alireda Aljaroudi and Ahmed Aljaroudi

Abstract

The failure of systems to meet the specified requirements may have adverse effects on their integrity and reliability. The systems could be mechanical, electrical, structural, telecommunications, or electronic that are designed and built to satisfy certain technical specifications and operational requirements. Failure does not necessarily mean the occurrence of a disaster or damage to the system, but also the degraded performance of such systems is considered a failure. One of the essential indicators of the performance and reliability of a system is the probability of failure which is computed by probabilistic methods. One of these methods is the first-order reliability method (FORM). Using FORM to estimate the probability of failure of systems having a nonlinear or a higher-order performance function may provide inaccurate results that may lead to misleading conclusions. To resolve this issue, the second-order reliability method (SORM) is recommended to estimate the probability of failure. This chapter presents commonly used probabilistic approximation methods to estimate the probability of failure for nonlinear performance functions. Illustrative examples to demonstrate the application of these methods are provided at the end of the chapter.

Keywords: second-order approximation, failure analysis, probabilistic methods in engineering, second-order reliability, reliability engineering

1. Introduction

Probability of failure can be calculated using analytical as well as simulation methods. Simulation methods such as the Monte Carlo technique can be used to calculate the probability of failure. It is an essential step in the calculation process to validate the analytical results. One of the analytical methods to compute the probability of failure is the first-order reliability method (FORM) which is based on the first-order expansion of the Taylor series. This method may provide inaccurate results when approximating nonlinear or higher-order performance functions. It should be noted that the performance function can be referred to as the limit state function (LSF).

The second-order reliability method, abbreviated as SORM, is used to resolve the nonlinearity issues of the performance function or the LSF. It uses the second-order expansion of the Taylor series to include the curvature of the LSF in the calculation to achieve better accuracy of the results.

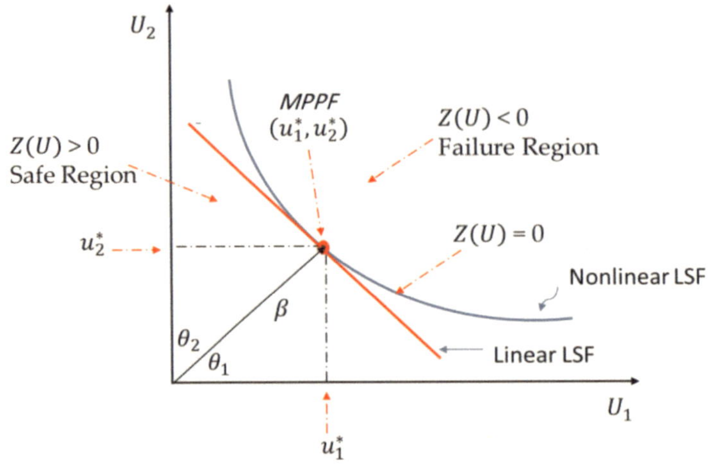

Figure 1.
Representation of the linear and nonlinear LSF.

Figure 1 shows the failure regions for linear and nonlinear LSFs. It also shows the most probable point of failure (MPPF) as the tangent point on the limit-state surface, $Z(U^*) = 0$, and the reliability index, β, is the shortest distance from the origin to the limit-state surface.

Breitung [1], Hohenbichler and Rackwitz [2], Tvedt [3, 4], and many other researchers came up with methods to calculate the probability of failure for nonlinear second-order LSFs. These methods use β obtained by FORM and the principal curvature, k_i, of the limit state surface in the calculation. Essentially, k_i estimates how much the curve diverges from the straight line, and it is an essential part of the calculation process. It is added to the calculation to provide a more accurate estimate of the probability of failure obtained by FORM. This chapter presents commonly used techniques that provide solutions to nonlinear second-order LSFs. The focus of this chapter is on SORM with the assumption that the random variables of the LSF are uncorrelated. A brief discussion about SORM and the commonly used methods for calculating the probability of failure are provided in the next section. Numerical examples are provided in Section 3 to demonstrate the calculation steps and provide in-depth understanding of these methods. Appendix A provides an overview of FORM and the steps involved in its calculation.

2. Second Order Reliability Method (SORM)

The LSF may exhibit nonlinearity when random variables have a non-normal distribution or a nonlinear relationship. Moreover, transforming uncorrelated variables to correlated variables may present nonlinearity in the LSF. SORM is used to deal with the nonlinearity issues to provide more accurate results. It uses the second-order Taylor series expansion of the LSF at the most probable point of failure (MPPF). The vectorized form of the second-order Taylor Series expansion is used to express the LSF around the MPPF as:

$$Z(U) \approx Z(U^*) + \nabla Z(U^*)^t(U - U^*) + 0.5(U - U^*)^t \nabla^2 Z(U^*)(U - U^*) \qquad (1)$$

where:

- *$Z(U)$: LSF.*
- *U: represents the vector of the input variables.*
- *U^*: is the MPPF.*
- *$Z(U^*)$: constant term.*
- *$\nabla Z(U^*)$: is the first-order partial derivative of the LSF evaluated at MPPF.*
- *$\nabla Z(U^*)^t(U - U^*)$: is the linear term, that is, the first-order derivative.*
- *$0.5(U - U^*)^t \nabla^2 Z(U^*)(U - U^*)$: is the quadratic term.*
- *$\nabla^2 Z(U^*)$: is a square matrix of the second-order partial derivative of the LSF evaluated at MPPF. It is called the Hessian matrix.*

Since U^* is located on the limit-state surface, $Z(U^*)$ equals zero.
$Z(U)$ is normalized by dividing Eq. (1) by $|\nabla Z(U^*)|$ which yields:

$$Z_{norm}(U) \approx \frac{\nabla Z(U^*)^t(U - U^*)}{|\nabla Z(U^*)|} + \frac{0.5(U - U^*)^t \nabla^2 Z(U^*)(U - U^*)}{|\nabla Z(U^*)|} \tag{2}$$

where $|\nabla Z(U^*)|$ is the length of the gradient vector of the LSF at the MPPF and is equal to:

$$|\nabla Z(U^*)| = \sqrt{\sum_{i=1}^{n} \left(\frac{\partial Z(U^*)}{\partial u_i}\right)^2} \tag{3}$$

The approximation of LSF at the MPPF is expressed as:

$$Z_{norm}(U) \approx \frac{\nabla Z(U^*)^t(U - U^*)}{|\nabla Z(U^*)|} + \frac{0.5(U - U^*)^t \nabla^2 Z(U^*)(U - U^*)}{|\nabla Z(U^*)|} = 0 \tag{4}$$

From the first-order reliability approximation (FORM approximation) we have:

$$\frac{\nabla Z(U^*)^t U^*}{|\nabla Z(U^*)|} = -\beta \tag{5}$$

$$\frac{\nabla Z(U^*)}{|\nabla Z(U^*)|} = \boldsymbol{\alpha} \tag{6}$$

α: the directional cosines, $\alpha_1, \ldots, \alpha_n$ the components of the unit gradient vector α. These are the cosines of the angles between the vector β and the axes.

Denoting $\frac{\nabla^2 Z(U^*)}{|\nabla Z(U^*)|}$ as ***H***:

$$\boldsymbol{H} = \frac{\nabla^2 Z(U^*)}{|\nabla Z(U^*)|} \tag{7}$$

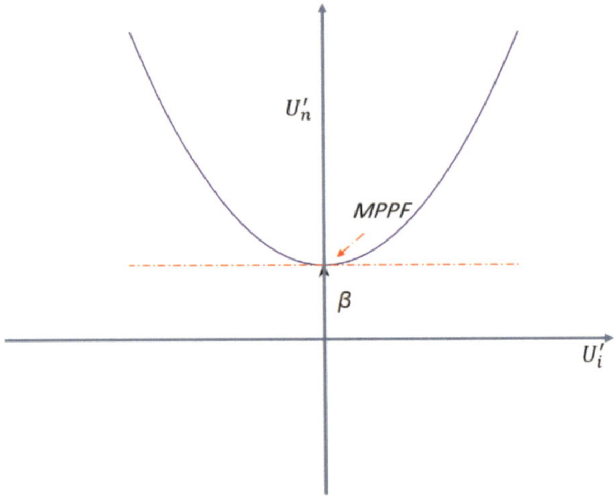

Figure 2.
Second-order failure surface shown in the rotated coordinates.

Using Eqs. (5) and (7), Eq. (4) yields:

$$Z_{norm}(U) \approx \frac{\nabla Z(U^*)^t U}{|\nabla Z(U^*)|} + \beta + 0.5(U - U^*)^t \boldsymbol{H}(U - U^*) = 0 \tag{8}$$

To go further with the solution of SORM, the U variables are transformed into a new set of standard normal random variables denoted as U'_i where the axis of the last variable coincides with the β vector as indicated in **Figure 2** [5–7].

This is an orthogonal transformation where:

$$\boldsymbol{U}' = \boldsymbol{RU} \tag{9}$$

$\boldsymbol{R}$: orthogonal n by n rotation matrix, where n is the number of variables. Since $\boldsymbol{R}$ is an orthogonal rotation matrix, $\boldsymbol{R}^{-1} = \boldsymbol{R}^t$

$$\boldsymbol{U} = \boldsymbol{R}^t \boldsymbol{U}' \tag{10}$$

The matrix $\boldsymbol{R}$ is constructed in two steps. First, define an initial matrix, $\boldsymbol{R}_{01}$ that consists of rows representing the unit vectors of the axes of the input variables. Since, we want to make β vector coincide with the axis of the last variable, we can simply substitute the last row of matrix $\boldsymbol{R}_{01}$ with the directional cosines of β. The following row vectors are the components of the $\boldsymbol{R}_{01}$ matrix:

$$\begin{aligned} r_{01} &= [1 \quad 0 \quad 0 \quad . \quad . \quad . \quad . \quad . \quad . \quad . \quad . \quad . \quad 0] \\ r_{02} &= [0 \quad 1 \quad 0 \quad . \quad . \quad . \quad . \quad . \quad . \quad . \quad . \quad . \quad 0] \\ &\vdots \\ r_{0n} &= \left[-\frac{\partial Z(U^*)/\partial U_1}{|\nabla Z(U^*)|} \quad . \quad . \quad . \quad -\frac{\partial Z(U^*)/\partial U_n}{|\nabla Z(U^*)|} \right] \end{aligned} \tag{11}$$

The Gram-Schmidt process is applied to orthogonalize the matrix [8]. The next step is to orthonormalize the matrix by dividing each element of the matrix by the length of its corresponding row vector.

Perform the Gram-Schmidt process for the matrix $\mathbf{R_{01}}$ using Eqs. (12) and (13), in reverse order:

$$r_n = r_{0n} \tag{12}$$

$$r_i = r_{0i} - \sum_{j=i+1}^{n} \frac{r_j r_{0i}^t}{r_j r_j^t} r_j \quad i = 1, 2, \dots, n-1 \tag{13}$$

Then orthonormalization is performed for each vector to produce the matrix $\boldsymbol{R}$. Using Eq. (10), Eq. (8) yields:

$$Z_{norm}(U') \approx -U'_n + \beta + 0.5(U' - U'^*)^t \boldsymbol{RHR^t}(U' - U'^*) = 0 \tag{14}$$

The $\boldsymbol{H}$ matrix is the second-order derivative of the LSF at the design point U^* divided by the length of the gradient vector, $|\nabla Z(U^*)|$, and is formulated as:

$$\boldsymbol{H} = \frac{1}{|\nabla Z(U'^*)|} \begin{bmatrix} \frac{\partial^2 Z(U'^*)}{\partial u_1^2} & \frac{\partial^2 Z(U'^*)}{\partial u_1 \partial u_2} & \cdots & \frac{\partial^2 Z(U'^*)}{\partial u_1 \partial u_n} \\ \frac{\partial^2 Z(U'^*)}{\partial u_2 \partial u_1} & \cdot & \cdots & \cdot \\ \cdot & \cdot & \cdots & \cdot \\ \cdot & \cdot & \cdots & \cdot \\ \frac{\partial^2 Z(U'^*)}{\partial u_n \partial u_1} & \cdot & \cdots & \frac{\partial^2 Z(U'^*)}{\partial u_n^2} \end{bmatrix} \tag{15}$$

Denoting $\boldsymbol{RHR^t}$ as $\boldsymbol{B}$, Eq. (14) becomes:

$$Z_{norm}(U') \approx -U'_n + \beta + 0.5(\boldsymbol{U'^t B U'}) = 0 \tag{16}$$

Then the solution of the second-order approximation is given as:

$$U'_n = \beta + 0.5(\boldsymbol{U'^t B U'}) \tag{17}$$

$\boldsymbol{B}$ is an $(n-1)$ by $(n-1)$ matrix and $\boldsymbol{U'}$ is a 1 by $(n-1)$ vector. The eigenvalues of matrix $\boldsymbol{B}$ are computed to obtain the main curvatures, $k_i's$, of the LSF at the MPPF. These $k_i's$ are used in the calculation of the probability of failure for nonlinear LSF as indicated in the next subsection [5–7]. The $0.5(\boldsymbol{U'^t B U'})$ term is simplified to $0.5\sum_{i=1}^{n-1} k_i U_i'^2$. Thus, Eq. (17) is expressed in terms of the curvatures, $k_i's$, as:

$$U'_n = \beta + 0.5\sum_{i=1}^{n-1} k_i U_i'^2 \tag{18}$$

If $k_i > 0$, the LSF will have a convex failure region, and if $k_i < 0$, it will have a concave failure region as shown in **Figure 3**. The figure shows the failure surfaces for an arbitrary LSF with three random variables.

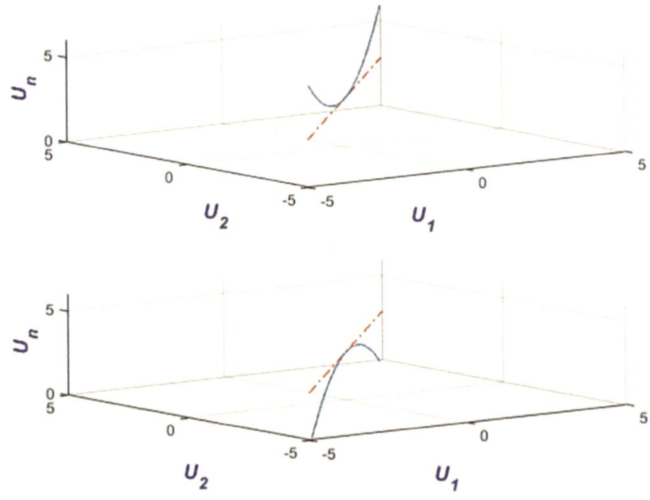

Figure 3.
Convex and concave failure regions for an arbitrary LSF with three random variables.

The calculation of $\nabla Z(U^*)$ and $\nabla^2 Z(U^*)$ can be performed in the X domain using the following equations:

$$\nabla Z(U'^*) = \frac{\partial Z(U'^*)}{\partial u_i} = \frac{\partial Z(X^*)}{\partial x_i}\sigma_{xi} \tag{19}$$

$$\nabla^2 Z(U'^*) = \frac{\partial^2 Z(U'^*)}{\partial u_i^2} = \frac{\partial^2 Z(X^*)}{\partial x_i^2}\sigma_{xi}^2 \tag{20}$$

where σ_{xi} is the standard deviation of the random variable x_i.

To summarize, the main objective of using SORM is to improve the estimate of the probability of failure that was obtained by FORM. The calculation includes three key steps. These are, the calculation of the reliability index and its associated design points either in the U domain or X domain using FORM, determining the principal curvatures, and determining the probability of failure using one of the commonly used methods for nonlinear LSFs.

2.1 Probability of failure

Breitung, Hohenbichler, and Tvedt methods use a correction factor which is expressed in terms of k_i and β to adjust the probability of failure obtained by FORM. The probability of failure for nonlinear LSF is then formulated as:

$$P_f = \Phi(-\beta)(CF) \tag{21}$$

The first term is the cumulative distribution function (cdf) of beta, CF is the SORM correction factor and β is the reliability index obtained by FROM.

2.1.1 Breitung formulation

Breitung formulation is expressed as:

$$P_{f\,Breitung} = \Phi(-\beta)\prod_{i=1}^{n-1}\frac{1}{\sqrt{(1+\beta k_i)}} \tag{22}$$

2.1.2 Hohenbichler formulation

Hohenbichler formulation is expressed in terms of the cdf of beta (the first term), probability density function of beta (pdf) which is the upper term in the denominator and the cdf of beta which is the lower term in the denominator as:

$$P_{f\,Hohenbichler} = \Phi(-\beta)\prod_{i=1}^{n-1}\frac{1}{\sqrt{\left(1+\frac{\phi(-\beta)}{\Phi(-\beta)}k_i\right)}} \tag{23}$$

2.1.3 Tvedt formulation

Tvedt formulation has three parts as indicated below:

$$P_{f\,Tvedt} = \Phi(-\beta)f1 + f2 + f3 \tag{24}$$

$$f1 = \prod_{i=1}^{n-1}\frac{1}{\sqrt{(1+\beta k_i)}}$$

$$f2 = (\beta\Phi(-\beta) - \phi(-\beta))\left(f1 - \left(\prod_{i=1}^{n-1}\frac{1}{\sqrt{(1+(1+\beta)k_i)}}\right)\right)$$

$$f3 = (1+\beta)(\beta\Phi(-\beta) - \phi(-\beta))\left(f1 - Real\left(\prod_{i=1}^{n-1}\frac{1}{\sqrt{(1+(j+\beta)k_i)}}\right)\right)$$

2.2 Calculation steps

- *Calculate the reliability index β using FORM.*
- *Calculate the design points U^* associated with the reliability index β following the steps in Appendix A.*
- *Determine the length of the gradient vector at the MPPF:* $|\nabla Z(U'^*)| = \sqrt{\sum_{i=1}^{n}\left(\frac{\partial Z(U^*)}{\partial u_i}\right)^2}$.
- *Construct the matrix $\mathbf{R_{01}}$ using Eq. (11).*
- *Perform Gram-Schmidt orthogonalization for the matrix $\mathbf{R_{01}}$ using* Eqs. (12) *and* (13) *and perform orthonormalization of each row vector to come up with the matrix* $\mathbf{R}$.
- *Compute the second-order derivative of the LSF at the design point, U^*, using Eq. (15) to obtain the **H** matrix.*
- *Compute the **B** matrix, $\mathbf{B} = \mathbf{RHR^t}$.*

- *Compute the eigenvalues of the matrix **B** to obtain the principal curvatures (k_i's).*
- *Calculate the probability of failure P_f using Eqs. (22)–(24).*

3. Examples

3.1 Example 1

The limit state function for a system has been formulated as:

$$Z(X) = 2.5\,x_1^2 - x_2^2 - 0.167 \text{ where } x_1 \sim N(2, 0.2),\ \ x_2 \sim N(2, 0.32).$$

A. Determine the reliability index (β) and the Most Probable Point of Failure (MPPF) in the X space and the U space. Use a tolerance of 0.0001 for convergence.

B. Determine the probability of failure using SORM by applying *Breitung*, *Hohenbichler,* and *Tvedt* methods.

C. Compare the results of parts A and B with results obtained by Monte Carlo simulation. Monte Carlo results are shown in **Table 1**.

Number of simulation cycles	P_f
2e5	0.00596500
1e6	0.00601600

Table 1.
MCS results – Example 1.0, Part C.

Solution

Part A

Determine β and the MPPF in the X space and the U space. FORM is used to solve part A. The computation continues until the solution converges with a tolerance error of 0.0001 $\left(\frac{|\beta_i - \beta_{i-1}|}{\beta_{i-1}} < 0.0001\right)$.

Assume the initial MPPF as the given mean of each variable.
Transform the LSF $Z(\boldsymbol{X})$ to $Z(\boldsymbol{U})$:

$$Z(\boldsymbol{U}) = 2.5(0.2u_1 + 2)^2 - (0.32u_2 + 2)^2 - 0.167$$

Compute the initial values of the design point in U space:

$$x_1 = \mu_{x_1} = 2$$
$$x_2 = \mu_{x_2} = 2$$

$$u_1 = \frac{x_1 - \mu_{x_1}}{\sigma_{x1}} = \frac{2-2}{0.2} = 0$$

$$u_2 = \frac{x_2 - \mu_{x_2}}{\sigma_{x2}} = \frac{2-2}{0.32} = 0$$

Compute the initial estimate of $Z(U)$:

$$Z(0,0) = 5.83300$$

Compute the partial derivatives of the LSF:

$$\frac{\partial Z(U^*)}{\partial u_1} = 0.2u_1 + 2 = 2$$

$$\frac{\partial Z(U^*)}{\partial u_2} = -0.2048u_2 - 1.28 = -1.28$$

Compute the standard deviation of the LSF:

$$\sigma_z = \sqrt{\sum_{i=1}^{n}\left(\frac{\partial Z(U^*)}{\partial u_i}\right)^2} = \sqrt{(2)^2 + (-1.28)^2} = 2.3745$$

Compute the initial reliability index β:

$$\beta = \frac{\mu_z}{\sigma_z} = \frac{5.8330}{2.3745} = 2.4565$$

Compute the directional cosines α_i:

$$\alpha_1 = -\frac{\left(\frac{\partial Z(U^*)}{\partial u_1}\right)}{\sqrt{\sum_{i=1}^{n}\left(\frac{\partial Z(U^*)}{\partial u_i}\right)^2}} = -\frac{2}{2.3745} = -0.8423$$

$$\alpha_2 = -\frac{\left(\frac{\partial Z(U^*)}{\partial u_2}\right)}{\sqrt{\sum_{i=1}^{n}\left(\frac{\partial Z(U^*)}{\partial u_i}\right)^2}} = -\frac{-1.28}{2.3745} = 0.5391$$

Iteration 1
Determine the new MPPF/design point in U space and X space:

$$u_1 = \beta\alpha_1 = 2.4565 * (-0.8423) = -2.0690$$

$$u_2 = \beta\alpha_2 = 2.4565 * (0.5391) = 1.3242$$

$$x_1 = \beta\alpha_1\sigma_{x_1} + \mu_{x_1} = u_1\sigma_{x_1} + \mu_{x_1} = -2.0690 * 0.2 + 2 = 1.5862$$

$$x_2 = \beta\alpha_2\sigma_{x_2} + \mu_{x_2} = u_2\sigma_{x_2} + \mu_{x_2} = 1.3242 * 0.32 + 2 = 2.4237$$

Compute the LSF in terms of the new design points:

$$Z(U^*) = Z(-2.0690, 1.3242) = 0.2485$$

Compute the partial derivatives and the standard deviation of the LSF at the new design points:

The partial derivatives:

$$\frac{\partial Z(U^*)}{\partial u_1} = 1.5862$$

$$\frac{\partial Z(U^*)}{\partial u_2} = -1.5512$$

The standard deviation of Z:

$$\sigma_z = \sqrt{(1.5862)^2 + (-1.5512)^2} = 2.2186$$

Compute the new β in terms of the new design points:

$$\beta = \frac{Z(\boldsymbol{U}^*) - \sum_{i=1}^{n} \frac{\partial Z(\boldsymbol{U}\ )}{\partial u_i}(u_i^*)}{\sqrt{\sum_{i=1}^{n}\left(\frac{\partial Z(\boldsymbol{U}^*)}{\partial u_i}\right)^2}}$$

$$= \frac{0.2485 - (1.5862 * (-2.0690) + (-1.5512 * 1.3242))}{2.2186}$$

$$\beta = 2.5171$$

We examine the tolerance:

$$\frac{|\beta_i - \beta_{i-1}|}{\beta_{i-1}} = \frac{|2.5171 - 2.45650|}{2.45650} = 0.02466$$

$$0.02466 > 0.0001$$

Since the error is greater than the established tolerance *(0.0001)*, the computation should continue until convergence is obtained.

The next step is to compute the directional cosines:

$$\alpha_1 = -\frac{1.5862}{2.2186} = -0.7150$$

$$\alpha_2 = -\frac{-1.5512}{2.2186} = 0.6992$$

The computation continues until β converges following the steps mentioned in Appendix A. Once convergence is obtained the probability of failure is calculated.

Table 2 shows the results for the next iterations. As the table indicates, convergence occurred at the third iteration with an error less than the tolerance.

$$\frac{|\beta_i - \beta_{i-1}|}{\beta_{i-1}} = \frac{|2.51171 - 2.51170|}{2.51170} = 0.000003981$$

$$0.000003981 \ll 0.0001$$

β is calculated to be *2.51171* and the MPPF is located at ***U****(-1.7758, 1.7762)* and ***X****(1.6448, 2.5684)*.

The probability of failure is computed in terms of the final value of β :

$$P_f = \Phi(-\beta) = 1 - \Phi(\beta) = 1 - \Phi(2.51171) = 0.006007$$

Iteration	0	1	2	3
u_1	0.0000	−2.0690	−1.7996	−1.7758
u_2	0.0000	1.3242	1.7599	1.7762
x_1	2.0000	1.5862	1.6401	1.6448
x_2	2.0000	2.4237	2.5632	2.5684
σ_1	0.2000	0.2000	0.2000	0.2000
σ_2	0.3200	0.3200	0.3200	0.3200
$Z(U^*)$	5.8330	0.2485	−0.0122	0.0000
$\partial Z/\partial u_1$	2.0000	1.5862	1.6401	1.6448
$\partial Z/\partial u_2$	−1.2800	−1.5512	−1.6404	−1.6438
α_1	−0.8423	−0.7150	−0.7070	−0.7073
α_2	0.5391	0.6992	0.7072	0.7069
σ	2.3745	2.2186	2.3197	2.3254
β	2.45648	2.51711	2.51170	2.51171
Pf	0.007015	0.005916	0.006008	0.006007

Table 2.
Summary of results – Example 1.0, Part A.

Part B

Determine the probability of failure using SORM by applying *Breitung*, *Hohenbichler,* and *Tvedt* methods.

Determine the length of the gradient vector at the MPPF:
The partial derivatives at the third iteration are:

$$\frac{\partial Z(U^*)}{\partial u_1} = 1.6448$$

$$\frac{\partial Z(U^*)}{\partial u_2} = -1.6438$$

$$|\nabla Z(U^*)| = \sqrt{\sum_{i=1}^{n} \left(\frac{\partial Z(U^*)}{\partial u_i}\right)^2} = \sqrt{(1.6448)^2 + (-1.6438)^2} = 2.3254$$

Construct the matrix $\mathbf{R_{01}}$ *using Eq. (11):*

$$\mathbf{R_{01}} = \begin{bmatrix} 1 & 0 \\ -\frac{\partial Z(U^*)/\partial u_1}{|\nabla Z(U^*)|} & -\frac{\partial Z(U^*)/\partial u_2}{|\nabla Z(U^*)|} \end{bmatrix} = \begin{bmatrix} 1 & 0 \\ -0.7073 & 0.7069 \end{bmatrix}$$

The second row is the directional cosines of the reliability index, β, at the MPPF.
Perform Gram–Schmidt orthogonalization for the matrix $\mathbf{R_{01}}$ *using Eqs.* (12) and (13), *and perform orthonormalization of each row vector to come up with the matrix* **R***:*

$$r_n = r_{02}$$

The last element of the matrix is calculated as:

$$r_i = r_{0i} - \sum_{j=i+1}^{n} \frac{r_j r_{0i}^t}{r_j r_j^t} r_j$$

$i = 2$

$$r_2 = r_{02} = [-0.7073\ 0.7069]$$

$i = 1$

$$r_1 = r_{01} - \sum_{j=i+1}^{n} \frac{r_j r_{02}^t}{r_j r_j^t} r_j = [1 \quad 0] - \frac{r_1 r_{02}^t}{r_1 r_1^t} r_1$$

$$= [1 \quad 0] - \frac{[-0.7073 \quad 0.7069]\begin{bmatrix} 1 \\ 0 \end{bmatrix}}{[-0.7073 \quad 0.7069]\begin{bmatrix} -0.7073 \\ 0.7069 \end{bmatrix}}[-0.7073 \quad 0.7069]$$

$$r_1 = [\ 0.4997 \quad 0.5000]$$

Normalizing the elements of the row vector, r_2 becomes:

$$r_2 = \frac{[0.4997\ 0.5000]}{\sqrt{0.4997^2 + 0.5000^2}} = [0.7073\ 0.7069]$$

The matrix $\boldsymbol{R}$ becomes:

$$\boldsymbol{R} = \begin{bmatrix} 0.7069 & 0.7073 \\ -0.7073 & 0.7069 \end{bmatrix}$$

Compute the second-order derivative of the LSF at the design point, U^*, using Equation (20) to obtain the $\boldsymbol{H}$ matrix:

$$\boldsymbol{H} = \frac{1}{|\nabla Z(U^*)|}\begin{bmatrix} \frac{\partial^2 Z(U^*)}{\partial u_1^2} & \frac{\partial^2 Z(U^*)}{\partial u_1 \partial u_2} \\ \frac{\partial^2 Z(U^*)}{\partial u_2 \partial u_1} & \frac{\partial^2 Z(U^*)}{\partial u_2^2} \end{bmatrix} = \frac{1}{2.3254}\begin{bmatrix} 0.2 & 0 \\ 0 & -0.2048 \end{bmatrix} = \begin{bmatrix} 0.0860 & 0 \\ 0 & -0.0881 \end{bmatrix}$$

Compute $\boldsymbol{RHR^t}$ matrix:

$$\boldsymbol{RHR^t} = \begin{bmatrix} -0.0011 & -0.0870 \\ -0.0870 & -0.0010 \end{bmatrix}$$

Compute the eigenvalues of the matrix **B** *to obtain principal curvatures (k_i's):*

The principal curvatures, $k_i's$, are computed by solving the eigenvalues of $\boldsymbol{RHR^t}$. The last column and last row are dropped from the above matrix, then the eigenvalues are obtained to determine the k_i values. From the above matrix, there is only one k at the MPPF, with a value of *(−0.0011)*.

Calculate the probability of failure P_f using Eqs. (22)–(24):

Breitung method

$$P_{f\,Breitung} = \Phi(-\beta)\prod_{i=1}^{n-1}\frac{1}{\sqrt{(1+\beta k_i)}}$$

$$= \Phi(-2.5117\,)\frac{1}{\sqrt{(1 + 2.5117\,(-0.0011))}}$$

$$P_{f\,Breitung} = 0.00601565$$

Hohenbichler method

$$P_{f\,Hohenbichler} = \Phi(-\beta)\prod_{i=1}^{n-1}\frac{1}{\sqrt{\left(1 + \frac{\phi(-\beta)}{\Phi(-\beta)}k_i\right)}}$$

$$= \Phi(-2.5117\,)\frac{1}{\sqrt{\left(1 + \frac{\phi(-2.5117\,)}{\Phi(-2.5117\,)}(-0.0011)\right)}}$$

$$P_{f\,Hohenbichler} = 0.00601671$$

Tvedt method

$$f1 = \prod_{i=1}^{n-1}\frac{1}{\sqrt{(1 + \beta k_i)}} = 1.00137$$

$$f2 = (\beta\Phi(-\beta) - \phi(-\beta))\left(f1 - \left(\prod_{i=1}^{n-1}\frac{1}{\sqrt{(1 + (1+\beta)k_i)}}\right)\right)$$

$$= (-0.00193)(\,1.00137 - (\,1.00192)) = 1.05670E - 06$$

$$f2 = 1.05670E - 06$$

$$f3 = (1+\beta)(\beta\Phi(-\beta) - \phi(-\beta))\left(f1 - Real\left(\prod_{i=1}^{n-1}\frac{1}{\sqrt{(1 + (j+\beta)k_i)}}\right)\right)$$

$$= (1 + 2.5117)(2.5117(\Phi(-2.5117)) - \phi(-2.5117))$$

$$\left(1.00137 - Real\left(\frac{1}{\sqrt{(1 + (j + 2.5117)(-0.0011))}}\right)\right)$$

$$f3 = -3.03430E - 9$$

$$P_{f\,Tvedt} = \Phi(-\beta)f1 + f2 + f3$$

$$= (6.00743\text{E} - 3)(1.00137) + (1.05670E - 06) + (-3.03430E - 9)$$

$$P_{f\,Tvedt} = 6.01671E - 03$$

See **Table 3**.

Part C

Compare the results of parts *A* and *B* with results obtained by Monte Carlo simulation:

The simulation was conducted using *2e5* and *1e6* simulation cycles as shown in **Table 1**. The results obtained by *Breitung*, *Tvedt*, and *Hohenbichler* methods are close to the Monte Carlo simulation results when using 1e6 simulation cycles.

Method	Reliability Index - β	P_f
FORM	2.5117077	0.00600743
SORM – Breitung		0.00601565
SORM – Hohenbichler		0.00601671
SORM – Tvedt		0.00601671

Table 3.
Summary of SORM results – Example 1.0 Part B.

3.2 Example 2

The performance function for a system has been formulated as:

$$Z(X) = 24x_1^2 + 13x_2^2 + x_3^2 - 100,\ x_1 \sim N(2, 0.2),\ x_2 \sim N(2, 0.32),\ x_3 \sim N(4, 0.4)$$

A. Determine the reliability index (β) and the most probable point of failure (MPPF) in the X space and the U space. Use a tolerance of 0.0001 for convergence.

B. Determine the probability of failure using SORM by applying *Breitung*, *Hohenbichler*, and *Tvedt* methods.

C. Compare the results of parts A and B with results obtained by Monte Carlo simulation. Monte Carlo results are shown in **Table 4**.

Number of simulation cycles	P_f
2e5	0.0018250
1e6	0.0018360

Table 4.
MCS results – Example 2.0, Part C.

Solution

Part A

Determine β and the MPPF in the X space and the U space. FORM is used to solve part A. The computation continues until the solution converges with a tolerance error of 0.0001. $\left(\frac{|\beta_i - \beta_{i-1}|}{\beta_{i-1}} < 0.0001\right)$.

Assume the initial MPPF as the given mean of each variable.
Transform the LSF $Z(\mathbf{X})$ to $Z(\mathbf{U})$:

$$Z(\mathbf{U}) = 24(0.2u_1 + 2)^2 + 13(0.32u_2 + 2)^2 + (0.4u_3 + 4)^2 - 100$$

Compute the initial values of the design point in U space:

$$x_1 = \mu_{x_1} = 2$$
$$x_2 = \mu_{x_2} = 2$$

$$x_3 = \mu_{x_3} = 4$$

$$u_1 = \frac{x_1 - \mu_{x_1}}{\sigma_{x1}} = \frac{2-2}{0.2} = 0$$

$$u_2 = \frac{x_2 - \mu_{x_2}}{\sigma_{x2}} = \frac{2-2}{0.32} = 0$$

$$u_3 = \frac{x_3 - \mu_{x_3}}{\sigma_{x3}} = \frac{4-4}{0.4} = 0$$

Compute the initial estimate of $Z(U)$:

$$Z(0,0,0) = 64$$

Compute the partial derivatives of the LSF:

$$\frac{\partial Z(U^*)}{\partial u_1} = 1.92u_1 + 19.2 = 19.2$$

$$\frac{\partial Z(U^*)}{\partial u_2} = 2.6624u_2 + 16.64 = 16.64$$

$$\frac{\partial Z(U^*)}{\partial u_3} = 0.32u_3 + 3.2 = 3.2$$

Compute the standard deviation of the LSF:

$$\sigma_z = \sqrt{\sum_{i=1}^{n} \left(\frac{\partial Z(U^*)}{\partial u_i}\right)^2} = \sqrt{(19.2)^2 + (16.64)^2 + (3.2)^2} = 25.6080$$

Compute the initial reliability index β:

$$\beta = \frac{\mu_z}{\sigma_z} = \frac{64}{25.6080} = 2.4992$$

Compute the directional cosines α_i:

$$\alpha_1 = -\frac{\left(\frac{\partial Z(U^*)}{\partial u_1}\right)}{\sqrt{\sum_{i=1}^{n} \left(\frac{\partial Z(U^*)}{\partial u_i}\right)^2}} = -\frac{19.2}{25.6080} = -0.7498$$

$$\alpha_2 = -\frac{\left(\frac{\partial Z(U^*)}{\partial u_2}\right)}{\sqrt{\sum_{i=1}^{n} \left(\frac{\partial Z(U^*)}{\partial u_i}\right)^2}} = -\frac{16.64}{25.6080} = -0.6498$$

$$\alpha_3 = -\frac{\left(\frac{\partial Z(U^*)}{\partial u_3}\right)}{\sqrt{\sum_{i=1}^{n} \left(\frac{\partial Z(U^*)}{\partial u_i}\right)^2}} = -\frac{3.2}{25.6080} = -0.1250$$

The initial calculation of the probability of failure is:

$$P_f = \Phi(-\beta) = 1 - \Phi(\beta) = 1 - \Phi(2.4992) = 0.00622$$

Iteration 1

Determine the new MPPF/design point in U space and X space:

$$u_1 = \beta\alpha_1 = 2.4992 * (-0.7498) = -1.8738$$

$$u_2 = \beta\alpha_2 = 2.4992 * (-0.64980) = -1.6240$$

$$u_3 = \beta\alpha_3 = 2.4992 * (-0.1250) = -0.3123$$

$$x_1 = \beta\alpha_1\sigma_{x_1} + \mu_{x_1} = u_1\sigma_{x_1} + \mu_{x_1} = -1.8738 * 0.2 + 2 = 1.6252$$

$$x_2 = \beta\alpha_2\sigma_{x_2} + \mu_{x_2} = u_2\sigma_{x_2} + \mu_{x_2} = -1.6240 * 0.32 + 2 = 1.4803$$

$$x_3 = \beta\alpha_3\sigma_{x_3} + \mu_{x_3} = u_3\sigma_{x_3} + \mu_{x_3} = -0.3123 * 0.4 + 4 = 3.8751$$

Compute the LSF in terms of the new design points:

$$Z(U^*) = Z\,(-1.8738, -1.624, -0.3123) = 6.8972$$

Compute the partial derivatives and the standard deviation of the LSF at the new design points:

The partial derivatives:

$$\frac{\partial Z(U^*)}{\partial u_1} = 1.92u_1 + 19.2 = 15.6022$$

$$\frac{\partial Z(U^*)}{\partial u_2} = 2.6624u_2 + 16.64 = 12.3163$$

$$\frac{\partial Z(U^*)}{\partial u_3} = 0.32u_3 + 3.2 = 3.1001$$

The standard deviation of Z:

$$\sigma_z = \sqrt{(15.6022)^2 + (12.3163)^2 + (3.1001)^2} = 20.1179$$

Compute the new β in terms of the new design points:

$$\beta = \frac{Z(\boldsymbol{U}^*) - \sum_{i=1}^{n} \frac{\partial Z(\boldsymbol{U}\ )}{\partial u_i}\left(u_i^*\right)}{\sqrt{\sum_{i=1}^{n}\left(\frac{\partial Z(\boldsymbol{U}^*)}{\partial u_i}\right)^2}}$$

$$= \frac{6.8972 - (15.6022 * (-1.8738) + 12.3163 * (-1.6240) + 3.1001 * (-0.3123))}{20.1179}$$

$$= 2.8384$$

We examine the tolerance:

$$\frac{|\beta_i - \beta_{i-1}|}{\beta_{i-1}} = \frac{|2.8384 - 2.4992|}{2.4992} = 0.1357$$

$$0.1357 > 0.0001$$

Since the error is greater than the established tolerance (0.0001), the computation should continue until convergence is obtained.

The next step is to compute the directional cosines:

$$\alpha_1 = -\frac{(78.011 * 0.2)}{20.1179} = -0.7755$$

$$\alpha_2 = -\frac{(38.488 * 0.32)}{20.1179} = -0.6122$$

$$\alpha_3 = -\frac{(7.7502 * 0.4)}{20.1179} = -0.1541$$

The computation continues until β converges following the steps mentioned in Appendix A. Once convergence is obtained the probability of failure is calculated.

Table 5 shows the results for the next iterations. As the table indicates, the convergence occurred at the third iteration with an error less than the tolerance.

Iteration	**0**	**1**	**2**	**3**
u_1	0.0000	−1.8738	−2.2013	−2.1911
u_2	0.0000	−1.6240	−1.7377	−1.7580
u_3	0.0000	−0.3123	−0.4374	−0.4478
x_1	2.0000	1.6252	1.5597	1.5618
x_2	2.0000	1.4803	1.4439	1.4375
x_3	4.0000	3.8751	3.8250	3.8209
$Z(U)$	64.0000	6.8972	0.1227	0.0007
$\partial Z/\partial u_1$	19.2000	15.6022	14.9735	14.9931
$\partial Z/\partial u_2$	16.6400	12.3163	12.0136	11.9596
$\partial Z/\partial u_3$	3.2000	3.1001	3.0600	3.0567
σ_1	0.2000	0.2000	0.2000	0.2000
σ_2	0.3200	0.3200	0.3200	0.3200
σ_3	0.4000	0.4000	0.4000	0.4000
σ_Z	25.6080	20.1179	19.4396	19.4208
α_1	−0.7498	−0.7755	−0.7703	−0.7720
α_2	−0.6498	−0.6122	−0.6180	−0.6158
α_3	−0.1250	−0.1541	−0.1574	−0.1574
β	2.49922	2.83840	2.84461	2.84463
Pf	0.006223	0.002267	0.002223	0.002223

Table 5.
Summary of the results – Example 2.0, Part A.

$$\frac{|\beta_i - \beta_{i-1}|}{\beta_{i-1}} = \frac{|2.84463 - 2.84461|}{2.84461} = 0.000007031$$

$$0.000007031 \ll 0.0001$$

β is calculated to be 2.84463 and the MPPF is located at $U(-2.1911, -1.7580, -0.4478)$ and $X(1.5618, 1.4375, 3.8209)$.

The probability of failure is computed in terms of the final value of β :

$$P_f = \Phi(-\beta) = 1 - \Phi(\beta) = 1 - \Phi(2.84463) = 0.002223$$

Part B

Determine the probability of failure using SORM by applying *Breitung*, *Hohenbichler*, and *Tvedt* methods.

Determine the length of the gradient vector at the MPPF:

The partial derivatives at the third iteration are:

$$\frac{\partial Z(U^*)}{\partial u_1} = 14.9931$$

$$\frac{\partial Z(U^*)}{\partial u_2} = 11.9596$$

$$\frac{\partial Z(U^*)}{\partial u_3} = 3.0567$$

$$|\nabla Z(U^*)| = \sqrt{\sum_{i=1}^{n}\left(\frac{\partial Z(U^*)}{\partial u_i}\right)^2} = \sqrt{(14.9931)^2 + (11.9569)^2 + (3.0567)^2} = 19.4208$$

Construct the matrix $\mathbf{R_{01}}$ *using Eq. (11):*

$$\mathbf{R_{01}} = \begin{bmatrix} 1 & 0 & 0 \\ 0 & 1 & 0 \\ -\frac{\partial Z(U^*)/\partial u_1}{\sqrt{\sum_{i=1}^{n}\left(\frac{\partial Z(U^*)}{\partial u_i}\right)^2}} & -\frac{\partial Z(U^*)/\partial u_2}{\sqrt{\sum_{i=1}^{n}\left(\frac{\partial Z(U^*)}{\partial u_i}\right)^2}} & -\frac{\partial Z(U^*)/\partial u_3}{\sqrt{\sum_{i=1}^{n}\left(\frac{\partial Z(U^*)}{\partial u_i}\right)^2}} \end{bmatrix}$$

$$\mathbf{R_{01}} = \begin{bmatrix} 1 & 0 & 0 \\ 0 & 1 & 0 \\ -0.77201 & -0.61581 & -0.15739 \end{bmatrix}$$

The last row is the directional cosines of the reliability index, β, at the MPPF.

Perform Gram-Schmidt orthogonalization for the matrix $\mathbf{R_{01}}$ *using Eqs.* (12) and (13), *and perform orthonormalization of each row vector to come up with the matrix* $\mathbf{R}$*:*

$$r_n = r_{0n} = r_{03}$$

$$r_i = r_{0i} - \sum_{j=i+1}^{n} \frac{r_j r_{0i}^t}{r_j r_j^t} r_j$$

i = 3

$$r_3 = r_{03} = [-0.77201 \quad -0.61581 \quad -0.15739]$$

The remaining elements of the matrix are calculated as:
i = 2

$$r_2 = r_{02} - \sum_{j=2}^{n} \frac{r_j r_{02}^t}{r_j r_j^t} r_j = [0\ 1\ 0] - \frac{r_3 r_{01}^t}{r_3 r_3^t} r_3$$

$$= [0 \quad 1 \quad 0] - \frac{[-0.77201 \quad -0.61581 \quad -0.15739]\begin{bmatrix}0\\1\\0\end{bmatrix}}{[-0.77201 \quad -0.61581 \quad -0.15739]\begin{bmatrix}-0.77201\\-0.61581\\-0.15739\end{bmatrix}}[-0.77201 \quad -0.61581 \quad -0.15739]$$

$$r_2 = [-0.47541 \quad 0.62077 \quad -0.096925]$$

Normalizing the elements of the row vector, r_2 becomes:

$$r_2 = \frac{[-0.47541 \quad 0.62077 \quad -0.096925]}{\sqrt{(-0.47541)^2 + 0.62077^2 + (-0.096925)^2}}$$

$$r_2 = [-0.6034\ 0.78789 - 0.12302]$$

Following the same procedures row vector r_1 becomes:

$$r_1 = [0.19977\ 0 - 0.97984]$$

The matrix $\boldsymbol{R}$ becomes:

$$\boldsymbol{R} = \begin{bmatrix} 0.19977 & 0 & -0.97984 \\ -0.6034 & 0.78789 & -0.12302 \\ -0.77201 & -0.61581 & -0.15739 \end{bmatrix}$$

Compute the second-order derivative of the LSF at the design point, U^*, using Equation (20) to obtain the ***H*** *matrix:*

$$\boldsymbol{H} = \frac{1}{|\nabla Z(U^*)|}\begin{bmatrix} \frac{\partial^2 Z(U^*)}{\partial u_1^2} & \frac{\partial^2 Z(U^*)}{\partial u_1 \partial u_2} & \frac{\partial^2 Z(U^*)}{\partial u_1 \partial u_3} \\ \frac{\partial^2 Z(U^*)}{\partial u_2 \partial u_1} & \frac{\partial^2 Z(U^*)}{\partial u_2^2} & \frac{\partial^2 Z(U^*)}{\partial u_2 \partial u_3} \\ \frac{\partial^2 Z(U^*)}{\partial u_3 \partial u_1} & \frac{\partial^2 Z(U^*)}{\partial u_3 \partial u} & \frac{\partial^2 Z(U^*)}{\partial u_3^2} \end{bmatrix}$$

$$= \frac{1}{19.4208}\begin{bmatrix} 1.9200 & 0 & 0 \\ 0 & 2.6624 & 0 \\ 0 & 0 & 0.3200 \end{bmatrix} = \begin{bmatrix} 0.0989 & 0 & 0 \\ 0 & 0.1371 & 0 \\ 0 & 0 & 0.0165 \end{bmatrix}$$

Compute $\boldsymbol{RHR^t}$ matrix:

$$\boldsymbol{RHR^t} = \begin{bmatrix} 0.0197650 & -0.0099306 & -0.0127056 \\ -0.0099306 & 0.1213464 & -0.0201427 \\ -0.0127056 & -0.0201427 & 0.1113186 \end{bmatrix}$$

Compute the eigenvalues of the matrix **B** *to obtain principal curvatures (k_i's):*

The principal curvatures, k_i's, are computed by solving the eigenvalues of $\boldsymbol{RHR^t}$. To do that, the last column and the last row are dropped from the matrix. The above matrix becomes:

$$\boldsymbol{RHR^t} = \begin{bmatrix} 0.0197650 & -0.0099306 \\ -0.0099306 & 0.1213464 \end{bmatrix}$$

Applying the eigenvalue method to obtain the principal curvatures:

$$k = [0.018803 \quad 0.122308]$$

Calculate the probability of failure P_f using Eqs. (22)–(24):

Breitung method

$$P_{f\,Breitung} = \Phi(-\beta) \prod_{i=1}^{n-1} \frac{1}{\sqrt{(1+\beta k_i)}}$$

$$= \Phi(-2.8446) \frac{1}{\sqrt{(1+2.8446(0.018803))(1+2.8446(0.122308))}}$$

$$P_{f\,Breitung} = 0.0018656$$

Hohenbichler method

$$P_{f\,Hohenbichler} = \Phi(-\beta) \prod_{i=1}^{n-1} \frac{1}{\sqrt{\left(1+\frac{\phi(-\beta)}{\Phi(-\beta)} k_i\right)}}$$

$$= \frac{\Phi(-2.8446)}{\sqrt{\left(1+\left(\frac{\phi(-2.8446)}{\Phi(-2.8446)}\right)(0.018803)\right)\left(1+\left(\frac{\phi(-2.8446)}{\Phi(-2.8446)}\right)(0.122308)\right)}}$$

$$= 0.0018363$$

Tvedt method

$$f1 = \prod_{i=1}^{n-1} \frac{1}{\sqrt{(1+\beta k_i)}} = 0.83918$$

$$f2 = (\beta\Phi(-\beta) - \phi(-\beta))\left(f1 - \left(\prod_{i=1}^{n-1} \frac{1}{\sqrt{((1+(1+\beta)k_i)}}\right)\right)$$

$$= ((2.8446)(0.0022231) - 0.0069786)(\ 0.83918 - (\ 0.9657)(0.82472))$$

$$f2 = -2.7977541E - 05$$

$$f3 = (1+\beta)(\beta\Phi(-\beta) - \phi(-\beta))\left(f1 - Real\left(\prod_{i=1}^{n-1}\frac{1}{\sqrt{(1+(j+\beta)k_i)}}\right)\right)$$

$$= (1+2.8446)(2.8446(\Phi(-2.8446)) - \phi(-2.8446))\left(0.83918\right.$$

$$\left. - Real\left(\frac{1}{\sqrt{(1+(j+2.8446)(0.018803))(1+(j+2.8446)(0.122308))}}\right)\right)$$

$$f3 = -7.5839622E - 06$$

$$P_{f\,Tvedt} = \Phi(-\beta)f1 + f2 + f3$$

$$= (0.0022231)(0.83918) + (-2.79775E - 05) + (-7.58396E - 06)$$

$$P_{f\,Tvedt} = 0.0018300$$

See **Table 6**.

Part C
Compare the results of parts A and B with results obtained by Monte Carlo simulation:

Method	Reliability Index – β	P_f
FORM	2.844633	0.0022231
SORM–Breitung		0.0018656
SORM–Hohenbichler		0.0018363
SORM–Tvedt		0.0018300

Table 6.
Analytical methods results – Example 2.0, Part B.

The simulation was conducted using 2e5 and 1e6 simulation cycles as shown in **Table 4**. The results obtained by *Breitung*, *Tvedt,* and *Hohenbichler* methods are close to the Monte Carlo simulation results when using 1e6 simulation cycles.

3.3 Example 3

Determine the second-order reliability index (β_{SORM}) for the probability of failure calculated in example 2.

Solution
The reliability index for the second-order LSF can be calculated by taking the inverse of the cumulative distribution of the probability of failure as:

$$\beta_{SORM} = \Phi^{-1}\left[1 - P_f\right]$$

$$\beta_{Breitung} = \Phi^{-1}[1 - 0.0018656\,] = 2.90004$$
$$\beta_{Tvedt} = \Phi^{-1}[1 - 0.0018300\,] = 2.90606$$
$$\beta_{Hohenbichler} = \Phi^{-1}[1 - 0.0018363] = 2.90499$$
$$\beta_{MC} = \Phi^{-1}[1 - 0.0018360\,] = 2.90505$$

The absolute difference between Monte Carlo method and the other methods is calculated below:

$$\text{Breitung method} = |2.90505 - 2.90004| = 0.00500$$
$$\text{Tvedt method} = |2.90505 - 2.90606| = 0.00101$$
$$\text{Hohenbichler method} = |2.90505 - 2.90499| = 0.00006$$

In this example, the reliability index obtained by *Hohenbichler* is the closest to the reliability index obtained by *Monte Carlo* method.

4. Summary and concluding remarks

This chapter presented commonly used analytical probabilistic methods for determining the probability of failure for nonlinear and higher-order systems. They included Breitung, Hohenbichler, and Tvedt methods. Furthermore, Monte Carlo Simulation was used to estimate the probability of failure to validate the results obtained by the methods mentioned above. It is concluded that the results obtained by Breitung, Tvedt, and Hohenbichler methods are closer to the results obtained by Monte Carlo Simulation when using 1e6 simulation cycles. Future work should include the analysis of multi-dimensional and higher-order failure functions having correlated variables. Moreover, it is recommended that future work apply finite element method to gain more insight into how the methods compare with each other.

A. An overview of FORM

A.1 Failure modeling

The performance function, also called the limit state function (LSF), is formulated in terms of a system's load and capacity. Both the load, L, and the capacity, C, have an impact on the performance of the system. The system fails when the load exceeds the capacity or when the LSF becomes less than zero. The LSF can be expressed as:

$$Z = C - L \tag{25}$$

The capacity C and the load L are formulated in terms of random variables, $x_1, x_2, \ldots x_n$:

$$Z(\boldsymbol{X}) = Z(x_1, x_2, \ldots \ldots \ldots \ldots, x_n) \tag{26}$$

These variables are assumed to be statistically independent random variables having a normal distribution. The probability of failure can be expressed as:

$$P_f = P(Z < 0) \tag{27}$$

This is the probability that the LSF becomes less than zero. In **Figure A1**, this is the region that exists above the straight line for linear LSF and above the curve for nonlinear LSF. The safe region or nonfailure region is below the curve, and this is the region where the LSF becomes greater than zero. The LSF becomes at the limit state when the LSF becomes equal to zero as **Figure A1** indicates.

Also, the figure shows the reliability index, β, which is the shortest distance from the origin to the surface. It is calculated as the ratio of the mean, μ_Z, to the standard deviation, σ_Z, of the LSF.

$$\beta = \frac{\mu_Z}{\sigma_Z} \tag{28}$$

$$\mu_Z \approx Z\left(\mu_{X_1}, \mu_{X_2}, \dots \dots \dots \dots, \mu_{X_n}\right) \tag{29}$$

$$\sigma_z = \sqrt{\sum_{i=1}^{n} \left(\frac{\partial Z(X^*)}{\partial x_i}\right)^2 \sigma_{x_i}^2} \tag{30}$$

Then, the probability of failure is expressed in terms of β as:

$$P_f = \phi(-\beta) = 1 - \Phi(\beta) \tag{31}$$

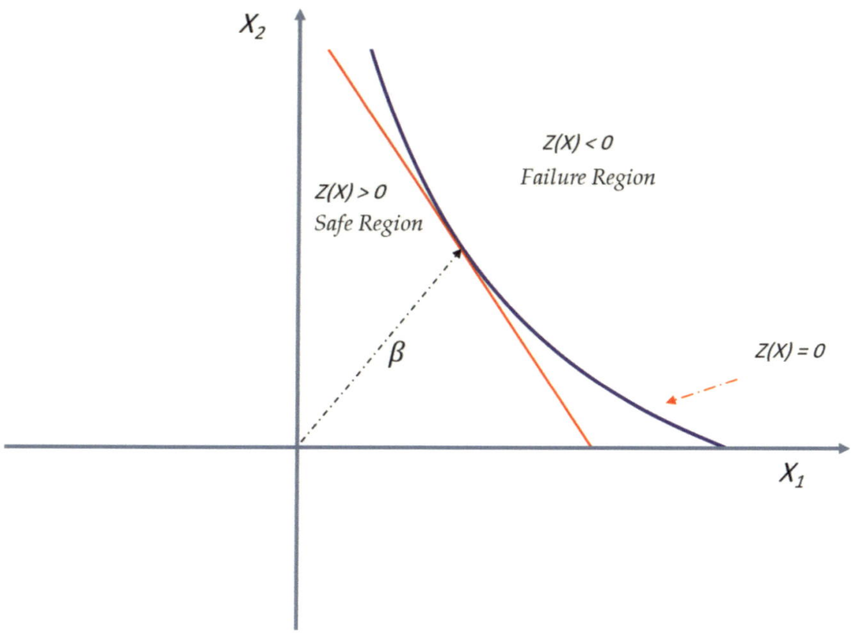

Figure A1.
Representation of linear and nonlinear LSF – original coordinates (X domain).

A.2 First Order Reliability Method (FORM)

This section will use the Hasofer and Lind method to calculate the probability of failure by FORM [9]. The first step is to transform the normal random variables into standard normal variables as:

$$u_i = \frac{x_i - \mu_{x_i}}{\sigma_{x_i}}, \quad i = 1, 2, \ldots, n \tag{32}$$

$$x_i = u_i * \sigma_{x_i} + \mu_{x_i}, \quad i = 1, 2, \ldots, n \tag{33}$$

where μ_{x_i} and σ_{x_i} are the mean and the standard deviation of the random variable x_i respectively, and u_i is the transformed standard normal variable. The LSF is then formulated in terms of the standard normal variables as:

$$Z(U) = Z\left(u_1 * \sigma_{x_1} + \mu_{x_1}, u_2 * \sigma_{x_2} + \mu_{x_2}, \ldots, u_n * \sigma_{x_n} + \mu_{x_n}\right) = 0 \tag{34}$$

Figure A2 shows the representation of linear and nonlinear LSF and the most probable point of failure, $(MPPF : u_1^*, u_2^*)$, or sometimes referred to as the design point. The next subsection outlines the steps for calculating the probability of failure using FORM.

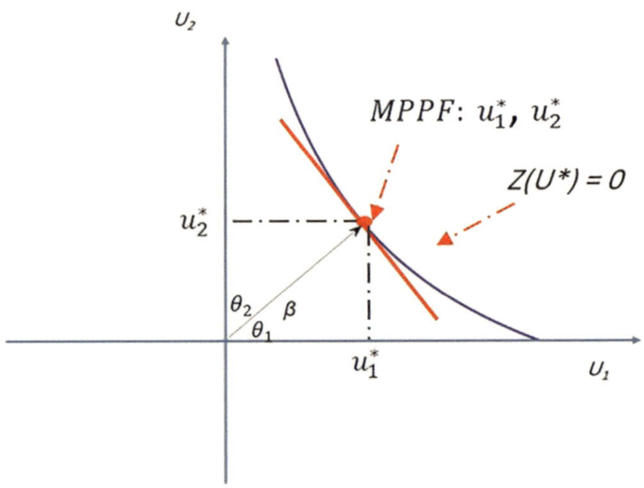

Figure A2.
Representation of linear and nonlinear LSF – transformed coordinates (U domain).

A.3 FORM calculation steps

The calculation steps for this method are listed below [10].

1. *Formulate the LSF/performance function in terms of the original random variables, x_i:*

$$Z(\mathbf{X}) = Z(x_1, x_2, \ldots \ldots \ldots \ldots x_n)$$

2. *Assume the initial MPPF/design point as the given mean of each variable.*

3. *Transform the LSF* $Z(\mathbf{X})$ *to* $Z(\mathbf{U})$ *using Eq. (34).*

4. *Compute the initial values of the design point in U space using Eq. (32). It should be noted that the initial values will be zero.*

5. *Compute the initial estimate of* $Z(U)$ *in terms of the initial values of design point U.*

6. *Compute the partial derivatives of the LSF:*

$$\frac{\partial Z(\mathbf{U}^*)}{\partial u_i} \tag{35}$$

7. *Compute the standard deviation of the LSF using the following Equation:*

$$\sigma_z = \sqrt{\sum_{i=1}^{n} \left(\frac{\partial Z(U^*)}{\partial u_i} \right)^2} \tag{36}$$

8. *Compute the initial reliability index* β*:*

$$\beta = \frac{\mu_Z}{\sigma_Z} \tag{37}$$

9. *Compute the directional cosines* α_i *using the following Equation:*

$$\alpha_i = -\frac{\left(\frac{\partial Z(\mathbf{U}^*)}{\partial u_i} \right)}{\sqrt{\sum_{i=1}^{n} \left(\frac{\partial Z(U^*)}{\partial u_i} \right)^2}} \tag{38}$$

10. *Determine the new MPPF/design point in U space using the following Equation and in the X space using Eq. (33):*

$$u_i^* = \beta \alpha_i \tag{39}$$

11. *Compute the LSF in terms of new design points.*

12. *Compute the partial derivatives, Eq. (35), and the standard deviation of the LSF, Eq. (36), at the new design points.*

13. *Compute the new* β *using the following Equation:*

$$\beta = \frac{Z(\mathbf{U}^*) - \sum_{i=1}^{n} \frac{\partial Z(\mathbf{U}^*)}{\partial u_i} (u_i^*)}{\sqrt{\sum_{i=1}^{n} \left(\frac{\partial Z(\mathbf{U}^*)}{\partial u_i} \right)^2}} \tag{40}$$

14. *Repeat steps 9 through 13 until β converges. A tolerance, ε, of 0.001 is usually used:*

$$\varepsilon = \frac{|\beta_i - \beta_{i-1}|}{\beta_{i-1}} < 0.001 \tag{41}$$

15. *Use Eq. (31) to calculate the probability of failure.*

Author details

Alireda Aljaroudi[1*] and Ahmed Aljaroudi[2]

1 Independent Engineering Research, Hamilton, ON, Canada

2 Engineering Freelancer, Hamilton, ON, Canada

*Address all correspondence to: aaa515@mun.ca

References

[1] Breitung K. Asymptotic approximations for multinormal integrals. Journal of the Engineering Mechanics Division. 1984;**110**(3): 357-366

[2] Hohenbichler M, Rackwitz R. Improvement of second order reliability estimates by importance sampling. Journal of Engineering Mechanics. 1988; **114**(12):2195-2199

[3] Tvedt L. Two Second-order Approximations to the Failure Probability: Section on Structural Reliability. A/S Vertas Research; 1984

[4] Tvedt L. Distribution of quadratic forms in normal space—Application to structural reliability. Journal of Engineering Mechanics. 1990;**116**: 1183-1197

[5] Choi S-K, Grandhi R, Canfie RA. Reliability-based Structural Design. London, UK: Springer London; 2007

[6] Haldar A, Mahadevan S. Probability, Reliability, and Statistical Methods in Engineering Design. New York, NY: John Wiley and Sons Inc; 2000

[7] Verma A, Ajit S, Karanki D. Reliability and Safety Engineering. London, UK: Springer-Verlag; 2010

[8] Rabenstein AL. Elementary Differential Equations with Linear Algebra. New York: Academic Press Inc.; 1982

[9] Hasofer A, Lind N. Exact and invariant second-moment code format. Journal of the Engineering Mechanics Division. 1974;**100**(1):111-121

[10] Aljaroudi A. Probabilistic modeling of failure. In: Huang Z, Hemeda S, editors. Failure Analysis. IntechOpen: London; 2019

Chapter 8

Fracture Mechanics Application in Soil Strength Identification: A Scientometric Analysis

Gobinath Ravindran, Vutukuru Mahesh, Herda Yati Binti Katman and A. Nikhil Kumar

Abstract

Fracture mechanics is a relatively new field that combines the study of fractures and mechanical characteristics. It focuses on understanding damage, fractures, and related events. Fracture mechanics helps quantify material properties, stress distribution, crack length, and the processes of crack propagation. Many academics are currently applying fracture mechanics in their analyses. In this study, we have examined the recent research progress (2000–2023) using a scientometric approach to assess its contribution. Our findings indicate a recent slowdown in research growth within this area. There have been a considerable number of publications (620) and a significant body of available research papers (1564). Moreover, a large number of authors (3985) are actively working in this field. International collaboration accounts for nearly 19% of the research output. On average, each document receives about 18.2 citations, reflecting the notable impact in this growing domain. This study serves as a valuable resource for new researchers interested in undertaking research within this field.

Keywords: scientometrics, research trend analysis, fracture mechanics, fracture toughness, soil mechanics

1. Introduction

Engineering material failure is an important concept engineers and designers look upon, material strength, if enhanced will give more durability and a quality product. Materials mostly fail due to loading (static, cyclic, fatigue) and avoiding failure through fracture is one of the key point we look upon. A fracture occurs when a body splits into two or more parts in response to a static stress applied at a temperature lower than the melting point of the material. There are two stages to a fracture: the crack starts, and then it spreads. There are two distinct fracture types which are brittle and ductile that varies according to the stress distribution in the member during loading.

1.1 Brittle failure occurs due to the following conditions that may occur in the material

- Too much high energy absorption and plastic deformation close to the crack precede fracture. "The spread of cracks is sluggish due to the high amount of energy absorption and plastic deformation that occurs in the material near the crack. More energy is needed to generate ductile fissures.
- In a tensile test, in general, failure comes after necking. Vacuum nucleation near the neck's midline is often the initial trigger. In time, these gaps will join together as the deformation continues. When the outside rim of the crack can no longer bear the weight, it shears off suddenly. A cup and cone fracture describes this type of global failure.
- Holes appear at inclusions when the inclusions themselves are brittle and/or the matrix's inclusion cohesion is inadequate [1].
- The broken surface is pitted with dimples. During the fracturing process, micro voids are generated and split in half, giving rise to the resulting dimples.
- The ductility is measured in terms of the percentage of length change or area decrease.
- Temperature, strain rate, and stress state all influence ductility. Propagation of cracks is stable [2].

1.2 Brittle failure in a material such as glass, rubber occurs due to the following reasons

- Crack As a result of little to no plastic distortion during propagation, very little energy is lost [3].
- Happens all of a sudden, with no prior notice. Cleavage is a possible mode of fracture (fracture on certain crystallographic planes by bond breaking). V-shaped chevron patterns or lines/ridges radiating from a crack are sometimes visible on a fracture's surface. Hard materials can have a relatively smooth surface. A fracture at a grain border, also known as an intergranular fracture, can occur without significant plastic deformation, making it difficult to detect prior to failure. Toughness is grain size dependent. Toughness and ductility improve with reduced grain size.
- Cracks spread in an unstable manner. A fracture can either be transgranular (between the grains) or intergranular (along the grain boundary) [4–6].

When a solid or structure has a geometrical discontinuity on the scale of the structure, its behavior can be described by the theories of fracture mechanics [2]. In two-dimensional media (such as plates and shells), the discontinuity features can take the shape of lines, whereas in three-dimensional media (such as spheres and cubes), the discontinuities can take the form of surfaces. Now a fully developed field of study, fracture mechanics has revolutionized our understanding of how engineering materials behave. Fracture mechanics has had a major influence by introducing a new design philosophy, damage tolerance design methodology, which is now the gold standard in aviation design.

2. Fracture mechanics

Fracture mechanics is a new field that integrates fracture mechanics with the study of mechanical characteristics [7]. Mechanical properties study and fracture mechanics have recently been combined to form the emerging topic of fracture mechanics. Quantifying the connection between material qualities, stress, crack length, and crack propagation mechanisms is the goal of fracture mechanics [8]. The study of fractures and other phenomena associated with breaking is the focus of fracture mechanics. The advancement of fracture mechanics is inextricably tied to certain recent, high-profile disasters. It all started during World War II, when hundreds of ships bringing citizens to safety on the high seas were heavily damaged, the reason for the same being unknown at that time. During World War II, the construction of Liberty ships transitioned from riveted to welded construction. However, this change resulted in numerous failures caused by subpar weld quality, stress concentrations, and the use of brittle materials. As a result, approximately 400 of the 2700 Liberty ships constructed during the war were severely damaged due to these factors. The investigation of these failures kickstarted a new area of study called "fracture mechanics", which aimed to understand the behavior of materials under stress and the mechanisms that lead to failure. The study of fracture mechanics has since become an important tool for identifying weaknesses and predicting the failure of materials in various applications, including soil strength identification. The Comet occurrences in 1954 led a thorough investigation into the causes, which greatly improved our understanding of fracture and fatigue. In July of 1962, the Kings Bridge in Melbourne suddenly collapsed after a car weighing 45 tons crossed one of the spans. Four girders failed because a crack extended from the bottom to the top of the girder, through the web and in some places the upper flange as well. Although no one was injured, the incident sparked a thorough investigation into the cause of the failure, which ultimately led to significant improvements in bridge safety and engineering practices. This kind of events become prelude for the initiation of Fracture mechanics study which progressed well and helped to avoid many disasters.

There are mainly two primary kinds of fracture mechanics approach: linear elastic fracture mechanics (LEFM) and elasto-plastic fracture mechanics (EPFM) (EPFM). LEFM excels at handling brittle-elastic materials such as high-strength steel, glass, ice, concrete, etc. In contrast, ductile materials, such as low-carbon steel, stainless steel, some aluminum alloys, and polymers, always exhibit plasticity before fracture. Linear fracture mechanics is still a good approximation to reality, but only for loads below a certain threshold. Hence it is evident that understanding the fracture properties of a material is necessary to produce a good quality product. In recent years, fracture mechanics has increasingly been applied in civil engineering to understand the properties and behavior of different materials. For example, stabilized soil and engineered soil are materials that have been extensively studied using fracture mechanics to analyze their cracking patterns and failure modes, among other factors.

2.1 Fracture mechanics approach towards soil strength

Soil always has some kind of defect, such as an inclusion, a vacuum, or a crack in any form. Environmental stress or mechanical strain could cause cracks to spread from these defects. Multi-layer pavement systems, buried pipes, and embankment dams use soil for sub-grade layers. Failure and fracture of this material can deform buildings and affect their long-term performance. Tensile type fracture (mode I fracture) of soil materials has been tested using several ways. Real-world loads can cause mixed mode tensile-shear deformations in clay materials. Among the failure pattern in soil, desiccation cracking [9–11] in cohesive soil, where cracks form due to volume changes with removal of water, has received a lot of attention from the scientific community. However, crack formation in the core of dams, the wall of tailings, and the liners of landfills, where the soil is saturated above the shrinkage limit, has received much less attention. Significant economic, environmental, and human losses have resulted from the collapse of these structures [12]. Over the course of the past three decades, linear elastic fracture mechanics (LEFM) has solidified its place as a key discipline within geotechnical engineering [13] Numerous geotechnical engineering literatures document the use of LEFM with great success in actual engineering projects. Because it is not possible to verify that failure criteria based on yield dominating failure of material, such as Tresca, Mises, or Coulomb, can be used to analyze the failure of brittle materials induced by fracture [14, 15] established that LEFM is a useful method for studying the fracture dominated failure or rupture of a wide variety of geomaterials, including stiff and over-consolidated soils, in particular those having cracks [16].

Correct knowledge of shear strength of the soil is necessary for the design of traditional geotechnical constructions including foundations, slopes, and retaining walls. But the void ratio, composition, friction angle, cohesion, stress history, temperature, strain, strain rate, and structure all have an impact on the shear strength of soil. When the tension exceeds the soil's bearing capacity, failure is expected [17, 18]. Owing to these factors influence, analyzing soil failure become cumbersome. When there are no fractures present, a brittle failure that happens quickly and is guided by linear elastic fracture mechanics (LEFM) [19, 20]. When it comes to design, traditional soil mechanics criteria like cohesive strength (c) and angle of internal shearing resistance (phi) fall short. If one want to prevent failure in brittle and quasi-brittle materials by employing LEFM, you need to pay attention to the parameters, especially the critical stress intensity factor K_{IC} [18, 21–24]. At low saturation, cohesive soil starts to operate like a quasi-brittle material, failing mostly in ways described by linear elastic fracture mechanics. In the field of fracture mechanics, the K_{IC} is a crucial mechanical characteristic that indicates a material's resistance to fracture failure under mode I stress circumstances. Numerous studies have focused on the best way to calculate K_{IC} and how it relates to other mechanical parameters like tensile strength t in geomaterials. Soil fracture toughness, defined as its resistance to crack initiation and propagation, is an important quality for determining the safe design limits of critical infrastructure including pipelines, bridges, and dams [25, 26]. Hence, it is found that fracture mechanics-based study on soil will help designers and engineers to develop better engineering soil [27]. In this work, we have tried to understand the real research works being undertaken in this domain.

Thus, research progress analysis is necessary to comprehend the research being done by different researchers and guide future research directions. This paper proposes quantitative visual representation and systematic analysis of existing studies to meet the requirements and provides a comprehensive and up-to-date review of fracture mechanics approach to engineered soil design. The scientometric analysis used here provides (a) a better understanding of the domain of knowledge encircling fracture mechanics approach towards engineered soil design, (b) more objectivity than prior reviews, and (c) a quantitative representation of the domain.

This study focuses on answering the following RQs:

1. How well-studied are Fracture mechanics methods?

2. What has changed in recent years?

3. How is this study and partnership distributed globally?

4. What areas are needed for these particular studies?

5. Where are researchers working?

This paper took a scientometric approach to answer the above-mentioned questions using most widely used bilbiometrics software for analysis. Scientometrics is widely used by many researchers to understand the research landscape and it can provide information related to important keywords, authors, affiliations, countries involved in research, organizations funding the research and other important information that can serve as a guideline for new researchers.

3. Methodology

To understand the research progress, we have taken a scientometric approach using Scopus database which is one of the widely used database for this purpose. Scopus is the most comprehensive index of scholarly articles, books, and proceedings from scientific conferences and journals. Scopus provides a global overview of research in the areas of science, technology, medicine, the social sciences, and the arts and humanities, with useful capabilities for monitoring, analyzing, and visualizing this data. We ran a search with keywords during December 2022 which is adopted by using trial and error method as mentioned below. Keywords and selection of them influence the outcome of bibliometric research and care should be taken in arriving at keywords. The whole methodology is divided into keyword finalization, database search, outlier removal, bibliometric study, analysis and recommendations which is provided in the following sections.

3.1 Method

Scientometric analysis was used to rank the papers written about fracture mechanics during a specific period (2000–2023). Publication trends in academic works such as research articles, conference papers, and other scholarly documents can be analyzed quantitatively and statistically with the help of a technique called scientometric. On December 3, 2022, researchers went through the Scopus database (www.scopus.com). By conducting the search on a single day, we avoided the potential for bias caused by the databases being updated on a daily basis. For this study, researchers looked back at the papers indexed by Scopus between the years 2000 and 2023. There are other databases available for the same study including Web of Science, Dimensions, Lens etc. which are also widely used by researchers. Scopus is the go-to database for scientometric and bibliometric investigations due of its large data set and its widely adopted in bibliometrics study [28–30]. The scholarly community agrees that Scopus is the best and largest database of its kind. It is now widely used as a bibliometric data source by many researchers. The authors of this study set out to create a bird's-eye view of the scientific landscape that has contributed to our knowledge of how fracture mechanics is used in engineered soil development. Both the intellectual (co-citation network) and social (collaboration network) structures of the gathered data were explored (co-occurrences of authors/keywords, theme progression).

3.2 Search query

To retrieve the scientometric data on finite element analysis and weak soil, the following search query was run in the main search interface of the Scopus database in the search field type:

(TITLE-ABS-KEY (fracture AND mechanics) AND TITLE-ABS-KEY (soil)) AND PUBYEAR >1999 AND PUBYEAR >1999

3.3 Inclusion/exclusion criteria

Authors ran the search using keywords in Scopus database and 1546 documents were found in the first search. A time frame of 2000–2023 is kept for publication only. Publications, conference papers, book chapters, reviews, and books were filtered out, leaving a total of 1558 results after narrowing the publication stage to "final." The top ten nations for publishing articles in the topic of Fracture mechanics of soil were also considered. All materials must be written in English to be considered. The remaining 1558 entries were published between the years 2000 and 2023 and include articles (N = 1079), conference papers (N = 402), book chapters (N = 14), review articles (N = 25), and books (N = 4). This paper introduces a novel methodology for conducting a comprehensive literature evaluation on soil fracture mechanics by combining scientometric and complex network analyses. Figure depicts a description of the research procedure and the incorporation of analytical techniques. In order to ensure that the items selected were consistent and that the results were accurate, two authors of this research followed the identical procedures.

The 1558 documents that were returned were analyzed using a variety of bibliometric techniques to extract useful information. Microsoft Office Excel (Version 2209) was utilized for several fundamental tasks, such as gaining insight into publication and citation trends. The authors used a variety of programmes to assist with data visualization, including VOS viewer (version 1.6.15), Origin Pro 2022b (64-bit), and Biblioshiny (version 4.1.1). Relationships in citations, bibliographic coupling, co-citation, and co-authorship were also discovered with the help of these technologies.

3.3.1 Analysis of the overall growth trend

Figure 1 below summarizes the main results of the bibliometric analysis to describe the collection size in terms of number of documents, number of authors, number of sources, number of keywords, timespan, references, and average number of citations. Furthermore, many different co-authorship indices are shown.

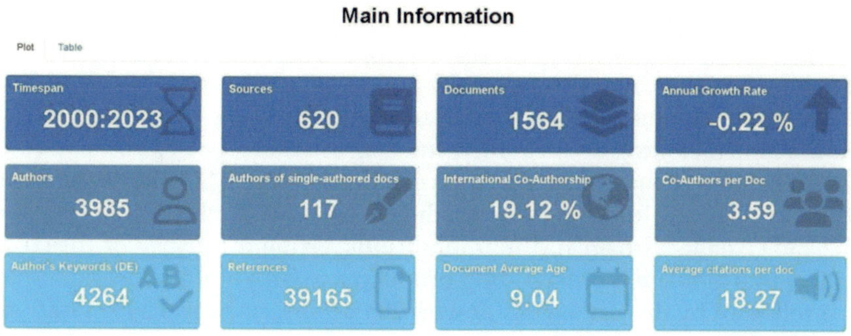

Figure 1.
Summary of the main results of the bibliometric analysis.

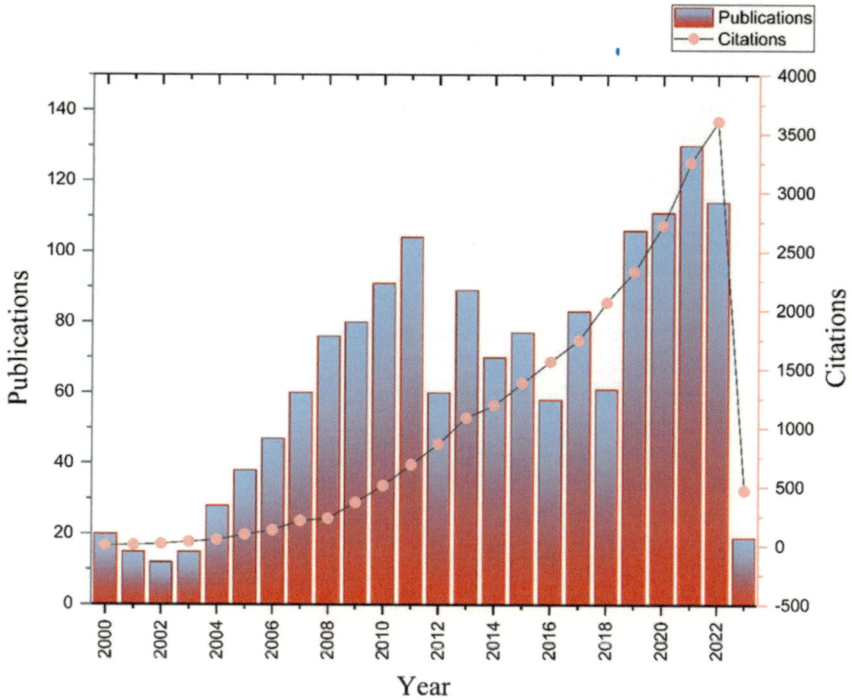

Figure 2.
Publication and citation trend on fracture mechanics.

Figure 2 shows the year-wise frequency of publications and citations of wave propagation through fracture mechanics literature from 2000 to 2023. The number of publications and citations have significantly increased over the years. The most successful publication year was 2021, in which the highest (130) number of articles were published. On the other hand, the highest number of citations (3607) were seen in the year 2022. More number of publications happening in an year is influenced by many factors including funding availability, increasing demand, increasing collaboration between authors, research tie up between institutions etc. It is found that the average publication in an year is well above many domains and it is showing the increasing awareness among authors.

3.3.2 Top ten publishing countries

The top ten highly productive countries are shown in **Figure 3**, which shows the interest generated among the country research which is in turn attributed to many other parameters including applicability, funding available, policy decisions etc. It can be seen that the most productive country, China had produced 604 publications, the United State and United Kingdom produced 284 and95 publications followed by Australia and Canada with 84 and 76 publications during the period of the study. The results show that China has outstanding productivity with 604 publications and 7068 citations; followed by the United States with 284 publications and 5999 citations.

3.3.3 Top ten high-research-producing institutions

The top 10 research-producing organizations are shown in **Figure 4**. The top 2 organizations had 80 and 78 publications, respectively. The Northwestern Polytechnic University emerged as the most productive organization with 80 publications, followed by the Not reported with 78 publications. The China University, Nanjing University and Tsinghua University produced 75, 57 and 55 articles, respectively.

3.3.4 Most prolific authors

Results from the top ten most prolific scholars in the field of fracture mechanics application in soil strength analysis from 2000 to January 2023 is obtained. These scholars have shown consistency by contributing to the research in this field. The highest number of documents are produced with no author name provided until 2000 to 2023. The results revealed that Wang Y from China had produced 24 documents and earned the highest citation count of 619. He also has the highest h-index, which

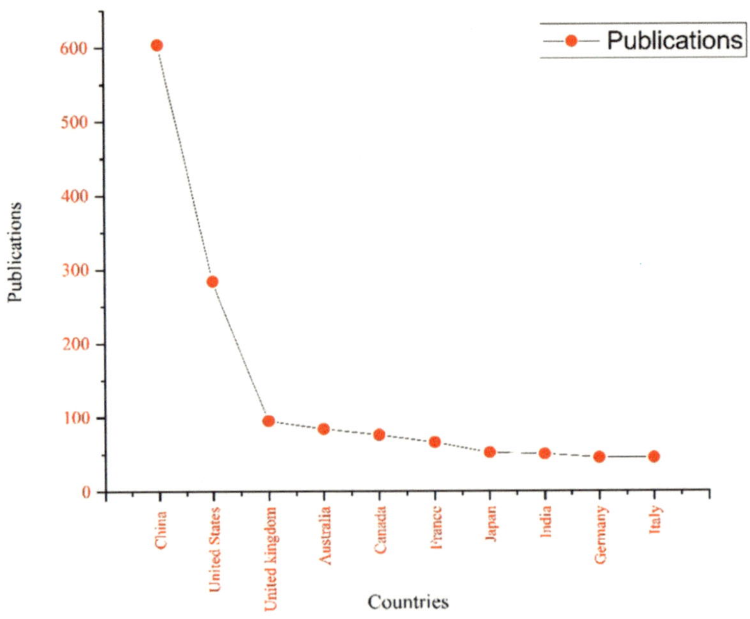

Figure 3.
Top ten highest- publishing countries and their corresponding citation.

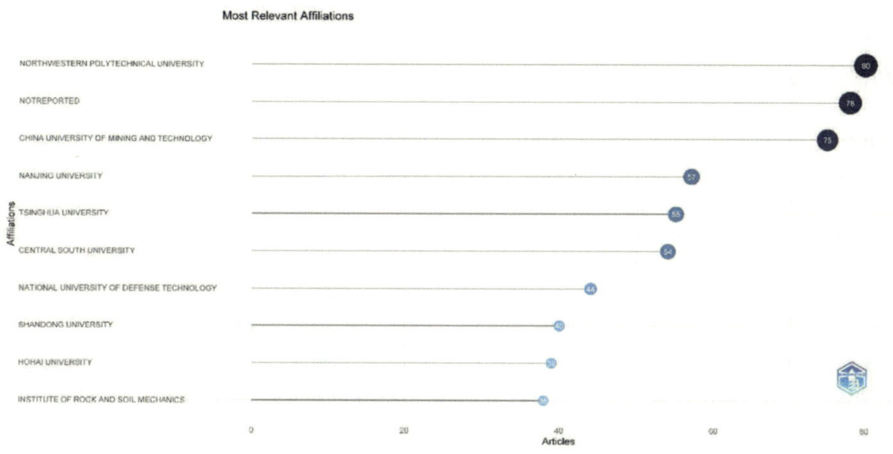

Figure 4.
Top ten most highly productive organizations.

suggests that Liux remains the most impactful author in the field of finite element analysis and weak soil. Liux's first article was published in 2011 with total citations per year of 6. Although our result shows that Liux has no publication in between 2012 and 2020, however, he has consistently published in this field between 2011 and 2021. The second most prolific scholar in this field is Zhang J from China. Zhang J has 15 publications. Zhang J began publishing in the field of finite element analysis and weak soil in 2010, where he had two publications and consistently published 1, 3, 2, and 2 papers in 2010, 2014, 2015, 2019, respectively, and yet to publish in the year 2022. Similarly, the results show that Shivashankar from India, Liu Y and Ylian S from China, have h-index of 6, 5, 4 and 4 respectively based on our dataset; hence, they have immensely impacted the field of finite element analysis. Other impactful scholars in this field may also be present but this work is confined to the database obtained during the study period only (**Table 1**).

Source	Impact measure (H – Index)
Yanshilixue Yu Gongcheng Xuebao/Chinese Journal of Rock Mechanics and Engineering	22
Journal of Rock Mechanics and Geotechnical Engineering	14
Computers and Geotechnics	13
Engineering Geology	12
Materials Science and Engineering A	11
Canadian Geotechnical Journal	10
Geotechnique	10
International Journal for Numerical and Analytical Methods in Geomechanics	10
International Journal of Geomechanics	10
Acta Geotechnica	9

Table 1.
Impact measure of the publications.

3.3.5 Three-factor analysis

3.3.5.1 Keywords, authors and countries

A more visualized representation of prolific scholars vis-à-vis their countries and the specific area of interest in the field of finite element analysis and weak soil is shown in **Figure 5**. This Figure is a three-field plot of article contributions by countries, authors, and themes within the field of fracture mechanics of soil. The rightmost column represents active countries, the rightmost column shows scholars' names contributing from those countries, and the middle column represents the most used keywords by the authors. The number of occurrences of these keywords forms what we refer to as 'themes' in this study. Note the height of the boxes and the thickness of the connecting lines. On the side of countries, China has more authors' affiliations, with 1250 authors connected to the country, followed by Australia, United States and United Kingdom with 65, 49, and 12 connections. Observing the thickness of the line leading from the countries to authors, we can see that Wang Y and Li X remain the giant contributors from China. In India, Kodikara J remains the prolific writer.

The emphasis is placed on the height of each box and thickness of the connecting lines; the taller the box, the more significant; and the thicker the lines' correlation, the more information or volume of work was produced.

3.3.5.2 Thematic evolution of keywords

Figure 6 demonstrates the evolution of keywords in two different stages (2000–2010 and 2011–2023). It is noticed that most of the research in the second stage relates to fracture, cracks and soil mechanics. This is supported by the height of the boxes and the thickness of the connecting lines during the period 2011–2023, which

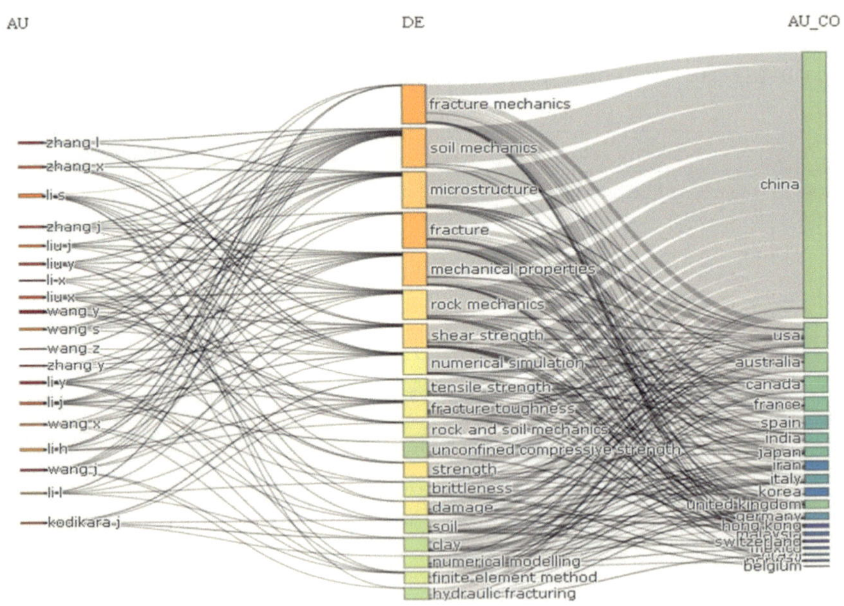

Figure 5.
A three-field plot of countries, authors and themes.

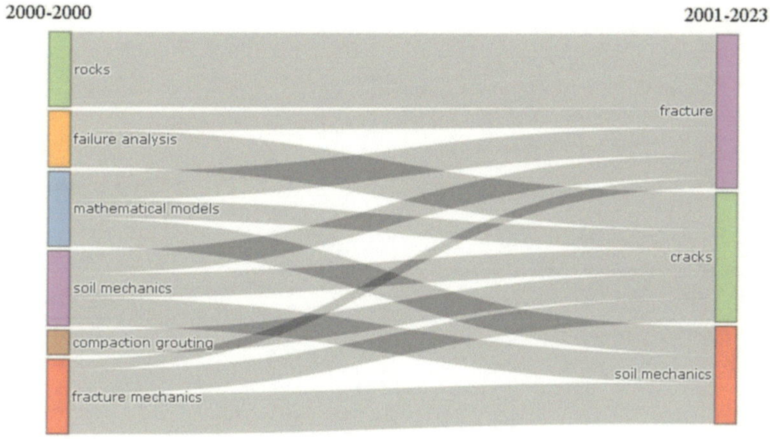

Figure 6.
Thematic evolution map of keywords from 2000 to 2023 with respect to fracture mechanics and soil.

manifests that the research focus on these three terms is relatively more. Most of the research works will start with a keyword and will evolve into multiple other keywords and progress via other keywords, hence understanding keyword variation dynamics is important to understand the research progress in a domain. Concerned with this study, keywords such as "rocks", "failure analysis", "mathematical models", "soil mechanics", "compaction grouting" and "fracture mechanics" defines the research work being undertaken by researchers across the globe.

3.3.6 Thematic map

Another analysis conducted in this study is the thematic map of fracture mechanics approach for engineered soil development. The aim of conducting a thematic map is to gain insight into the field's current status and what its future sustainability holds. This analysis is useful in providing knowledge to researchers and stakeholders regarding the potential of future research development of thematic areas within a field.

Thematic analysis takes clusters of keywords plus and their interconnections to obtain themes. These themes are characterized by properties (density and centrality). The density is represented on the vertical axis, while centrality takes the horizontal axis. Centrality is the degree of correlation among different topics; density measures the cohesiveness among the nodes. These two properties measure whether certain topics are well developed or not, important, or not. The higher the number of relations a node has with others in the thematic network, the higher the centrality and importance, and it lies within the essential position in the network. Similarly, cohesiveness among a node, which represents the density of a research field, delineates its capability to develop and sustain itself. In **Figure 7**, we provide the thematic map of the field studied in this paper, which is divided into four quadrants (Q1 to Q4) and sustain itself. In **Figure 7**, we provide the thematic map of the field studied, which is divided into four quadrants (Q1 to Q4).

The upper right quadrant (Q1) represents the driving themes, the lower right quadrant (Q4) denotes the underlying themes, the upper left quadrant (Q2) represents the very specialized themes, and the lower left quadrant (Q3) denotes emerging or disappearing themes. Notably from the figure, a theme such as "Fracture Mechanics,"

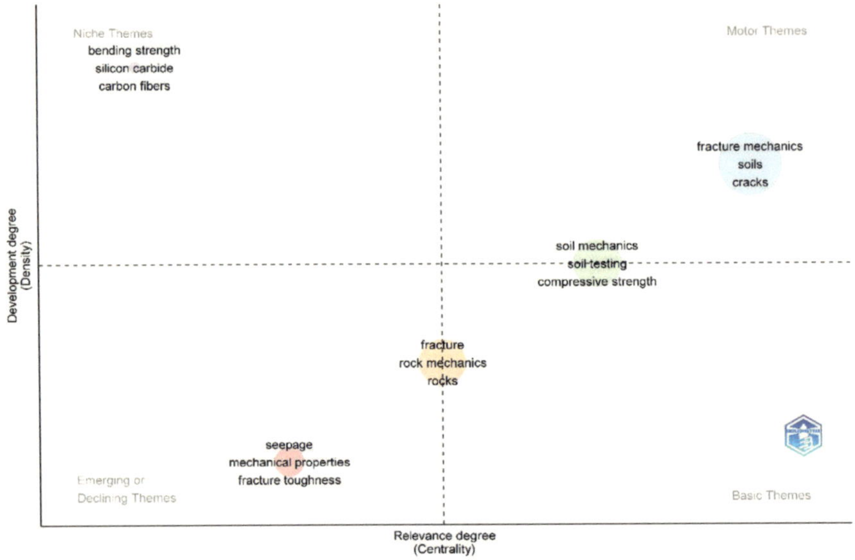

Figure 7.
Thematic map: Q1 contains the main theme, Q2 contains highly developed and specialized themes building ties with the leading theme, Q3 contains disappearing or emerging themes, Q4 consists of foundational and transversal themes.

placed in Q1, is well developed and capable of structuring the research field. In other words, fracture mechanics remains the leading theme within the field. Themes such as "soil mechanics," "soil mechanics", "soil testing" and "soils compressive strength" seen in Q4 are the basics and are critical for the field's development. Themes in Q2 have developed internal bonds but still of marginal contribution to the development of the field of fracture mechanics. This finding suggests that themes in Q2 such as bending strength, silicon carbide and carbon fibers are potential topics that need to be more connected to fracture mechanics. Scholars in this field may explore these themes to provide twenty-first-century solutions in the field of finite element analysis.

The theme in Q3, "seepage", "mechanical properties" and "fracture toughness" appears to be emerging but shows cohesiveness towards the themes of Q4, indicating that some of its components are basic and necessary for developing the field of Fracture Mechanics of soil. The thematic analysis suggests that more efforts are needed to develop themes deformation," to establish more ties with "Fracture mechanics of soil."

3.3.7 Thematic focus of the field of fracture mechanics of soil

3.3.7.1 Keywords analysis, co-occurrence network, and trend topics

This section investigates the themes that dominate the research landscape of fracture mechanics of soil and areas that scholars have focused on during the period 2000–2023.

Besides, the study also attempts to gain insight into whether there is a shift in the topic of discussion among scholars within the field. We first began by analyzing keywords plus and their frequency of occurrences. Next, we carried out an analysis of keywords dynamics, trending topics, co-occurrence network, and thematic areas of the field.

DOI: http://dx.doi.org/10.5772/intechopen.112451

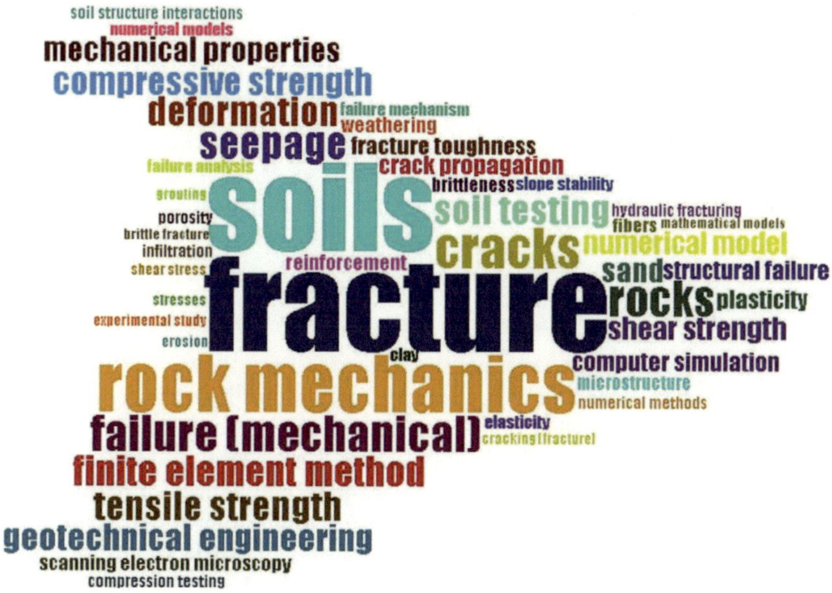

Figure 8.
A visualized word-cloud of frequently used keywords in the finite element analysis and weak soil, these are among the highest number of repetitive keywords within the field.

Analysis of keywords used by authors in publications is an essential tool for investigating trending topics and scholars focus on the field [7]. This analysis is so because publication keywords help us to identify the topic and focus of that publication quickly. The word-cloud in **Figure 8** shows frequently used keywords in this work. Using this word cloud, one can understand the various keywords authors used to define their research related to fracture mechanics approach on soil development, thickness of the keywords is related to the high frequency of its usage among the articles analyzed in this research.

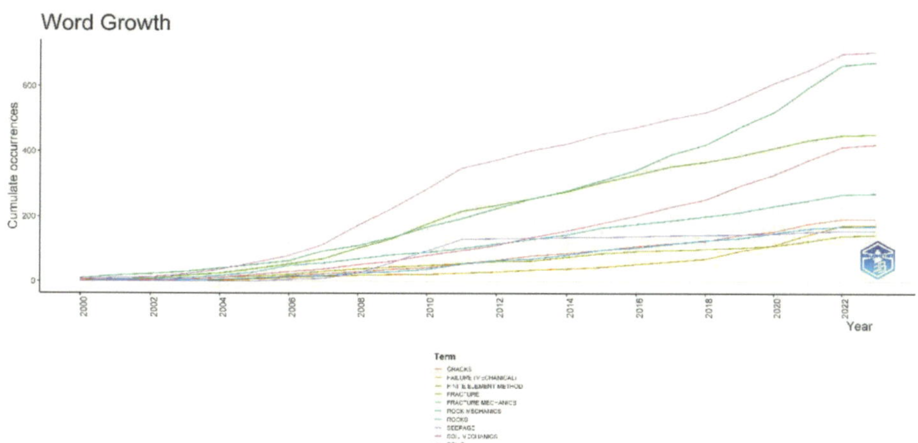

Figure 9.
Visualized word dynamics of keywords.

Specifically, **Figure 9** shows the visualized word dynamics of the used keywords plus. Soils became one of the most used keywords from 2000 and grew quickly until 2023. This finding signified that *fracture* became the most discussed topic as an aspect of fracture mechanics among scholars within those years. However, frequently used keywords such as *rock mechanics, cracks, soil testing, failure (mechanical) and finite element method* are also continued to increase (see **Figure 9**). This finding suggests that the field of wave propagation will continue to be researched these prevailing aspects.

4. Limitations of the study

This study is done for a limited period between 2000 and 2023 owing to several attributes and hence careful understanding is solicited while interpreting results. Authors have considered only articles, conference papers, book chapters, reviews and books only and there may be other avenues where the research publications may happen which is not included in this study. Also, we have considered only publications in English language excluding other language publications. This studies primary focus is on engineering application of weak soil, finite element analysis and other applications was ignored in this work. We had used Scopus as the database, further studies using Web of Science, Lens, Dimensions and other databases also needed, each database has unique indexed content and the results may vary according to the database used.

5. Conclusion

In this work we have performed research trend analysis for Fracture mechanics applications in soil, through this study we have found the following important points that will give directions to future researchers.

1. Concerned with scientific output there is a negative growth of 0.22% in the study period which shows that the publication progress and in turn research outcome is reducing which may be owing to many attributes.

2. The main keywords surrounding which the work goes on includes “fracture mechanics”, “soil”, “finite element analysis”, “fracture” “clay”, “computer simulation”.

3. Concerned with this work done by countries, China has more authors’ affiliations, with 1250 authors connected to the country, followed by Australia, United States and United Kingdom with 65, 49, and 12 connections.

4. Some of the emerging topics include “seepage”, “fracture toughness”, “rock mechanics” and thus its evident that recent research revolves surrounding this topic.

5. Average citations per document in this domain is above average and its around 18 per paper, this shows that considerable interest exists among research fraternity.

6. Last five years had seen good number of publications done in these keywords and 2021 had maximum number of papers.

Hence, we suggest further study in this area using other databases and we are sure that this work will serve as a point of interest.

Author details

Gobinath Ravindran[1*], Vutukuru Mahesh[1], Herda Yati Binti Katman[2] and A. Nikhil Kumar[1]

1 SR University, Warangal, Telangana, India

2 Institute of Energy Infrastructure, Universiti Tenaga Nasional, Kajang, Malaysia

*Address all correspondence to: gobinathdpi@gmail.com

References

[1] Petroleum Related Rock Mechanics. Developments in Petroleum Science. 1992. DOI: 10.1016/s0376-7361(09) x7008-1

[2] Xie H. Fractals in Rock Mechanics. 2020. DOI: 10.1201/9781003077626

[3] Eppes M-C, Keanini R. Mechanical weathering and rock erosion by climate-dependent subcritical cracking. Reviews of Geophysics. 2017;**55**(2):470-508. DOI: 10.1002/2017rg000557

[4] Hallett PD, Newson TA. Describing soil crack formation using elastic-plastic fracture mechanics. European Journal of Soil Science. 2005;**56**(1):31-38. DOI: 10.1111/j.1365-2389.2004.00652.x

[5] Hu T, Guilleminot J, Dolbow JE. A phase-field model of fracture with frictionless contact and random fracture properties: Application to thin-film fracture and soil desiccation. Computer Methods in Applied Mechanics and Engineering. 2020;**368**:113106. DOI: 10.1016/j.cma.2020.113106

[6] Hallett PD, Dexter AR, Seville JPK. The application of fracture mechanics to crack propagation in dry soil. European Journal of Soil Science. 1995;**46**(4):591-599. DOI: 10.1111/j.1365-2389.1995.tb01355.x

[7] Schnaid F, de Spinelli L, Iturrioz I, Rocha MM. Fracture mechanics in ground improvement design. Proceedings of the Institution of Civil Engineers – Ground Improvement. 2004;**8**(1):7-15. DOI: 10.1680/grim.2004.8.1.7

[8] Rizvi Z, Amin A, Arp JCC, Wuttke F. Fracture toughness mode I of glass fibers improved soil. Materials Today: Proceedings. 2022;**62**:3276-3281. DOI: 10.1016/j.matpr.2022.04.230

[9] Tang C-S, Zhu C, Cheng Q, Zeng H, Xu J-J, Tian B-G, et al. Desiccation cracking of soils: A review of investigation approaches, underlying mechanisms, and influencing factors. Earth-Science Reviews. 2021;**216**:103586

[10] Einav I. Breakage mechanics—Part II: Modelling granular materials. Journal of the Mechanics and Physics of Solids. 2007;**55**(6):1298-1320. DOI: 10.1016/j.jmps.2006.11.004

[11] Aliha MRM, Jafari Haghighatpour P, Tavana A. Application of asymmetric semi-circular bend test for determining mixed mode I + II fracture toughness of compacted soil material. Engineering Fracture Mechanics. 2022;**262**(108268):108268. DOI: 10.1016/j.engfracmech.2022.108268

[12] Xu J-J, Zhang H, Tang C-S, Cheng Q, Tian B-G, Liu B, et al. Automatic soil crack recognition under uneven illumination condition with the application of artificial intelligence. Engineering Geology. 2022;**296**:106495

[13] Sture S, Alqasabi A, Ayari M. Fracture and size effect characters of cemented sand. International Journal of Fracture. 1999;**95**:405-433

[14] Chudnovsky A, Saada A, Lesser AJ. Micromechanisms of deformation in fracture of overconsolidated clays. Canada Geotechnical Journal. 1988;**25**:213-221

[15] Morris PH, Graham J, Williams DJ. Cracking in drying soils. Canadian Geotechnical Journal. 1992;**29**:263-277

[16] Krishnan JC, Zhao XL, Zaman M, Roegiers JC. Fracture toughness of a soft sandstone. International Journal of Rock Mechanics and Mining Sciences. 1998;**35**:695-710

[17] Mitchell JK, Soga K. Fundamentals of Soil Behavior. 3rd ed. John Wiley & Sons; 2005

[18] Amarasiri AL, Costa S, Kodikara JK. Determination of cohesive properties for mode I fracture from compacted clay beams. Canadian Geotechnique Journal. 2011;**48**(8):1163-1173

[19] Hallett PD, Newson TA. A simple fracture mechanics approach for assessing ductile crack growth in soil. Soil Science Society of America Journal. 2001;**65**(4):1083-1088. DOI: 10.2136/sssaj2001.6541083x

[20] Kienle D, Aldakheel F, Keip M-A. A finite-strain phase-field approach to ductile failure of frictional materials. International Journal of Solids and Structures. 2019;**172-173**:147-162. DOI: 10.1016/j.ijsolstr.2019.02.006

[21] Lakshmikantha MR, Prat PC, Ledesma A. Experimental evidence of size effect in soil cracking. Canadian Geotechnique Journal. 2012;**49**(3):264-284

[22] Yang J, Li L, Lian H, Wang J. Experimental investigation of the effects of water content on the anisotropy of mode I fracture toughness of bedded mudstones. PLoS One. 2020;**15**(8):e0237909

[23] Hallett PD, Newson TA. Describing soil crack formation using elastic-plastic fracture mechanics Eur. Journal of Soil Science. 2005;**56**(1):31-38

[24] Xu J-J, Tang C-S, Cheng Q, Xu Q, Inyang HI, Lin Z-Y, et al. Investigation on desiccation cracking behavior of clayey soils with a perspective of fracture mechanics: A review. Journal of Soils and Sediments. 2021;**22**(3):859-888. DOI: 10.1007/s11368-021-03082-y

[25] Li H, Tang J, Lu Y, Zhou L, Han S, Dai R. Experimental measurements of shale fracture conductivity under cyclic loading. Arabian Journal for Science and Engineering. 2017;**43**(11):6315-6324. DOI: 10.1007/s13369-017-3032-y

[26] Juárez-Luna G, Ayala G. Application of fracture Mechanics to cracking problems in soils. The Open Construction and Building Technology Journal. 2014;**8**(1):1-8. DOI: 10.2174/1874836801408010001

[27] Wang J-J, Zhu J-G, Chiu CF, Zhang H. Experimental study on fracture toughness and tensile strength of a clay. Engineering Geology. 2007;**94**(1-2):65-75. DOI: 10.1016/j.enggeo.2007.06.005

[28] Mahesh V, Gobinath R, Kırgız MS, Shekar RPV, Shewale M. Global trends of research productivity in natural fibre reinforced composites: Comprehensive scientometric analysis. Journal of Natural Fibers. 2022;**19**(16):13088-13105. DOI: 10.1080/15440478.2022.2085227

[29] Salunkhe AA, Gobinath R, Vinay S, Joseph L. Progress and trends in image processing applications in civil engineering: Opportunities and challenges. Advances in Civil Engineering. 2022;**2022**:1-17. DOI: 10.1155/2022/6400254

[30] Srivastav P, Vutukuru M, Ravindran G, Awad MM. Biofortification—Present scenario, possibilities and challenges: A scientometric approach. Sustainability. 2022;**14**(18):11632. DOI: 10.3390/su141811632

Chapter 9

Perspective Chapter: Analysis of the Operational Reliability of Forest Equipment

Ján Kováč

Abstract

In the given chapter called Research of operational reliability of forest machines we deal with the reliability of forest machines in the field of logging, handling and removal of wood by world famous manufacturers. The main content of the work is the definition of the issue of operational reliability of forest logging machines, considering that this is very important for the high-quality and economical functioning of maintenance and diagnostics operations. The first part of the chapter focuses on an overview of current topics and thematic analysis. It describes the basics of operational reliability, maintenance and provide basic information about forest machines and their distribution. The second part deals with the reliability of specific forest machines. It discusses the methods and forms of maintenance, control of technological equipment and evaluation of operational reliability of forest harvesting machines.

Keywords: operational reliability, maintenance, means of transport, forest machines, transportation

1. Introduction

In recent years, forestry in Slovakia has seen a very rapid development trend in all its aspects. With increasing demands on the quality of care in forestry, emphasis is constantly placed on the quality and reliability of forest machinery and technological equipment. At present, forestry is based on the widespread use of forest machinery and equipment.

High requirements for care and processing are closely related to the requirements for the quality and reliability of forest machinery and technological equipment. These are closely related to the care of the facilities. The care of equipment used in forestry consists of daily operation, treatment and supervision of their operation, activities aimed at putting new equipment into operation, to eliminate faults and defects, to improve the technical condition, to technical modernization, to store, conservation temporarily decommissioned forest machinery and technological equipment and for the disposal of decommissioned and finally also the replacement of decommissioned forest machinery and technological equipment with new ones.

The above description of the life cycle of forest machinery shows that the operational reliability and care of forest machinery play an important role in forestry. The current and future degree of mechanization and automation of production in forestry, as well as the tendency to increase the efficiency of forest technology, establish as one of the primary tasks to ensure its operational reliability.

Costs related to the operational reliability of forest equipment can be affected by both the manufacturer and the user of the machine or equipment; the manufacturer with the correct conception and construction of the machine or the equipment and the user by placing the machine or an equipment in such conditions for which it is intended and by quality of maintenance, operation, and repair. Tasks with ensuring the operational reliability of forest stands are complicated by some specific factors e.g., assortment of machines, number of users, differences in the complexity of construction of different machines, seasonality in the use of forest machines, high demands on the qualification of operators, dispersion of technology over a large area, lack of storage space, operating and weather conditions, etc.

The main content of the chapter is to define the purpose of operational reliability of forest harvesting machines, given that this is very important for the quality and economical operation of maintenance and diagnostic operations.

2. Forest machines

There are two interrelated equivalent production processes in forestry, namely: the biological process (organic forest production such as establishment, education, and protection of forest stands) and the logging process [1].

There are many companies in Slovakia which primary activity is logging, concentrating, transporting, storing, and processing wood and wood. The role of Slovak forestry is to ensure the development of the forests to fulfill ecological, social, and economic functions. Regarding the optimization of economic results, it is important to focus on the transport of timber for export and removal by motor means of transport [2]. At present, machinery and equipment are used mainly to facilitate human labour in almost all sectors of the economy [3]. High acquisition costs for securing machines and equipment compensate for the resulting effect of the activity and that is labour productivity and the final profit. To achieve the goal, which currently in the forestry sector is also the care of machinery and equipment used for reforestation and logging, maintenance of machinery and equipment used in this process [4].

For effective provision of the tree felling and transportation process, it is necessary to ensure maintenance, repairs and modernization of machinery, equipment and vehicle fleet of vehicle operators, provided that the equipment meets the strictest safety and environmental criteria [5]. This is aided by the ubiquitous Fourth Industrial Revolution.

2.1 Timber harvest and transportation process

The tree felling and transportation process has a long history. Slovakia, as a mountainous country in the past as well as in the present, was an ideal place for forestry to flourish. As a conscious queen, Maria Theresa sought to provide education to her subjects to ensure the well-being of her country and the protection of the environment through the care of the forest. Welfare through the extraction of mineral wealth and environmental protection through the care of the forest and our mountains [6]. For this reason, she founded the Mining and Forestry Academy in Banská Štiavnica which began to educate experts in the field of mining and forestry. Graduates of this academy contributed their knowledge about the expansion of forestry in Slovakia [7]. Therefore, one of the sectors that has achieved an unprecedented boom is logging and wood concentration.

Logging is one of the activities performed in the care of the forest. It is one of the basic production processes in forestry. Logging consists of activities such as shading trees, their processing, sorting, approaching, concentrating, removal and shipping of individual wood assortments [8].

The transport of timber is important activity in the logging process. Timber transport consists of two stages. The primary transport of timber and timber is referred to as concentration (timber export). Secondary transport of timber is held on the modified roads (timber removal) [4]. Concentration of wood and wood mass is performed in several ways using different forces: manual, gravitational, animal and partially or complex mechanized concentration of wood [5]. The predominant means of concentration in Slovakia are special wheeled tractors. Other types of tractors that are used are universal and tracked (FOREST ACTIVITIES, online).

The transport of wood and timber also includes the removal of wood. The removal of timber follows on from the concentration of timber (export of timber and timber). It is the transport of wood from the forest stock or from the transport point in the forest to the main handling or shipping stock, or directly to the customer. Motor vehicles and trailers are used for these purposes. When transporting wood, transport sets with wood transport in lengths over 6 m predominate (ACTIVITIES IN THE FOREST, online).

The technologies used to remove timber from a hauling site by public road are very diverse. Road haulage vehicles for motor transport are divided according to the mode of propulsion into motor vehicles (self-propelled) and trailers (they do not have their own engine and are attached to motor vehicles). They depend on the form of transported wood (whole trees or their sections, shortened trunks, medium length cut-outs, short cut-outs, etc.), vehicle design and loading equipment [9].

Functions of forest machines Reliability of an object is the object's property which expresses the measure of capability to fulfill stated objectives of the object. It can be expressed by means of time in which the object fulfills the defined objectives until the time when it does not fulfill these objectives (downtime) [10].

Breakdown is a phenomenon resulting from the transition of an object from its working condition into non-working condition e.g., a phenomenon resulting from ending the working condition of an object to fulfill required functions.

Idle time can be defined as a total sum of times when an equipment is out of operation due to a defect, e.g., from the moment of the stop up to the moment of renewed operation. Downtime presents a longer time than the net time required for the equipment repair (**Figure 1**) [11].

Duration of idle time can vary with equal type of a defect, for various reasons. The causes and their consequences are important for the evaluation of the seriousness of down-times but often the downtimes are recorded which represent only the so called "typical worst cases". In real life situation it is common that the down-time duration in night shift during the weekend, caused by the same defect as during the week, lasts longer.

In case that defects affect the defect, it is important to record the downtime, for two basic reasons, as the time before the defect. Many people mistake the word "repair time" with idle-time duration, which can cause a false assessment of the break consequence only from the point of view of interruption of equipment operation, and the estimate of consequences is limited to only so called "typical worst cases".

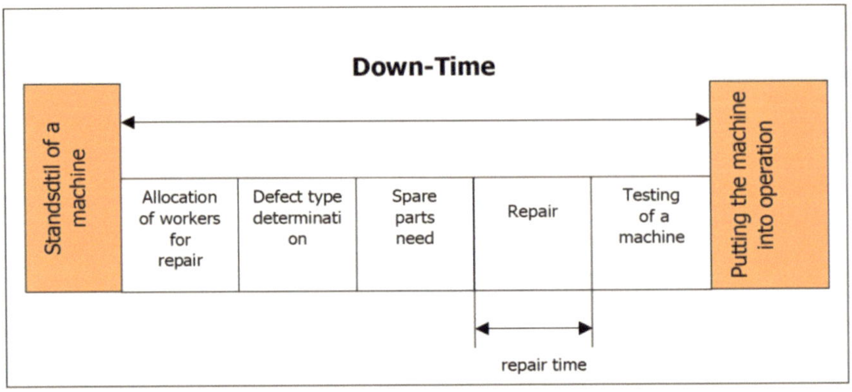

Figure 1.
Comparing down-time duration with the time needed for repair [1].

Due to idle time in production further influences occur which affect:

- production quality, e.g., customer services which can manifest themselves in a form of financial penalization,
- stopping or performance reduction of other equipment or activities,
- total production cost increase e.g., direct maintenance costs (due to energy cost increase, etc.),
- development of secondary damages due to the defect.

The parameters of idle times characterize the effectiveness of the conducted maintenance. The idle times can be divided into two basic groups:

- technological idle time –planned repairs, object exchange, necessary inspections, current maintenance, care, and equipment checking, etc.,
- idle time due to a negative event development – defect, accident, injury, etc.

The downtimes can also be divided into:

I.planned

- standstill (work rest),
- repairs,
- technological,

II.not planned

- defects,
- technological,
- others.

The losses that are caused by the downtime include following costs:

- material costs,
- energy costs,
- damaged parts repair costs,
- costs due to production drop out during the downtime.

Repair is a set of activities by executing of which the object turns from the state of non-operable to the operable condition (**Figure 2**).

Downtime has always affected the productive capability of physical assets by reducing output, increasing operating costs and interfering with customer service. By the 1960's and 1970's, this was already a major concern in the tree felling, manufacturing, and transport sectors. In manufacturing, the effects of downtime are being aggravated by the world-wide move towards just-in-time systems where reduced stocks of work-in-progress mean that quite small breakdowns are now much more likely to stop a whole plant. In recent times, the growth of mechanisation and automation has meant that *reliability* and *availability* have now also become key issues in sectors as diverse as health care, data processing, telecommunications and building management.

Reliability: Definitions Reliability was defined as a general property of an object (e.g. a vehicle), consisting in the ability to perform the required functions while maintaining the values of the set operating indicators within the given limits and in time according to the set technical conditions. It was therefore an objective, general and complex property. However, most experts considered this definition of the basic term reliability to be less appropriate. According to the newly introduced standards ISOP 8402 and, in more detail, IEC 50(191), **reliability** is understood more narrowly as a term for describing readiness and the factors that influence it: failure-free, maintainability and assurance of maintenance.

While the definition of fault-free has not changed much, the term maintainability is understood more generally and corresponds to the original definition of general maintainability under the legislation in force until 1993 in **Table 1**.

Maintenance is generally understood as a combination of all management, technical, organizational, control and administrative activities, aimed at maintaining or returning the object to a condition in which it can perform the required function

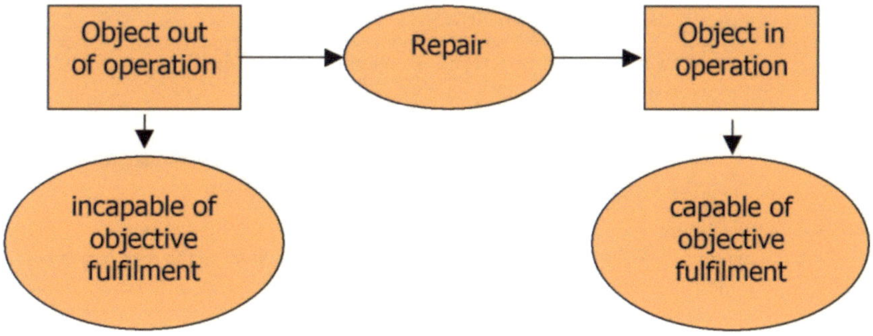

Figure 2.
A repair characteristics [1].

(generally defined maintenance in this way = preventive maintenance + maintenance after a failure); is performed by the user, operator, supplier, manufacturer, etc.

a. Reliability quantification

Reliability theory examines the patterns of occurrence of failures and methods of predicting them, looks for ways to increase the reliability of products, in the period of their design, projection, construction and production, but also methods of achieving inherent reliability (internally given) and operational reliability during the period of operation and storage. The concept of operational reliability includes the technical side of reliability and the reliability of people - operators, operators, etc. Reliability theory also elaborates methods of checking and testing reliability. The brief definitions of reliability properties given in **Table 1** have only a **qualitative** character - they provide data on the content of each property.
Reliability quantification involves two steps:

- the choice of quantitative measures (numerical, functional, etc.), i.e. indicators and characteristics of failure-free, maintainability, repairability, readiness, or next,
- determination of their required values based on specifications, requirements, or contracts, market research, etc.

Determining reliability indicators (safety, durability) for a specific type of product depends on its functional complexity, purpose of use, etc. It generally takes place in the following steps:

Termin	Definice
safety	a state in which the risk of endangering health, life of persons, the environment or property is limited to an acceptable level
durability	the ability of the object to perform the required function under the given conditions of use and maintenance to a limit state that can be characterized by the end of its useful life, economic or technical unsuitability, or other serious reasons
dependability	collective term for the description of readiness and the factors that influence it, i.e. fault-free, maintainability and maintenance assurance
availability	the ability of an object to be in a state capable of performing a required function under given conditions, at a given point in time or interval, provided that the required external conditions are provided
reliability	the ability of the object to perform the required function in the given conditions and in the given time period
maintainability	the ability of an object under the given conditions of use to remain in a state or to return to a state in which it can perform the required function, if maintenance is carried out in the given conditions and the specified procedures and means are used
maintenance support	the ability of the organization providing maintenance services to provide, according to the requirements in the given conditions (they apply both to the own object and to the conditions of use and maintenance), the means needed for maintenance according to the given maintenance concept

Table 1.
Overview of basic terms and definitions according to the new legislation [2].

- it is determined by which indicators reliability (as well as safety and service life) will be described; this task is referred to as the choice of indicator nomenclature,

- numerical values are assigned to the chosen nomenclature of indicators - in this phase, target values are determined based on the identified requirements for specified technical parameters and conditions of use,

- a preliminary analysis of the attainability of the set goals (including economic efficiency) is carried out; In this phase, the product is usually considered as a system and the purpose of the analysis is:

    1. divide requirements into components, i.e. system elements (subsystems, modules, identified critical or newly developed parts, etc.),

    2. establish requirements for ensuring the quality of supply (i.e. materials, services, software, etc.),

    3. evaluate the capabilities and capabilities of the manufacturer and suppliers (technical and personnel level, technological equipment, control equipment, people's qualifications, etc.).

b. Analysis of reliability objectives

   It is implemented in the defining, preparatory phase of the product life cycle from the viewpoints of:

   - fault-free requirements, or lifetime of the product (its elements, systems, critical parts, etc.),

   - requirements for maintainability and provision of maintenance (scope and conditions of provision of preventive maintenance, diagnosability, need for operational diagnostics, repairability, general scope of spare parts needs during operation, etc.).

c. Economic evaluation

   One of the basic aspects of assessing a newly developed or innovated product is the consideration of expected costs and achieved benefits for its required (or assumed) reliability, or lifetime while respecting safety requirements. To optimize reliability costs, two related methods are usually used: the LCC qualitative assessment method "Life Cycle Cost" and the LSC quantitative assessment method "Life Support Cost", which according to IEC recommendations have be part of the technical documentation.

The LCC analysis broadly includes the following steps:

- drawing up a list of cost items that will be used,

- the formulation of mathematical relationships for the expression of cost items based on product characteristics and operating conditions, the essence of the LSC method,

- cost correction (instability of the value of money, valorization, inflation, increase in input prices, investments and depreciation, current annual expenses, etc.),
- collection of input data from the manufacturer (design data, fault-free and maintainability data, price lists and other financial data, etc.) and from the user (data on implementation, installation, costs of ensuring conditions of use, maintenance, supply of spare parts, etc.), possible computer processing.

The level of reliability as a whole and the level of individual sub-properties is created, shaped and maintained in two basic spheres:

1. In the sphere of design, projection, construction and production, the so-called inherent reliability, internally, genetically given to the product by creative workers.

2. In the sphere of users, where the initial inherent reliability properties are developed and formed (mainly, however, maintained and also reduced by the action of the human factor) in the manner of operational use and operational care of the product.

Operational care of an object (vehicle) is an integrating term that includes all measures and activities carried out in the sphere of users (in the ACR, e.g. in units, departments, etc.) and in the ACR logistics units and departments. Operational care mainly includes preventive care, i.e. system of inclusion and use, maintenance, technical diagnostics, repair system, etc.

Maintenance is an activity carried out for the purpose of maintaining the object in an operable condition for a period determined by the technical conditions; it consists in checking the condition of the object, carrying out preventive interventions and maintenance after a fault. Maintenance includes washing, cleaning, adding fuel, lubricants and operating fluids, lubrication, adjusting, adjusting, checking parameters, troubleshooting, repairs and more.

Technical diagnostics is a field dealing with methods and means of determining the technical condition of objects. Technical diagnostics means non-dismantling and non-destructive diagnostics. Its purpose is to evaluate the technical condition and draw conclusions for further operation, repair, etc.

Repair is a set of activities carried out after a fault in order to return the object to an operational state. Includes disassembly, replacements, adjustments, partial repairs, assembly, etc.

The entire complex system of the mentioned activities, which make up the operational care of the vehicle, can significantly influence the achieved parameters of reliability properties. E.g. the level of maintainability depends not only on the level of construction and manufacturing, but also on the quality of maintenance, the level of people performing maintenance (driver, crew, workshop specialists), workshop and mechanization equipment, diagnostic technology and its adoption, tools, devices, workshop equipment, etc.; the same applies to other reliability properties.

2.1.1 Failure rate and product life characteristics curve

In practice, reliability is determined by the *number of failures per unit time during the duration under consideration* (called the **failure rate**).

In considering the failure rate of a product, suppose that a large group of items is tested or used until all fail, and that the time of failure is recorded for each item. Plotting the cumulative percent of failures against time results in a curve such as the one shown in **Figure 3**.

It can be seen from **Figure 3** that 34% of items failed within the period from 0 to 500 hours and 86% of items have not survived more than 4500 hours operating time. The latter data can be re-interpreted as follows: when the product is placed in service, then, as time goes on, 66 items of 100 probably continue to meet specification after 500 hours, while 14 items of 100 are expected to survive more than 4500 hours operating time under given operating conditions. Hence, the **cumulative failure curve** can be also used to estimate the reliability of the tested product.

Both` reliability and unreliability vary with time. Reliability $R(t)$ decreases with time; an item that has just been tested and shown to meet specification has a reliability of 1 when first placed in service, 1100 hours later this may have decreased to 0.5. Unreliability $F(t)$ increases with time; an item that has just been tested and shown to meet specification has an unreliability of 0 when first placed in service, increasing to 0.5 after 1100 hours. Since, at any time t, the product has either survived or failed, the sum of reliability and unreliability must be 1, i.e.:

$$R(t) + F(t) = 1. \tag{1}$$

The situation is shown in **Figure 4**.

For example, knowledge of a product's reliability is useful in developing warranties.

The **instantaneous** failure rate (failures per unit time) λ at any point in time t is defined by eq. (2):

$$\lambda(t) = \frac{\left(\frac{dF}{dt}\right)}{R(t)} \tag{2}$$

The second phase of the life characteristics curve describes the normal pattern of random failures during a product's **useful life**. This period usually has a low, relatively constant failure rate caused by uncontrollable factors, such as sudden and unexpected

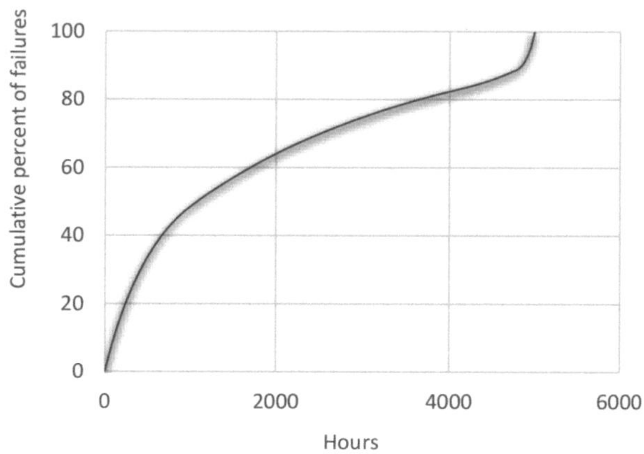

Figure 3.
Cumulative failure curve of the product.

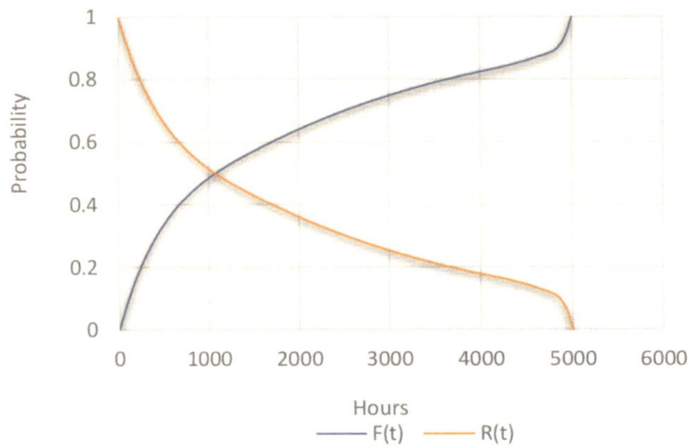

Figure 4.
Unreliability (F) and reliability (R) of the product.

stresses due to complex interactions in materials or the environment. These factors are usually impossible to predict on an individual basis. However, the collective behavior of such failures can be described statistically.

Finally, as age takes over, the **wear-out period** begins, and the failure rate increases, a common experience with automobile components or other consumer products [12].

It is very easy for a machine manufacturer to declare "our machines are reliable", but behind that statement hides complex verification and reliability analyses. Each analysis contains a detailed knowledge function and all possible available failures of the analyzed equipment, failure rates of individual components, etc.

The curve in **Figure 5** is typical only for some types of simple devices. The course of the period of life-threatening disorders is e.g. often affected by wear and tear.

The bathtub curve theory also describes the course of breakdowns and maintenance from a historical perspective. In general, today's devices are much more complex than they were twenty years ago. It follows that the curves of life indicators change – **Figure 6**. The curves in the figure show the failure rate of various electrical and mechanical elements depending on the time of operation (curve A describes the behavior of about 4% of objects, B - 2%, C - 5%, D - 7%, E - 14% and curve F - about 68%). Although the relative representation of objects with different behavior (curves of type A to F) is not the same in industrial sectors, the failure rate curves of equipment are increasingly approaching curves of type E and F as their complexity increases [13].

Knowing the product life characteristics curve for a particular product helps engineers predict behavior and make decisions accordingly. Though many research institutes and large manufacturers conduct extensive statistical studies to identify distinct patterns of failure over time, gathering enough data about failures to generate as smooth a curve as shown in **Figure 5** is not always possible.

If limited data is available, the **average failure rate** is computed using the following formula:

$$\bar{\lambda} = \frac{total\ number\ of\ failures}{(number\ of\ identical\ items\ tested) \times (test\ duration\ in\ hours)} \tag{3}$$

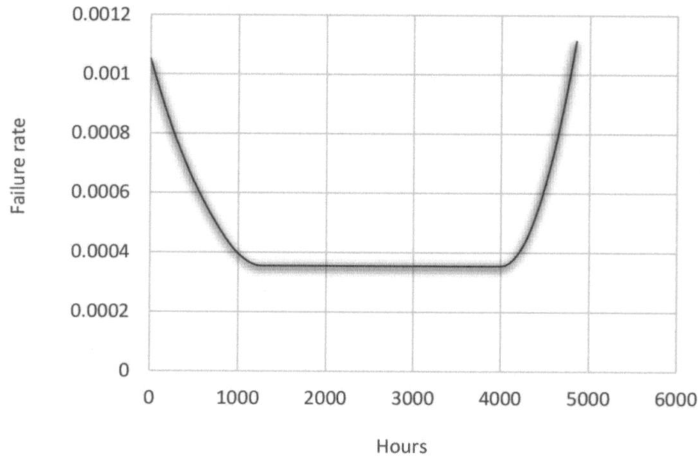

Figure 5.
Failure rate curve (a "bathtub curve").

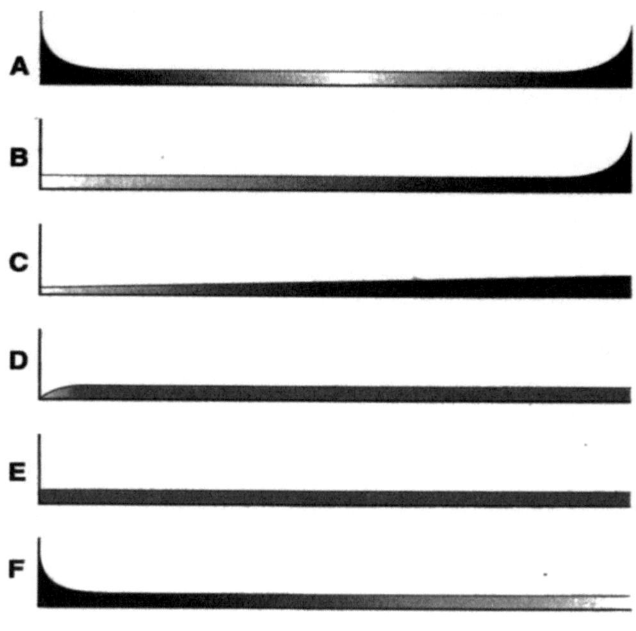

Figure 6.
Six failure rate vs. operating age curves.

3. Maintenance systems

The maintenance kinds include activities by which the maintenance is conducted. The maintenance kinds are as follows:

1. Running maintenance consists of

    - cleaning,

- lubrication,
- set up.

2. Inspection activities are:
   - revision,
   - diagnostics,
   - prophylaxis.
3. Repair activities are defined as:
   - small repair,
   - medium repair,
   - general overhaul.
4. Renovation is known as:
   - modernization,
   - reconstruction,
   - module replacement (exchange),
   - upgrading,
   - machine replacement.

Running maintenance means a regular care of objects. It has a prevention character. For this activity, the terms care-upkeeping are used.

It is a running, regular machines, and equipment maintenance. By this maintenance a premature wear and defects can be prevented. The running maintenance includes cleaning, set up of machines and equipment, tightening of loosened connections, refilling of recording (writing) parts of measuring and recording apparatuses. But it is the measurement which has the greatest significance in the whole maintenance process. To the maintenance also small repairs belong, which are conducted during the maintenance. The difference between maintenance and repairs lies also in the fact which workers are conducting them. Whereas specialists conduct the repairs, especially operatives who are also appointed with other duties conduct the maintenance. The operatives of machines and equipment conduct the running maintenance.

Inspection activity in maintenance presents a revision and control activity regarding the condition, maintenance and repair of machines and equipment within the framework of maintenance. From the point of view of a professional, the inspection concerns all kinds and ways of maintenance. It is appropriate to create professional inspection departments (e.g., machinery, electrical, construction, etc.) and to record the results of these controls and revisions into inspection books.

The monitoring of defects belongs to the important duties of an inspection. The inspection is dependent on the maintenance organization. It is not uniform, and the organizational incorporation is not uniform.

The inspection work content results from its basic objective.

Diagnostics presents the detecting of the object condition, e.g., its monitoring and registration of magnitude values which define the condition and development of a subject [14].

By introducing the subject maintenance on the base of the diagnostics of their conditions it is necessary to solve the following tasks:

- subject selection for diagnostics,
- selection of measurable magnitudes on an object that can identify the defect occurrence,
- selection of means and the diagnostics.

The diagnostics of a subject is conducted on the base of:

- a consecutive defect,
- inaccuracy,
- expenses.

Prophylaxis means a regular control of the condition of electrical equipment, information technology appliances, instruments in medicine, etc.

General Overhaul represents the repair of the whole basic equipment. By the general overhaul, the effects of wear and damage are removed. The aim here is to renew the original.

Performance, technical properties, operational quality, and profitability of the basic equipment. It is of advantage to improve its properties based on the technological progress. This can be achieved through the modernization. The general overhaul must ensure that the basic equipment will be able to work until the next general overhaul. It is convenient to conduct the general overhauls in specialized shops. General overhauls belong to the biggest, most important but also most costly repairs.

Medium repair is a repair performance of a greater extent, in which a larger number of machine or equipment parts is replaced or repaired. The extent of a medium repair is determined on the base of inspection results and findings detected during small repairs. Medium repairs represent a significant activity in the operation. Their extent is large, and they are conducted often.

Small repair includes all remaining repair performances, the extent of which is smaller in comparison with medium repairs and general overhauls. Small repairs present a large number of repairs. A part of small repairs can be done during preventive inspections by workers who conduct an inspection.

Renovation of an Object.

An intensive technical development requires that during repairs (especially during general overhauls) not only the physical but also the moral wear of equipment be removed. The total dismantling of a machine is to be used for the removal of its technical obsolescence. In this way a modernization is conducted. It includes various

Maintenance is set of activities to secure the determined operational state of an object by technical conditions, as well as to define and consider its actual state.			
	Care Examination Checking	**Inspection**	**Repairs**
Definition	Set of activities for maintaining the determined operational state of an object by technical conditions	Set of activities for determination and consideration of the actual state of a technical object	Set of activities for the restoring of a given operational state of a technical object by technical conditions
Example	Cleaning Lubrication Setting	Measurement Testing Diagnostics	Replacement of object parts Adjustment Adaptation of an object part

Table 2.
The characteristics of maintenance kinds [15].

technical improvements of a machine, by which a higher performance can be achieved, easier handling (operation), higher precision, longer life, higher safety at work, mechanization of operational manipulation, etc [13].

The modernization as a sustainable process acting against the moral wear, becomes an organic component of a repair. In association with the modernization the term reconstruction is to be specified. A modernization can represent such an improvement, which changes its original character. A reconstruction is such a performance, by which the original character of the reconstructed object is (permanently) changed (e.g., when general purpose machines are adopted to a single purpose, special machines).

Current trend of machines and equipment development, new trends in IT- equipment maintenance, when the manufacture of a module in a large series production is cheaper than the repair of a faulty one and it results in module exchange or when the downtime due to the repair, and the loss due to this downtime, and repair costs are comparable with the value of a new module.

In IT, where the replacement of an original is enlarged by a new one which ensures the original function or extends it, is called upgrading. In case that any of this kind of activity does not put the object into an operable condition, or if the costs for its maintenance are inadequate, the exchange of the machine takes place (**Table 2**).

4. The objective of maintenance and repair function

The basic function of the maintenance and repair activity is to ensure work capability of HIM and NIM in production. The significance of this function increases simultaneously with the increase of production technical level, and specially with the increase of automatic machines, lines, and integrated production processes.

The sum of work, the task of which is to secure the capability of machine and equipment operation, and their total effective running, is called maintenance-repair activity, sometimes only a maintenance. In a narrower sense, under this term we understand cleaning, lubrication.

and other regular protection of basic means prior to repair e.g., current maintenance. In a broader sense, under the term maintenance we understand any care of basic means, by which their working capability is ensured. Here belongs current maintenance of machines during operation, as well as all kinds of planned and non-planned repairs, by which the consequences of wear are removed.

In real life situation it is difficult to draw the exact border line between maintenance and repairs because these activities often overlap. During current maintenance small repairs are done, and vice versa, during repairs, sometimes current maintenance must be conducted.

Principally the removal of wear consequences is called repair. From the economic point of view, it is execution of additional work on the fixed assets, by which their wear is to be removed. The inner differentiation of maintenance-repair work is being derived from the level of functioning of the fixed assets (basic means). In trouble-free functioning the supervisory-inspection activity will be enough. It must avoid undesired conditions. If the functioning worsens, it is necessary to conduct repairs. According to their extent there are various degrees of maintenance/repair activities:

Current maintenance consists of cleaning, lubrication, inspection, and other pre-repair activity.

Small repair consists of such work as repair and replacement of smaller parts, which are subject to wear, replacement of sealing of taps, valves, cleaning and exchange of oil and cooling systems, oil cups, etc.

Medium repair is characterized by the fact, that during its execution the opening and inspection of all boxes and closed mechanisms is done, larger parts and aggregates are being repaired and replaced, the whole lubrication system is being flushed. According to circumstances machine slide ways are being repaired and the whole set-up of a machine is conducted, including precision checking. The medium repairs are conducted on the site, where the machines are installed. Only exceptionally they are conducted in repair shops, specially determined for repairs. In any case, after the medium repair a control protocol is written down, where there is the record of handing over the machine and the record on precision control results.

General overhaul (GO) presents the biggest repair performance in maintenance. In this kind of repair, the whole machine undergoes the repair, not just its parts. For this reason, it is being released from its base, it is disassembled to its individual parts and repaired in a special workshop. The extent and quality of GO must ensure that the machine will obtain the original technical properties and its planned life will be obtained. The machine or other fixed asset undergoes a test according to standards after the general overhaul and is handed over to operation by means of a protocol.

The intensive technical development requires still more that in repairs (specially in general overhauls) not only physical, but also moral wear is removed. The complete disassembly of a machine must be also used for eliminating its technical obsolescence. In such a way modernization takes place. It includes various technical improvements of the machine by which its higher performance, easier handling (operation), higher precision, longer life, improved safety at work, mechanization of operation manipulation, etc. can be achieved. In such a way the modernization, as a continuous process acting against the moral wear, becomes an organic part of repairs.

In association with the modernization the term reconstruction must be specified. Modernization can represent such a machine improvement which does not change its original character. Reconstruction, on the other side, presents such a performance, by which the original character of the reconstructed machine has been changed permanently (e.g., when multi-purpose machines are adapted to single-purpose, special machines) (**Tables 3** and **4**).

Maintenance activities	Maintenance kinds	Stages
Cleaning Lubrication Set up	Current maintenance	I.
Testing Diagnostics Prophylaxis	Inspection	II.
Small repair Medium repair General overhaul	Repairs	III.
Modernization Reconstruction Module replacement(exchange) Upgrading Machine replacement(exchange)	Renovation	IV.

Table 3.
The objective of maintenance and repair function [16].

Assessment criteria	Central form	Decentral form	Integrated form	Combined form
Operation know-how	satisfactory	good	very good	satisfactory up to very good
Roads, ways	long	negligible	negligible	negligible to long
Decision acts	many	negligible	negligible	negligible to many
Capacity balancing	very good	good	insufficient	good to very good
Rationalization possibility	very good	negligible	almost none	negligible to good
Investment for machine deployment	average	high	very high	average to high
Quality	very good	good	satisfactory	good
Cooperation within the company	very good	negligible	none	negligible to very good
Organizational demand	high	average	negligible	average
Qualification of workers	average	average	very high	average to high
Production kind sequence	none	negligible	very strong	negligible to strong
Experience exchange	very good	satisfactory	negligible	satisfactory to very good
Number of workers required	negligible	high	average	negligible to average
Worker's motivation	negligible	good	high	good to negligible

Table 4.
Organizational maintenance forms [17].

Decentralized organization form results from the fact that workers are assigned on the base of their qualification and work-legal relations to individual lower organizational units, e.g., plant. With consequent implementation of this form of maintenance the incorporation of maintenance operation into the structure of each production unit is assumed. This unit works independently and is linked directly to the organizational structure of the production unit. The workers have been specialized in the equipment of the given operation. This fact is of advantage specially both at work within current maintenance, as well as with equipment inspection.

Integrated organizational form – is based on the supposition that service workers (service men) conduct, beside the maintenance activities, also current operational work. For this work universal specialists with broad knowledge and skills are required.

5. Research into the operational reliability of forest machines

This methodology of operational reliability research is intended for monitoring the actual condition of the machine in operating conditions and evaluation of selected indicators of operational reliability of selected forest harvesting machines.

There are several methods of processing information on reliability indicators, but some of them are very complex and difficult to apply under normal operating conditions [18]. The information processing method listed below is simpler but accurate enough which in turn facilitates its implementation in real business conditions.

The methodological procedure of mathematical processing of empirical information on the reliability indicators of forest machines is as follows:

1. Create a Machine Card for each device separately

   The machine card contains:

   - CV of the machine, i.e. its current use,

   - empirical values of machine reliability indicators arranged in an arithmetic series.

2. Using mathematical statistics and software, calculate individual characteristics for failure times (standard deviation *S*, coefficient of variation *V*, arithmetic mean, left confidence interval limit, right confidence interval limit, median, minimum, maximum, lower quartile, upper quartile, variation range, quartile range, variance, standard deviation, standard error, asymmetry, point).

3. Construction of histogram and curve of cumulative relative number of failure rate indicator (resp. time between failures).

4. Selection of theoretical division and determination of its parameters.

5. Determination of fault intensity $\lambda\ (t)$ on the basis of theoretical distribution.

6. Determination of the mean time to *TS* failure.

Combined organizational form is composed of a combination of central, de-central and integrated organizational form of maintenance activities, while the principles of management are changing within hierarchical levels [19]. This form is most convenient for companies in the metallurgy industry, with a broad production program and with a large number of workers with various structure of qualification.

The examined forest machines of various manufacturers are necessary for the needs of data collection from operation and for the needs of subsequent analysis and evaluation divided into the following subgroups [20]:

1. Cabin, control units, cabin equipment (on-board computer/control unit, tachograph, outside thermometer)

2. Sensors - electrical (safety) equipment of the vehicle

3. Engine (fuel system with tank, block and lower engine cover, cylinders, pistons, pins, connecting rods, bearings, shaft, flywheel, wiring)

4. Engine and its cooling and lubrication

5. Gearboxes and transmission mechanisms (gearboxes, clutches, shafts, joints, gearboxes, differentials, distributions)

6. Chassis (body, frame, shock absorbers, suspension, wheels, tires, axles, steering, brakes)

7. Hydraulic system of the machine

8. Hydraulic crane with log grab

9. Trailer/semi-trailer

10. Trailer poles, trailer equipment [21].

6. Conclusions

The main reason for paying attention to this area of reliability and maintenance is that for every company, creating the most reliable system possible is a challenge and nowadays a common need. It is therefore necessary to be able to assess the reliability of all machinery and equipment and, in the event of deterioration in the characteristics of the means of transport monitored or stagnation, to be able to take appropriate steps to remedy this situation over time. The dependence of companies and people on technology is growing, and it is therefore necessary to ensure that the failure rate of used machinery and equipment is kept to a minimum or that machinery and equipment have maximum controlled maintenance based on real operating conditions and conditions [22].

Regarding the electronic systems used in forest machinery, we can say that this area of the logging and transport process has already been affected by Industry 4.0. All manufacturers try to implement their own know-how, their own technologies, which they try to address their customers and operators of the vehicles. Most of the mentioned systems of manufacturers are usable only with a certain limitation and therefore the manual collection of operational data is currently used for the evaluation of operational data. As advances in software and hardware for vehicles as such are growing very rapidly, there is a real need for tools to be available to collect and evaluate in-service data, thus increasing the efficiency of data evaluation and the operational reliability of these vehicles. In the future, the use of Industry 4.0 for the collection of operational data using computer technology as well as its subsequent evaluation is interesting. At present, it is possible to collect operational data only partially through the control unit and then their laborious analysis.

Operational reliability prediction grows with the introduction of software and hardware innovations. For this reason, it is clearly necessary to say that Industry 4.0. has also affected this area of industry, which greatly affects the use of these machines and equipment in real operating conditions [23].

The maintenance organization system is an important internal source for increasing the reliability of machinery and equipment. A positive result can be achieved through planning, management, improvement of work organization and data recording up to the management of spare parts [24]. In other words, a maintenance system is a means of maintaining or restoring equipment to good technical condition for the duration of its technical life or for a period which its operator can maintain in view of the amount of costs incurred.

The current state of the study of operational reliability in real conditions still carries a great deal of ambiguity and it can be said that the electronic area (software as well as hardware) has not yet matured, even with the current progress in the field of forest machinery. The high requirements for care and processing are closely related to the requirements for the quality and reliability of forestry machines and technological equipment. And these, in turn, are closely related to the care of the devices in question. The care of equipment used in forestry consists of daily operation, treatment and supervision of their operation, activities aimed at putting new equipment into operation, eliminating malfunctions and defects, improving technical condition, technical modernization, storage, preservation of temporarily decommissioned forestry machines and technological equipment and for the liquidation of decommissioned forestry machines and technological equipment and finally for the replacement of decommissioned forestry machines and technological equipment with new ones. The current and prospective degree of mechanization and automation of production in forestry, as well as the tendency to increase the performance of forestry technology, establish as one of the primary tasks the provision of its operational reliability. The system of organizing maintenance represents an important internal source of increasing the reliability of machines and equipment. A positive result can be achieved through planning, management, improving the organization of work and recording data up to the management of spare parts. In other words, the maintenance system is a means for equipment to keep them in good technical condition or restore this good technical condition for the duration of their technical life or for the period that their operator is able to maintain considering the amount of costs incurred.

Acknowledgements

This publication is the result of the project implementation - VEGA project No. 1/0364/21 "Research of forest machines working mechanisms regarding the new constructional parameters and working principles".

Conflict of interest

If you have any conflicts of interest, please declare them here. If no conflict exists, please put the text: "The authors declare no conflict of interest." or delete this entire section.

Author details

Ján Kováč
Faculty of Technology, Department of Environmental and Forest Machinery,
Technical University in Zvolen, Zvolen, Slovakia

*Address all correspondence to: jan.kovac@tuzvo.sk

References

[1] Neruda J, Simanov V, Klvač R, Skoupý A, Kadlec J, Zemánek T, et al. Technika a technologie v lesnictví – díl druhý. Brno: Mendelova univerzita v Brne; 2013. p. 300

[2] Kováč J. Teória mobilných strojov: riešené úlohy a príklady [elektronický zdroj] 1. Vyd. Elektronické dokumenty (listy, články, monografie). Zvolen: TU vo Zvolene; 2013 Cd-rom. p. 175

[3] Kováčová K, Marko J. Metóda RCM a jej implementácia pri údržbe a prevádzke lkt. In: coyous 2008: konference mladých vědeckých pracovníků, 2.4.2008. Praha; 2008. pp. 126-134

[4] Štollmann V, Slugeň J. Lesnícke mechanizačné prostriedky. Zvolen: TU vo Zvolene; 2009. p. 213

[5] Kováčová K. Research of forest machines. In: problemy inżynierii rolniczej i leśnej. Problems of argo and forestry engineering, xix międzynarodowa konferencja naukowa studentów, warszawa, 26 maja 2010. Warszawa: Szkoła glówna gospodarstwa wiejskiego w warszawie; 2010

[6] Mikleš MA, kol. Teória a stavba lesných strojov II. Zvolen TU vo Zvolene; 1993. p. 273

[7] Mikleš M, Holík J. Teória navrhovania zemných strojov pre stavebníctvo a lesníctvo. Vedecké štúdie 11/1999/a. Zvolen: TU vo Zvolene; 1999

[8] Mikleš M, Marko J. Teória a stavba lesných strojov I. Zvolen TU vo Zvolene; 1992. p. 243

[9] Mikleš M, Kučera M, Mikleš J. Cestné motorové vozidlá. Zvolen: TU vo Zvolene; 2007. p. 244

[10] Bentley JP. An Introduction to Reliability and Quality Engineering. Longman; 1993

[11] Ireson WG, Coombs CF, editors. Handbook of Reliability Engineering and Management. McGraw-Hill; 1988

[12] Pačajová H. Príklady aplikácie teórie rizík Pri plánovaní údržby (Examples of Application of Risk Theories in Maintenance Planning). Košice: TU-Košice; 1999

[13] Dummer GWA, Winton RC. An Elementary Guide to Reliability. 4th ed. Pergamon Press; 1990

[14] Lebder G. Teória údržby a opráv (Maintenance and Repair Theory), Strojnícka Fakulta (Faculty of Mech. Eng). Košice: TU, Košice; 2000

[15] Smith DJ. Reliability, Maintainability and Risk. 4th ed. Butterworth Heinemann; 1993

[16] Moubray J. Reliability-Centetred Maintenance. Butterworth-Heinemann Oxford; 1997

[17] Ziegler J. Údržba zařízení (Equipment Maintenance), Fakulta strojní (Faculty of Mechanical Engineering). Ostrava: VŠB-TU Ostrava; 1993

[18] O'Connor PDT. Practical Reliability Engineering. 3rd ed. John Wiley & Sons; 1994

[19] Weck M. Werkzeugmaschinen–Alles Konstruktionsbeispiele aus Werkzeug-Maschinenbau, Grundwerk 1. VBI Verlag Duesseldorf; 1991

[20] Boroška J, Krištin Š. Údržba Strojov a Mechanizmov (Machines and

Mechanism Maintenance). Košice: Edičné stredisko VŠT Košice; 1986

[21] Mooren R. Instandhaltungsgerechtes Konstruiren und Projektieren. Grundlagen, Methoden und Checklisten fuer die Maschinen-und Apparatenbau. Berlin: Springer; 1991

[22] Davis RK. Productivity Improvements through TPM. Prentice Hall; 1955

[23] Evans JR, Lindsay WM. The Management and Control of Quality. 3rd ed. West Publishing Company; 1996

[24] Hamelin B. Údržba a její nové pojetí (Maintenance and its New Understanding). SNTL Praha; 1983

Chapter 10

Root Cause Failure Analysis of Castings: A Case Study of a Brake Rotor

Samuel A. Awe

Abstract

A premature failure of a vehicle brake rotor was investigated and reported. The brake rotor was manufactured from a grey cast iron material and had covered about 10 670 miles before it failed. The failure-generated tremendous concern for the autoparts manufacturer due to the warranty claims from the Original Equipment Manufacturer (OEM). This chapter, however, intends to describe the methodical approach used to identify the failure's main cause using a root cause failure analysis technique and offer suggestions to prevent a similar failure from re-occurring. The results of this investigation showed that the disc's early failure was caused by oxide inclusions that were accidentally entrapped into the disc's neck region. The eventual disc failure was initiated by micro-cracks developed within the inclusion particles and propagated through the weakest interface between flaky graphite and the pearlitic matrix. To ensure that nonmetallic inclusions are kept out of cast components, several solutions for improving casting quality were proposed.

Keywords: failure analysis, gray cast iron, brake rotor, root cause analysis, fractography, nonmetallic inclusions

1. Introduction

A failure occurs when an engineering system, a mechanical component, an engineering material, or a process cannot correctly perform its intended or design function. When a system or component fails, it causes unimaginable disruption of operations and services; sometimes, it attracts or incurs a legal tussle and a stringent warranty claim. On several levels, failure can be described as (i) a loss of function, implying that the component or system works but cannot accomplish the designed function; (ii) a loss of operational life, that is, when the component or system performs its function but is unsafe or unreliable to operate; and (iii) an inoperable, meaning that the system or engineering component is completely unusable [1]. The above definitions categorize the failure of an engineering system, component, material, machine, assembly, or process. However, an engineering system or components can experience a mechanical failure in service due to design insufficiencies, maintenance deficiencies, manufacturing and material imperfections, service abuse and overload, and a hostile service atmosphere [2–4].

Structural and engineering components are generally used to produce automobiles, aircraft, buildings, power generation plants, jet engines, military equipment, manufacturing plants and marine equipment. These components are made from different materials, such as gray cast iron (GCI), steel, aluminum alloys, titanium alloys, etc. Engineering components are typically exposed to various externally applied shear and perpendicular stresses [5]. These engineering components habitually function under changeable amplitude recurrent loadings in service, although they are designed considering the level of static stress or constant amplitude fatigue strength [6]. In many cases, failure of components may lead to fatalities, property loss, degrading of the company image or loss of credibility, and many legal issues; therefore, it is of great concern to the industry [4]. Failure of some engineering components and accessories used in various engineering applications, such as railway wheels, automotive brake discs, engine cylinder heads, crankshafts, foundation bolts, chains and hooks, cranes, conveyors, and excavators, have been studied to establish the root cause of failure [6–14].

An automotive brake rotor is a critical component of a brake system designed to reduce the acceleration of or stop vehicles in motion. The fundamental comprehensive responsibilities of vehicle braking systems are to decrease the car's speed, bring the vehicle to a standstill, prevent spontaneous acceleration during downhill driving, and keep the car stationary when stopped [15, 16]. Automotive brake rotors have been generally produced from GCI material for several decades because of their inherent properties. Besides the automotive, many other industrial sectors, such as wind power, agriculture, machines, and tools, use cast iron parts extensively. Several engineering components of complex geometries and thin sections that require a combination of mechanical and thermal properties to achieve the desired function are commonly produced from cast iron materials. The existence of flaky graphite in their microstructure offers good thermal conductivity, while the pearlitic matrix is usually accountable for its mechanical properties. GCI material has the essential attributes required for the functioning of an automotive disc brake. It is a material of choice for brake rotors due to its excellent friction properties, good thermal conductivity and castability, retain strength at elevated temperatures, relative ease of manufacture and thermal stability with excellent damping capacity [11, 17–19]. However, the significant disadvantages of using GCI brake rotors include high density, susceptibility to corrosion and the propensity to noise and vibration issues. Despite these drawbacks, GCI brake rotors remain the most sought in the automotive industry to manufacture brake discs rotors due to their relatively cheap cost, inherent properties, and well-known production process.

The automotive brake discs are usually produced through a metal casting process technique. A sand mold casting is used to manufacture brake rotors, where four to six brake discs are simultaneously cast in a mold. Cast components may look perfect visually, but occasionally they may contain concealed inadequacies which might not be noticed until the part fails in service. Different factors could influence the failure of an engineering component; these factors may include (i) hidden manufacturing defects/imperfections, (ii) poor design, (iii) the type and size of load the component is exposed to, (iv) inappropriate raw materials, (v) improper repair or maintenance, and (vi) the environmental conditions under which the component served. However, cast components' functionality strongly depends on the soundness of the castings. Cast components are consequently required to be virtually free of imperfections that can impair their quality and lead to early failure of castings, failed machine tools, or poor mechanical properties of cast components. According to the International Committee of Foundry Technical Association, casting defects are classified into seven categories - metallic projections, cavities, discontinuities, defective surfaces, incomplete casting, incorrect dimension, and inclusions or structural anomalies [20]. The operation of the foundry process is a complex multi-step process with varying technical levels. Therefore, the final quality of the cast products can be influenced by the operator's skills, defective pattern design, improper metal melting, adopted quality management system and equipment, defective molding material, incorrect quality of raw materials used, improper mold venting, casting processing problems, and improper service condition and maintenance of the available equipment. If the defects in cast products are not effectively inspected, it will impair the quality of the manufactured products and hence lead to unsatisfied customers due to a faulty product. Generally, the most severe defects that can serve as stress raisers or crack promoters in cast components include pre-existing cracks, internal voids, and nonmetallic inclusions.

Casting inclusions defects can be defined as nonmetallic and occasionally intermetallic phases embedded in a metallic matrix. They are frequently simple oxides, sulfides, nitrides, or their complex compounds in ferrous alloys and can include intermetallic phases in nonferrous alloys. In almost all instances of metal casting, they are considered detrimental to the cast component's performance [11, 21]. Casting inclusions are subsurface defects, which may sometimes be detected during the machining operations or even, in many cases, remains undetected until the component fails in service. Nonmetallic inclusions can adversely influence the mechanical properties of castings because they act as a stress raiser. Some mechanical properties are more sensitive to inclusions than others; for instance, elongation or reduction in area is very sensitive to the presence of inclusions and generally adjusted more significantly than ultimate tensile strength. Therefore, cast products use ductility specifications as standard quality-control indices. Inclusions defects have different chemical roots and negatively influence the cast products' mechanical properties, such as machinability, corrosion resistance, fracture toughness, and formability. Inclusions are classified into indigenous or exogenous, depending on their source [11, 21]. The inclusions derived from external sources, such as slag, dross, ladle lining, eroded and entrapped mold and refractories materials, are classified as exogenous; at the same time, those that are native, innate, or inherent in the molten metal treatment process are known as indigenous inclusions. Exogenous inclusions are also derived from ferroalloys, flux materials, and other starting materials that do not float to the surface of the liquid metal or dissolve in it [22]. Sometimes, exogenous inclusions are visible to the naked eye at the casting surface. They may be seen beneath the peripheral casting surface when the casting is sectioned if they have had insufficient time to float out or settle due to density differences concerning the molten metal. Indigenous inclusions, however, are of micro size and can easily be identified under the microscope, and they are frequently distributed uniformly within the casting's microstructure. These inclusions are the products of the liquid melt reactions with deoxidizers such as silicon, manganese, and aluminum particles or during desulphurization, leaving some residual oxide inclusions in the casting. The distribution of indigenous inclusions in a grain boundary in the microstructure of the cast components can severely impair their mechanical properties [21] and reduce the components' service life.

To unravel how a cast product failed correctly, the failure analysis techniques adopted for the castings must include a systematic understanding of how the component was made and processed. However, the scope of the failure examination depends on the problem's definition. Still, in all failure analyses, the understanding of the reasons (how and why) for a casting failure can only be determined if the pertinent background information is collected, thorough examinations are conducted, representative castings are inspected, the proper material assessments are accomplished, and the service conditions to which the cast components are subjected to are undoubtedly understood [23]. The brake rotor under investigation was newly manufactured and installed in a brand-new car. The rotor had been in service for about one year and three months and covered 10,670 miles before its failure. However, under normal circumstances, a quality set of automotive brake rotors should have an average life expectancy ranging from 30,000 to 70,000 miles traveled, which depends primarily on the size of the vehicle, the way it is driven, and the quality of the brakes (including pads and discs) [11]. The above information suggests the rotor had failed prematurely in service. There are several reasons for undertaking a comprehensive understanding of why and how a cast product failed prematurely. The reasons may include – i) evaluating the effectiveness of an in-process quality-control system, ii) assessing the efficiency of the installation or assembly plant where the part was assembled, iii) improving product design and manufacturing processes; iv) preventing similar problems with identical components, v) absolving a company of liability, and vi) determining and understanding the cause(s) of a failure to resolve financial warranty claims where applicable [23–25].

A failure analysis investigation generally involves collecting and analyzing failure information, identifying the root causes, improving product design, ensuring product compliance and assessing the liability of product failure. Besides, the data collected from the failure analysis can improve component design, regulate maintenance plans, improve maintenance processes where necessary, and improve asset reliability. However, this chapter describes the practical application of root cause analysis in automotive brake rotor failure, details the step-by-step approach to unraveling failure root cause, documents the inspection and characterization of the failed brake rotor, and provides recommendations to avert the recurrence of such a failure in the future.

2. Practical application of root cause failure analysis in brake rotor failure

Nowadays, automakers and automotive part manufacturers are under intense pressure to avoid warranty claims/costs and litigation by elevating the quality of their products to ensure safe operation. The component that fails earlier than expected usually attracts financial liability and challenges the manufacturers to scrutinize the root cause(s) of failure to evade future recurrences. However, in some cases, if the cause of the damage cannot be recognized and resolved, it increases the service and warranty costs until the appropriate solution is accomplished.

A failure is when a component or machine cannot satisfactorily perform its intended functions correctly. A metallurgical failure analysis technique is usually adopted for determining the next step to resolving the problem. Failure analysis is a systematic, methodical process to determine the physical causes of failures. Sometimes, the failure analysis process can be complex, draws upon many different technical disciplines, and employs various observation, inspection, and laboratory techniques [25]. Performing a failure analysis is tedious and complex, and it does not necessarily conclude when the physical causes of the failure are recognized.

Consequently, a well-organized, comprehensive, and straightforward way of resolving failure analysis problems is required to achieve enhanced product quality and failure prevention. A reasonable failure analysis approach first requires a clear understanding of the problem definition and the distinction between an indicator, a cause, a failure mechanism, and a consequence [25]. However, a complete understanding of the conditions associated with the failure would enhance the knowledge of its causes and significantly improve the ability to specify appropriate corrective measures.

The root cause failure analysis (RCFA) procedure is commonly used to discover more profound contributors to failures, such as human and hidden root causes. RCFA is a process for identifying the actual root cause of a particular failure and applying the information to set a pathway for corrective and/or preventive measures. RCFA, in combination with physical analysis, are essential steps in the general problem-solving process and are vital constituents for amending and preventing failures, achieving higher levels of quality and reliability, and ultimately enhancing customer satisfaction [25].

In solving failure or quality-related problems, it is better to visually structure the information relating to the problem because it can sometimes be challenging to see how different aspects of the issue interact. The fishbone diagram is an applicable technique for visualizing causal factors and their effect on the quality or failure of a product. This diagram is also known as the cause-and-effect diagram (**Figure 1**). The chart is a visual method for root cause analysis that organizes cause-and-effect relationships into categories [26]. The idea behind this diagram is that several factors may be responsible for the poor quality or failure of a product/component, which is hidden but can be categorized and viewed clearly from the chart. This diagram has been employed by aerospace, information technology, and medical industries for process and product quality improvement. Cause-and-effect graphs allow the visualization and organization of potential causes of a problem into an applicable framework for solving it. The charts are also helpful for incorporating cross-functional influences.

The damaged rotor in this investigation was produced by sand casting, machined to the specified dimensions, and installed in a new vehicle. The rotor had operated for approximately fifteen (15) months and covered 10,670 miles before it failed. However, the potential causal factors that possibly led to the failure of the rotor are classified and identified, as illustrated in **Figure 1**. The "fish" head represents the

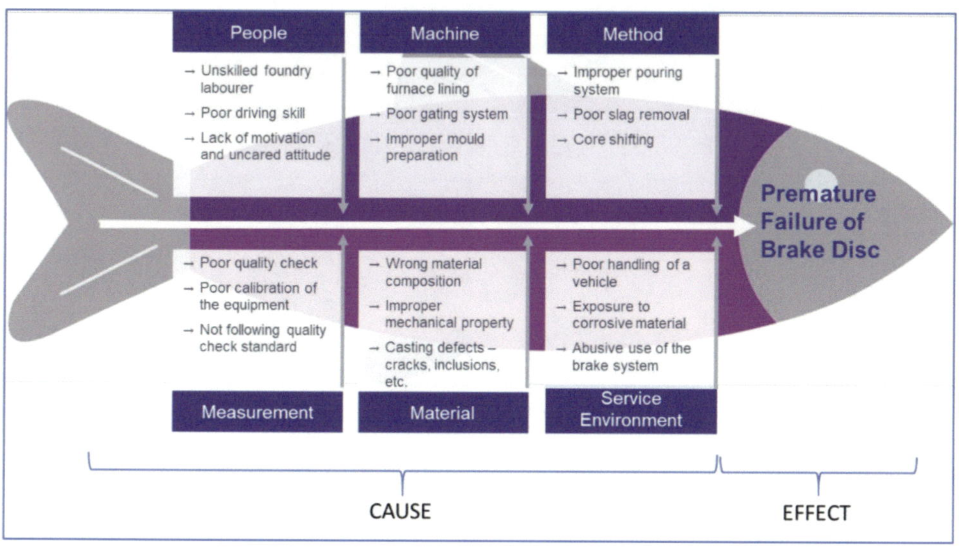

Figure 1.
Fishbone diagram illustrating the potential causes of brake rotor failure.

problem to be studied, and each critical branch denotes a specific functional area. Based on the information presented in **Figure 1**, the failure root cause analysis was begun by critically evaluating each operational unit of the diagram. The background information provided at the beginning of the investigation assisted in eliminating the need to investigate the other branches further. The root cause investigation was focused on the manufacturing process and the application of the brake rotor.

However, in this investigation, a systematic approach of root-cause failure analysis was undertaken to understand the factors contributing to the brake rotor's failure. In addition, this chapter presents some recommendations that will improve the quality of automotive brake rotors during manufacturing and inspection and prevent future failure reoccurrence.

3. Methodology

Firstly, the "as-received" failed GCI brake rotor (**Figure 2a** and **b**) was critically inspected to observe any abnormalities or predominant features on the disc before samples were extracted from it. The rotor was cautiously handled to avoid further

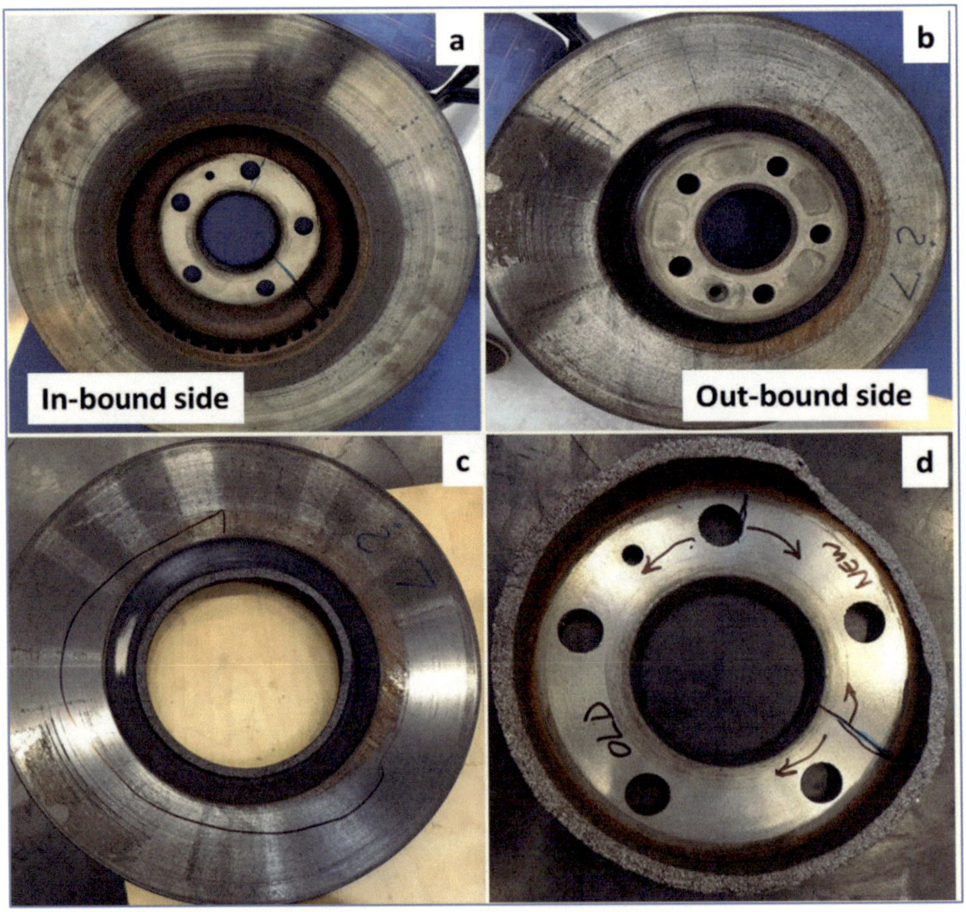

Figure 2.
The as-received brake rotor (a) in-board (b) out-board sides of the disc, and (c) and (d) represent the split rotor.

damage to the cracked surface throughout the investigation. To inspect the fracture surface, the as-received rotor was cautiously divided into two (**Figure 2c** and **d**) along the fracture line. The broken surface was critically examined to identify abnormalities and the origin of the fracture. Visual inspection was done using a stereo microscope (SZ-CTV Olympus, Japan). Samples for chemical composition and microstructure analyses were obtained from the swan-neck region of the as-received rotor using a spark optical emission spectrometer (OES) and scanning electron microscope, respectively. The rotor swan neck thickness was determined using a TI-007X Precision Ultrasonic Wall Thickness Gauge.

The metallographic samples were extracted from the damaged section of the rotor and prepared to identify any microstructural irregularities. The samples were etched with a 2% Nital solution to reveal the pearlitic structure of the GCI rotor. For microstructural investigation, the samples were observed under a light optical microscope (LOM) and a scanning electron microscope coupled with energy dispersive X-ray microscopy (SEM-EDS). A scanning electron microscope (JEOL JSM-IT200) and

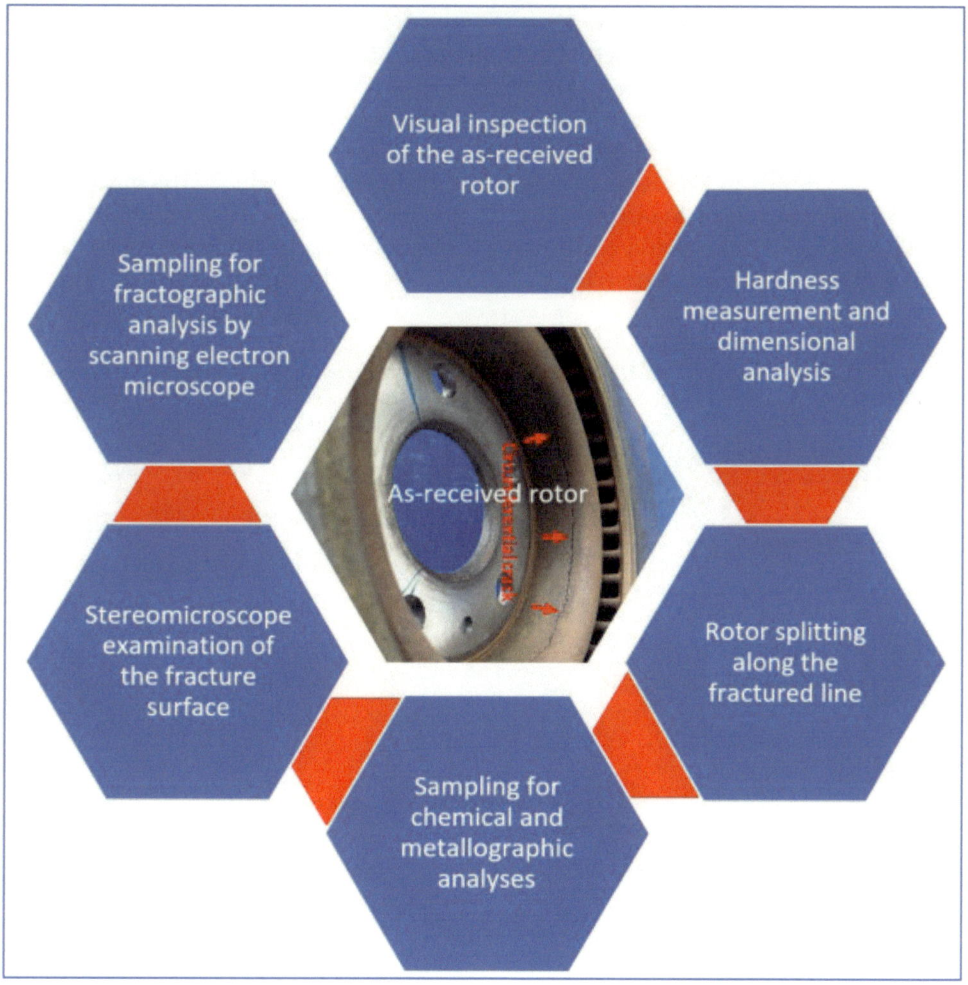

Figure 3.
Chart illustrating the step-by-step approach to finding the cause of the rotor failure.

Olympus optical microscope (Leica, DM2500M, Germany) were used for the microstructure analysis. A Brinell hardness tester was employed to measure the rotor's hardness on the friction surface and the disc hat. The average value of the hardness was reported. The fractographic investigation was accomplished using SEM-EDS to determine the constituent phases found on the cracked surface. **Figure 3** illustrates the systematic method adopted to understand the root cause of the rotor failure.

4. Results and discussion

4.1 Visual inspection

The failed "as-received" brake rotor was carefully examined before cutting samples out of it to observe any signs of aggressive use or inappropriate installation of the rotor with other brake system components, such as pads and calipers. As shown in **Figure 4**, a circumferential crack was noted at the swan neck region of the rotor, indicated by the arrows. The damage covered over 75% of the disc neck perimeter. It is also observed that the in-board area and the rotor's vents have been rusted, while the out-board region is still coated with graphite black paint that protects it against corrosion. A thorough examination of the brake surfaces (in-board and out-board) does not suggest any symptoms of abusive use of the rotor (**Figure 2a** and **b**). The friction planes do not show a blue discolouration marking. Blue discolouration of the friction surface emanates from excessive heat generated from braking due to the misalignment of brake calipers or inadequate heat dissipation by typical brake components [27]. This discolouration could occur due to extreme pressure mounted on the brake pads even when the brake system is not used. However, the brake surface blue marks

Figure 4.
A circumferential crack around the swan neck of the rotor.

can suggest faulty calipers and brake pads, disc cracks, and non-uniform brake pad wear. This analysis does not indicate any of the abovementioned issues as the cause of the failure, implying that the rotor was installed correctly and operated normally.

4.2 Analysis of brake disc material

4.2.1 Dimensional analysis

During the casting operation, a core shifting may occur, which could alter the dimensional correctness of the disc swan neck thickness. Therefore, inspecting the correctness of this thickness is sometimes crucial to avoid a reduction in load-bearing capacity. A core is a pre-determined shape of the mold, which provides internal cavities, recesses, or projections in the casting. The tendency for core instability during the casting operations is high due to pressure build-up or operational error during the mold filling. A core shift is a defect due to the buoyancy of liquid metal that causes the core of the cast to move from its correct position. If a core shift occurs, it causes a dimensional inconsistency in the casting. **Figure 5** shows a schematic illustration of the brake rotor suggesting the design requirement of the rotor swan neck thickness. As indicated in the figure, the specified thickness is 6.5 ± 0.25 mm. The failed disc swan neck thickness was determined as 6.62 ± 0.02 mm, which is still within the design specifications. This analysis suggests that the failure was not caused by core shifting.

4.2.2 Hardness measurement

Hardness is an easy method to assess the mechanical property of castings, and it is the measure of material resistance to deformation. Brinell hardness test is one of the standard techniques used for hardness measurement. The hardness was determined on the friction surface, and the hub of the failed rotor, and the measured value ranges from 163 to 180 HB. However, the design hardness specification is 160–220 HB. It can be said that the hardness of the disc is within the specification when the measured value is compared to the required hardness.

4.2.3 Chemical composition

It is essential to check the chemical composition of any failed cast component to approve that it was manufactured from the standard material. A sample was extracted from the neck region of the rotor and investigated for its elemental composition. The measured value was verified against the standard material composition. The results are presented in **Table 1**. The chemical analysis of the failed rotor confirmed that the disc was produced from the appropriate GCI material and that there were no deficiencies or abnormalities in the composition.

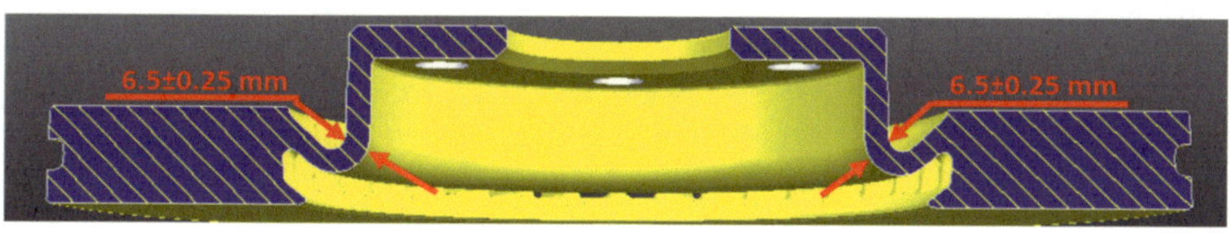

Figure 5.
A schematic representation of the rotor showing the specified swan neck thickness.

Element	Failed rotor	Material standard
C	3.79	3.60–3.90
Si	1.84	1.80–2.20
Mn	0.57	0.50–0.80
P	0.05	0.00–0.10
S	0.07	0.00–0.10
Cr	0.19	0.20–0.40
Mo	0.04	0.00–0.10
Cu	0.26	0.60–1.00
Fe	Bal.	Bal.

Table 1.
Elemental composition of the fractured brake disc (wt.%).

4.2.4 Microstructure analysis

The microstructure of a material depends on its elemental composition and the processing technique used. Even at the same nominal constituents, the effect of the processing method can produce several other microstructural features in different metallic materials. Light optical and scanning electron microscopes were used to examine the brake rotor samples for their microstructure constituents. **Figure 6a** displays the LOM micrograph of the etched sample, and the SEM micrograph of the disc is presented in **Figure 6b**. The microstructure, shown in **Figure 6a**, displays no traces of free cementite, but 0.15% free ferrite (by phase volume fraction) was observed. It is also evident in **Figure 6** that the GCI disc contains graphite flakes, which conforms with the requirements according to the European Standard (EN ISO 945–1:2008) for GCIs rotor [28]. Based on this standard, the rotor microstructure should primarily consist of pearlite in its matrix and a maximum of 5% and 2% free ferrite and free cementite, respectively, by phase volume fractions. Also, the graphite morphology should be mainly Type-A with flakes size 3–4 according to the SS-EN ISO 945–1:2008 standard (**Figure 6b**). It is apparent in **Figure 6** that the matrix is principally pearlite

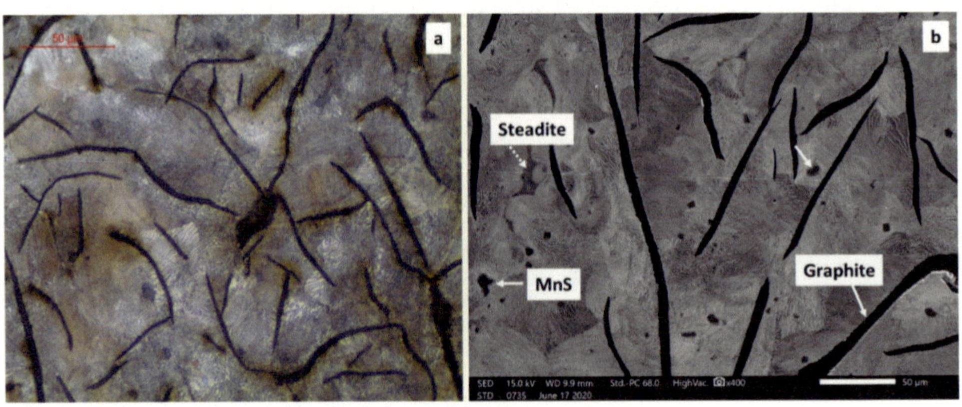

Figure 6.
Light optical (a) and SEM (b) micrographs of the rotor.

with a few manganese sulphide (MnS) particles and traces of steadite. It can be concluded from this study that there were no inconsistencies in the microstructure constituent of the failed disc, and therefore conforms with the microstructural specification for the disc material.

4.3 Fractographic examination

The aims of conducting fractographic analyses are to identify the origin and the propagation direction of a crack. This analysis is often performed using a combination of low and high-magnification microscopes. With the aid of a stereomicroscope, the origin/cause of a failure can be detected. At the same time, a scanning electron microscope can reveal detailed features and the identity of the failure cause. As shown in **Figure 7a**, the region identified as "Old fracture" is the fracture surface created during the in-service of the rotor, while the one marked as "New fracture" is the surface made during the splitting process. The "Old" fractured surface was critically inspected using a stereomicroscope to identify the failure's origin and root cause, as illustrated in **Figure 7a**. A thorough examination of the cracked surface showed some microstructural features variation from the primary matrix microstructure. The broken arrows in **Figure 7a** indicated some whitish gray phases, presumed to be corrosion

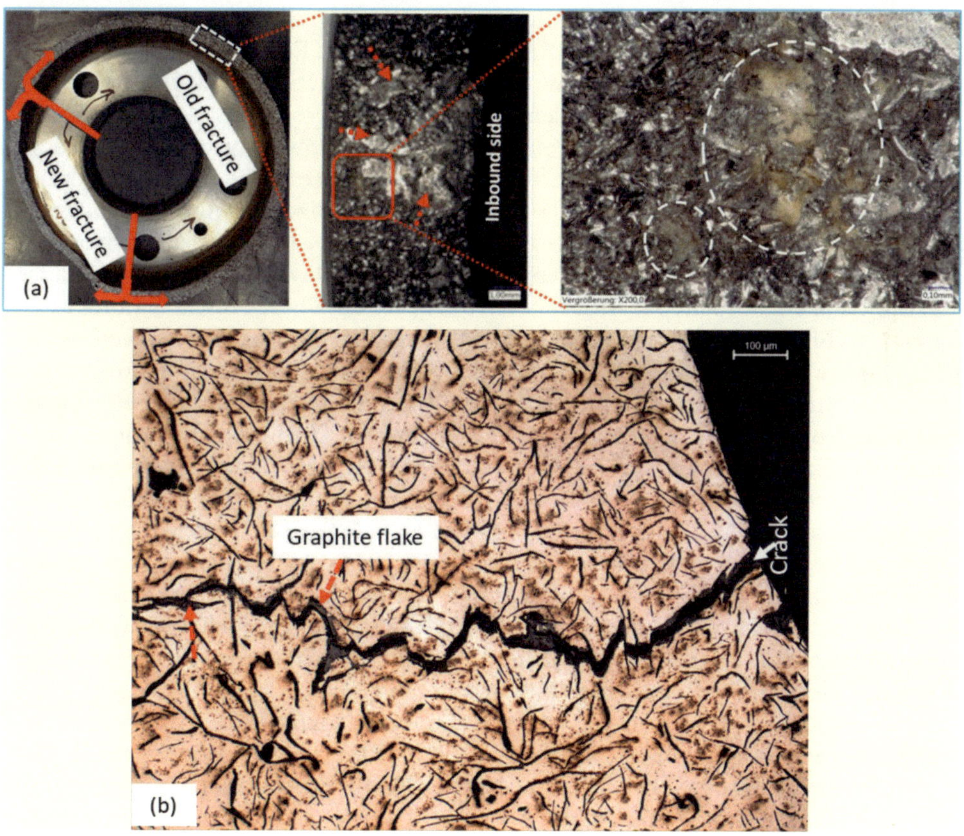

Figure 7.
(a) Stereomicroscope analysis showing fracture surface features. The broken arrows indicate corrosion products; the circled areas indicate foreign structures; and (b) a micrograph showing the propagation of cracks along the graphite-pearlite interface.

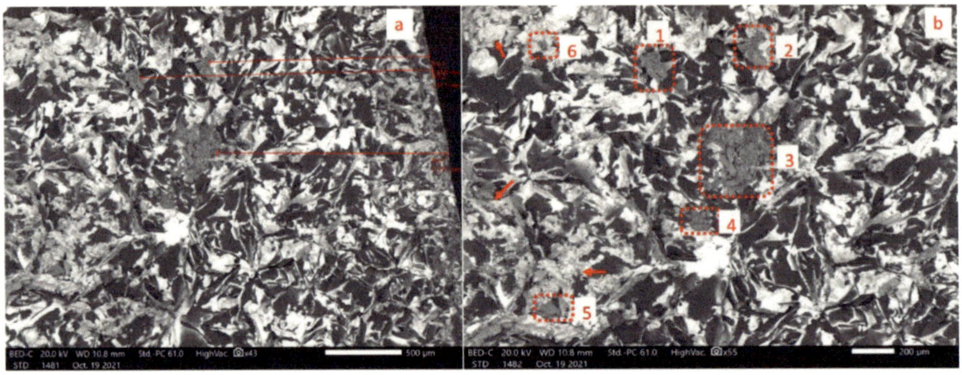

Figure 8.
SEM micrographs of the rotor's fracture surface.

products. The region marked with a solid rectangle was further magnified to reveal its microstructural make-up. However, some distinct phases were noticed in the magnified micrograph (areas highlighted in broken circles), which were further examined through an SEM-EDS analysis for identification. Additionally, a metallographic sample was extracted from the region around the fracture and observed under the optical microscope, revealing that the crack, having initiated, propagated circumferentially along the flaky graphite-pearlite interface, as illustrated in **Figure 7b**.

Figure 8 displays the SEM micrographs of the rotor's fracture surface. Different phases with varying textures can be identified from the micrographs. An inspection of the crack surface showed a relatively rough texture with a significant amount of oxidized products and foreign bodies inclusion, as displayed in **Figure 8**. The dark gray is the graphite phase, while the whitish-colored areas are the oxidized pearlite matrix due to the penetration of corrosive agents into the fractured surface leading to the corrosion of the cracked surface. Besides, some areas were recognized with a distinct appearance from the major constituent phases, as shown in **Figure 8b**. These critical features are labeled 1–6 (**Figure 8b**) and further investigated at high magnification to reveal their details. **Figure 9** presents the high magnification of some of the distinct microstructure features identified on the cracked surface of the rotor. A high number of foreign inclusions were also identified, some highlighted in **Figure 8b**. The foreign inclusions occasionally appear as a blocky (**Figure 9**, images 3–6) or lumpy (**Figure 9**, images 1, 2 and 7) structure. The location of the inclusions is approximately 1.620 ± 0.237 mm away from the disc in-board radial perimeter, as indicated in **Figure 8a**. **Figure 8** also displays that the cracked surface portrays layered structures suggesting brittleness, a characteristic fracture behavior exhibited by GCI materials [11, 29]. A network of fine microcracks can be seen on the fracture surface and even in the vicinity of the inclusions (**Figure 8**). Numerous microcracks are observed around these inclusions (**Figure 9**), suggesting potential nucleation sites for crack initiation.

However, an SEM-EDS spot and elemental mapping analysis were employed to identify the origin and composition of the foreign inclusions shown in **Figure 9**. The results of these analyses are presented in **Figures 10–13**. Furthermore, a close examination of the EDS analysis of Spots a and c in **Figure 10** reveals that the foreign materials in micrographs 3 and 5 (**Figure 9**) are silica sand (SiO_2) inclusions. The shape of the sand inclusions is comparatively regular, as seen in the figure. However, some other oxides can be observed admixed with the silica as represented in the EDS microanalysis of **Figure 10d** and the SEM-EDS elemental mapping of micrograph 3 (**Figure 11**).

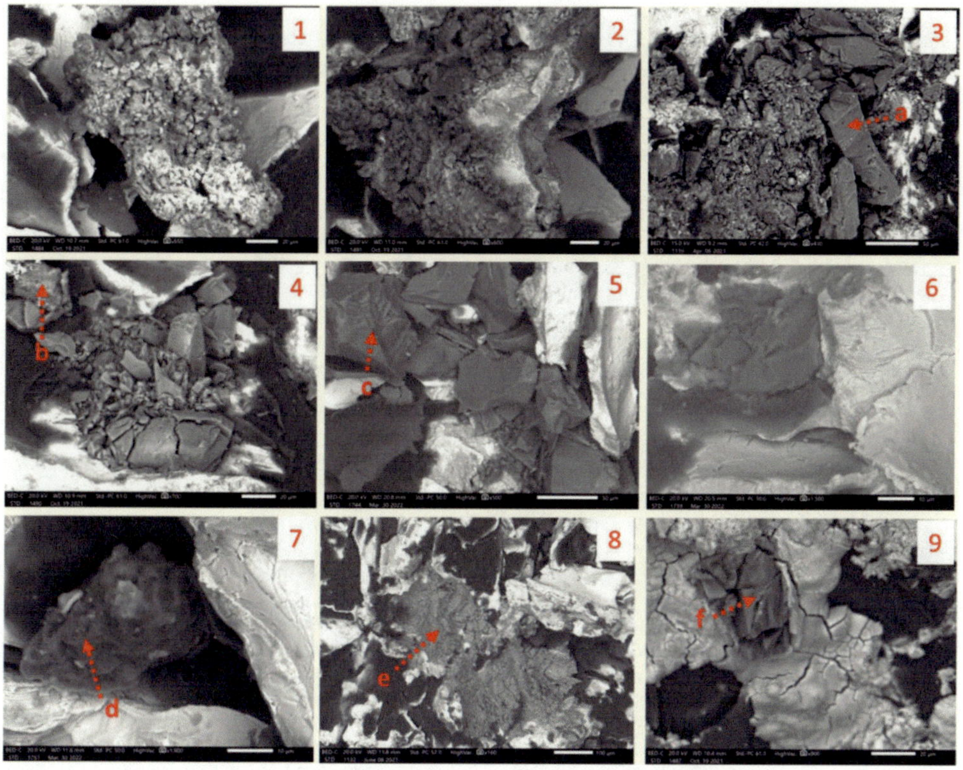

Figure 9.
SEM micrographs of the inclusions found on the fracture surface of the rotor.

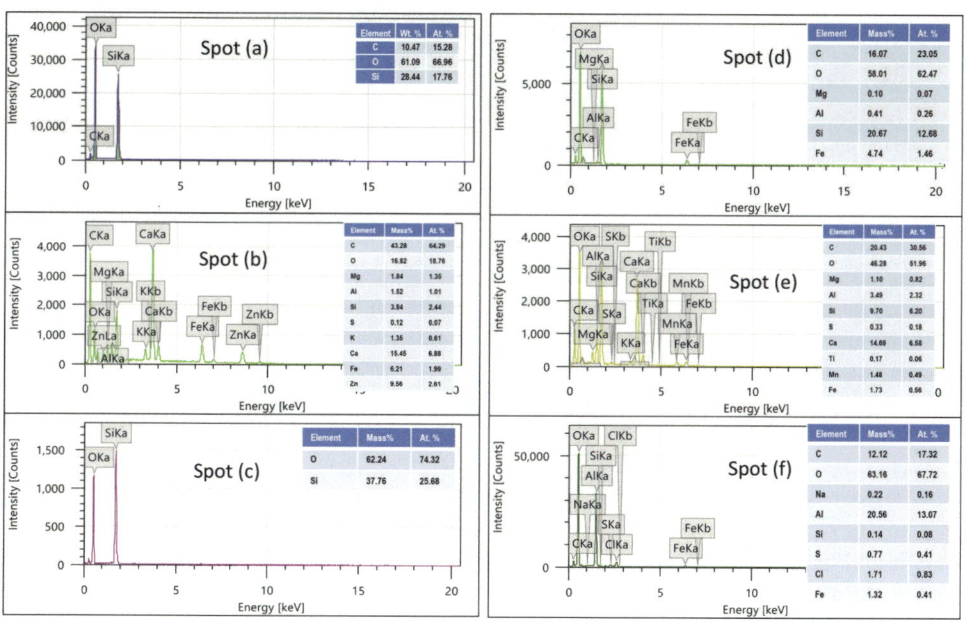

Figure 10.
SEM-EDS spectra of spots (a-f) indicated in ***Figure 9****.*

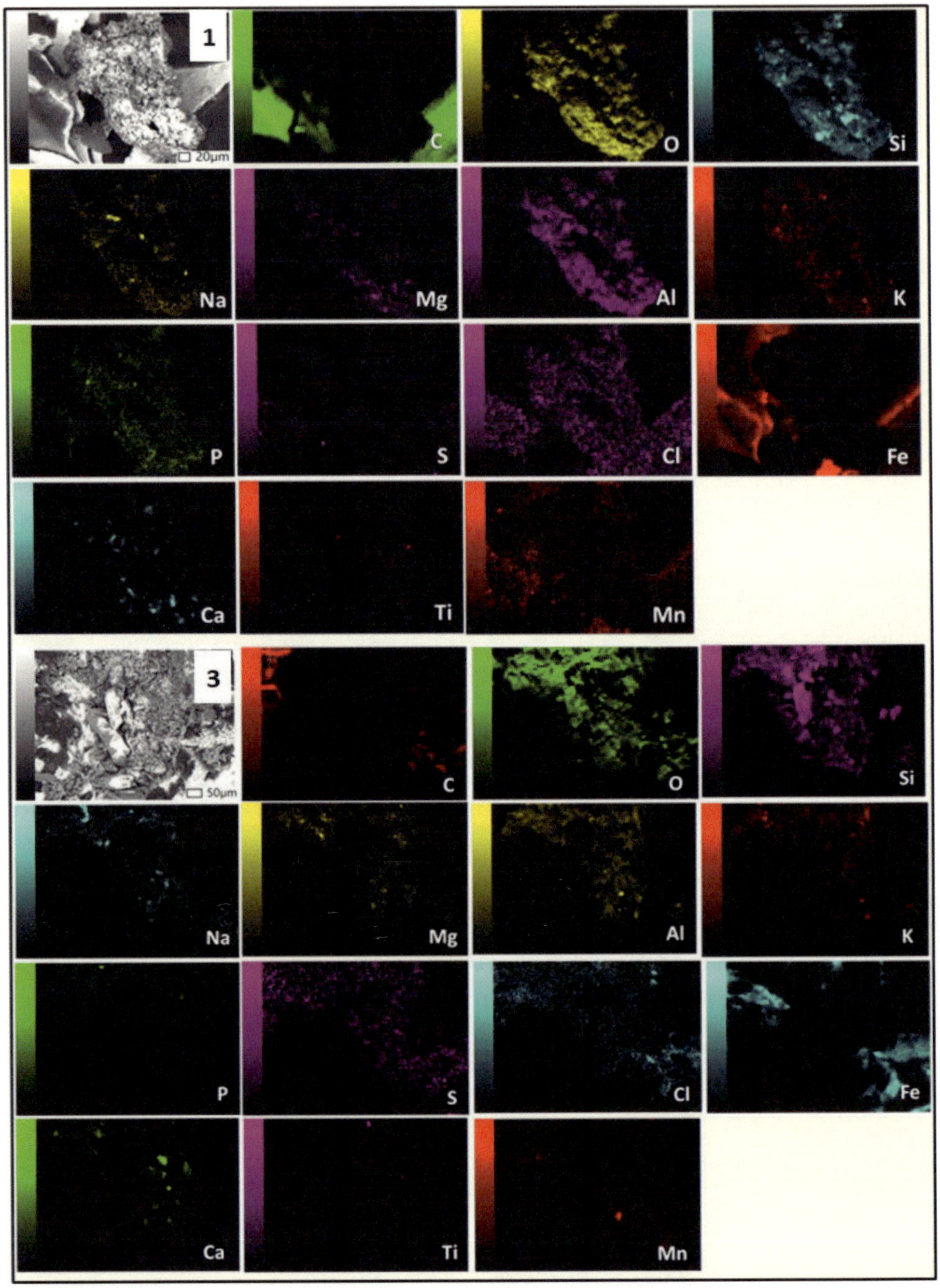

Figure 11.
SEM-EDS elemental mapping of micrographs 1 and 3 of ***Figure 9****.*

Silica sand with binders is generally used for making mold and core in the iron casting process and can be differentiated from slag because it comprises a single phase (**Figure 10a** and **c**). The presence of silica inclusions can be attributed to the entrapment of mold material during metal pouring due to the erosion of or loose mold

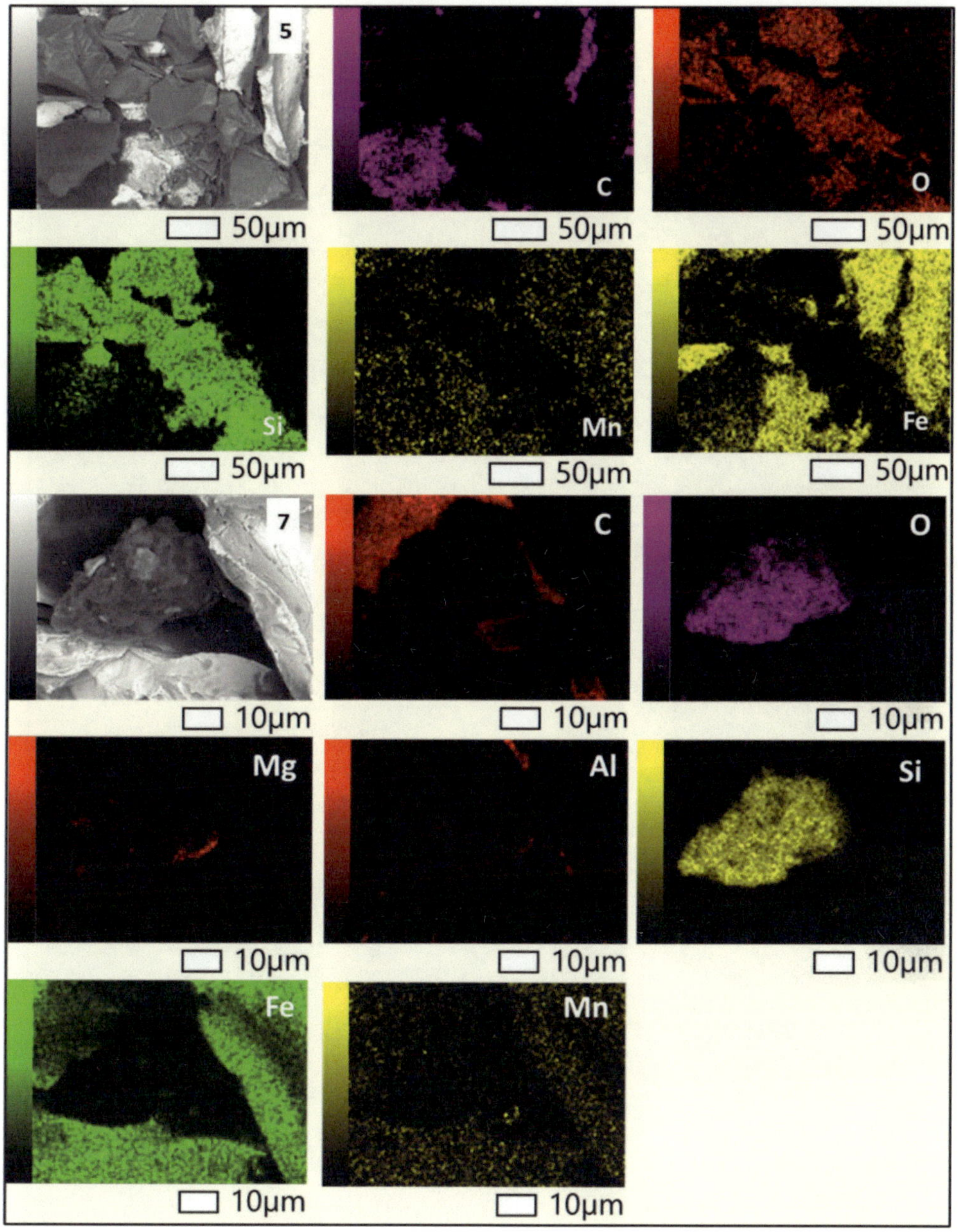

Figure 12.
*SEM-EDS elemental mapping of micrographs 5 and 7 of **Figure 9**.*

material. By analyzing the EDS microanalysis of spots b, e-f (**Figure 10**) and comparing it with the elemental mappings shown in **Figure 11** (micrograph 1) and 12 (micrograph 7), it can be seen that some of these lumpy inclusions are the oxides rich in calcium, silicon, zinc and aluminum with traces of admixture oxides of Mg, Ti, Mn, Na, K and Fe. The analysis also reveals minor phosphorous, sulphide and chloride concentrations (**Figure 10b, e, f, 11** and **13**).

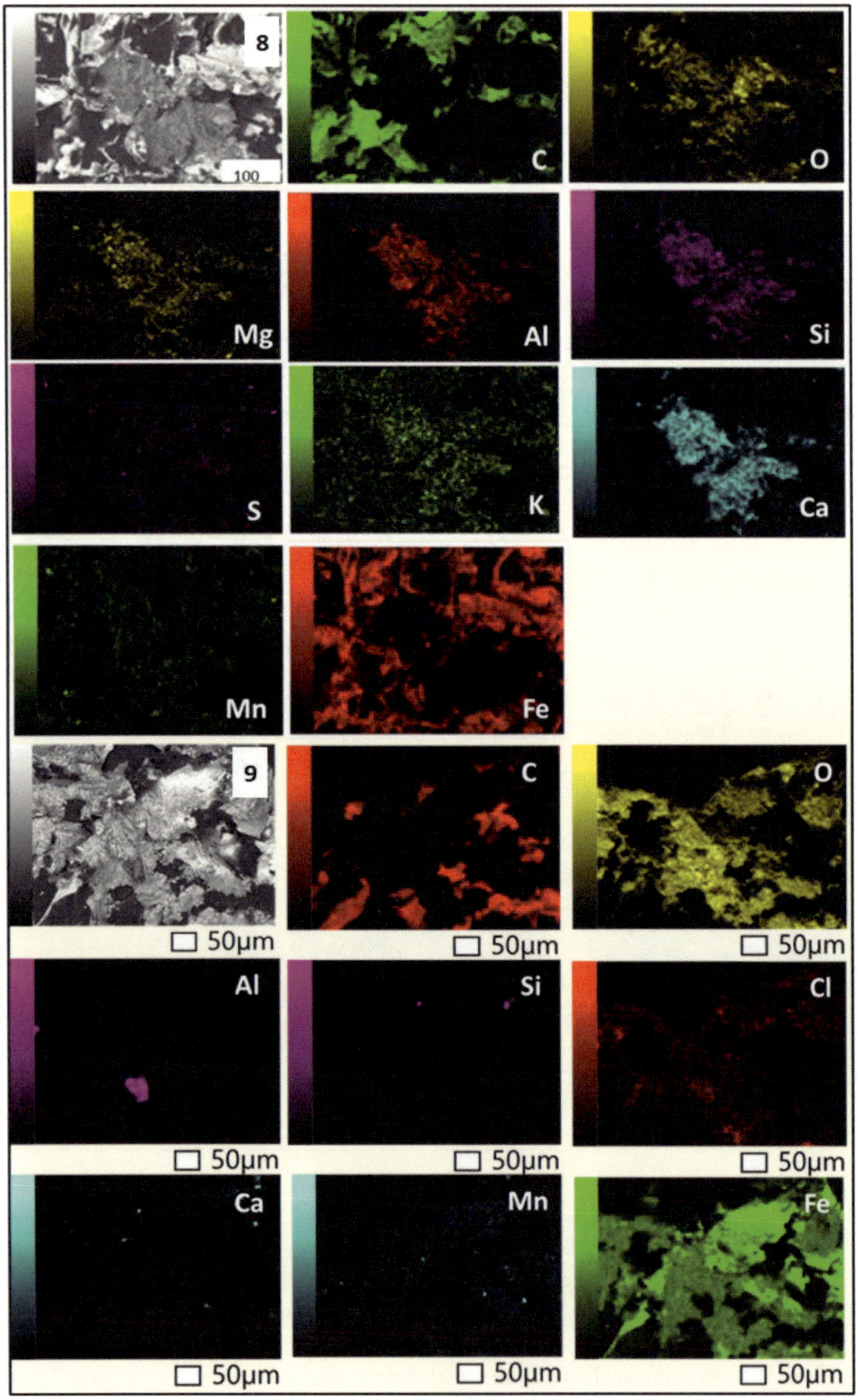

Figure 13.
*SEM-EDS elemental mapping of micrographs 8 and 9 of **Figure 9**.*

The fractographic examination conducted on the failed rotor has identified indigenous and exogenous inclusions on the disc's cracked surface. The inclusions from external sources like slag, dross, ferroalloys, flux materials, ladle lining, and entrapments from eroded mold, core and refractory materials are categorized as exogenous. Those native or inherent in the molten metal treatment process are known as indigenous inclusions. Indigenous inclusions are the products of the liquid melt reactions with deoxidizers such as silicon, manganese, and aluminum particles or during desulphurization, creating some residual oxide inclusions in the casting [22]. However, previous studies have recognized ferroalloys as a significant source of inclusions in steel and cast-iron castings [11, 30]. Ferroalloys are pre-alloyed raw materials bonded with iron used in the treatment of molten iron and steel to produce castings with desired chemical composition. In general, the primary applications of ferroalloys in iron and steel makings include (i) alloying sources to enhance the mechanical properties and functional characteristics of iron and steel products (e.g., FeCr, FeMo, FeTi, FeMn, FeW), (ii) deoxidizers such as FeSi, FeMn, SiMn and FeAl, and (iii) reducing agents such as FeSi which can be used as a reducing agent to produce FeMo, FeV and other alloys [30]. Unfortunately, several impurities accompanied ferroalloys production, including H, N, O, S and P and other trace elemental impurities such as Mg, Al, Ti, V, Ca, etc. In the cast-iron melting process, FeSi ferroalloys are the common source of silicon addition to the molten iron to decrease its melting point, improve fluidity, and promote graphitization. These FeSi alloys are the source of Al and Ca impurities, which can significantly affect the quality of the castings by forming oxide inclusions. Unfortunately, these inclusions provide excellent nucleation sites for microcracks during cooling, which is why they were consistently found within microcracks [30]. Earlier studies [31] have shown that the predominant composition of slag formed from spheroidal and lamellar irons processing consists of several oxides, including FeO, MnO, SiO_2, Al_2O_3 and MgO. In a similar investigation, Jonczy [32] concluded that the dominant component of the cast iron slag is silica (62.04%) in addition to 11.03% of Al_2O_3, 10.38% of MnO, 6.32% of MgO, 5.37% of CaO with the admixture of iron, sodium, potassium, barium and sulfur oxides. The chemical composition of slag can vary greatly depending on the level of impurities in the starting raw materials (e.g., pig iron and scraps) and other additives (e.g., flux agents and ferroalloys). For instance, the slags formed using steel scrap-based charges showed the highest zinc and aluminum contents and created a crystalline $ZnAl_2O_4$ (gahnite) phase in addition to other admixture oxides (e.g., SiO_2, Al_2O_3, CaO, MgO and MnO). These high zinc and aluminum contents are due to the use of galvanized steel scrap as raw material (Zn and potentially Al) and of FeSi and SiC as additives (Al) [31]. This probably explains the high concentrations of Zn, Al and Ca found in some of the inclusions identified in **Figure 10(b, e** and **f)**.

The SEM-EDS elemental mappings shown in **Figures 11** and **13** depict the existence of chloride ions on the cracked surface of the rotor, which is also corroborated by the EDS microanalysis presented in **Figure 10f**. It can be observed that the chloride ions covered the entire disc's fractured surface, suggesting that the chloride ions possibly originate from an outside source, such as road salt. During winter, calcium and sodium chloride salts are generally used for roadway de-icing, thereby creating a high concentration of chloride ions on the motorways. Salt solution on the road can easily penetrate the crack surface of the rotor (**Figure 4**) and hasten the corrosion of the fracture surface (**Figure 8**). Chloride ions are a highly corrosive agent that serves as a catalyst and exacerbates the deterioration of steel and cast-iron components [11]. Analysis of the disc's fractured surface showed that the rusty products were only noticed in the pearlitic matrix areas, as displayed in **Figure 9**, micrographs 6–9, since graphite does not corrode.

4.4 Failure root cause and prevention

Failures of components are frequently activated by defects introduced during manufacturing [33]. Due to the prevalence of manufacturing defects, critical components are usually subjected to a thorough inspection to prevent defective parts from entering service, and sometimes this effort is unsuccessful. Several shortcomings are associated with the casting operations, which are potential causes of product failure. For example, it is well known that core or subsurfaces discontinuous, including voids, blowholes, shrinkage cavities, pipes and porosities created during ingot solidification, are identified sources of imperfections during succeeding manufacturing steps or in service as these weak points initiate cracks in the components that eventually lead to their failure [34]. The oxide inclusions in castings could indirectly induce crack initiation by enhancing local stress concentration, promoting cleavage fracture and detrimental to the fracture toughness of the components [12]. Nonmetallic inclusions also create discrepancies such as thermal expansion mismatch, stiffness mismatch, chemical mismatch, and ductility mismatch [22, 35], which ultimately impair the cast products' mechanical properties and service performance. It is therefore imperative to understand the impact of inclusion defects on the structural integrity and the fracture toughness of cast parts about their interaction with the main characteristics of fracture mechanics, including defect crack size, loading and material toughness.

As illustrated in **Figure 14**, nonmetallic inclusions in the rotor's swan neck reduce the disc's adequate thickness. The design thickness in the rotor neck region is 6.62 ± 0.01 mm, but the inclusions defects were positioned at approximately 1.620 ± 0.237 mm away from the rotor inner radial perimeter. During the braking regime, the disc swan neck can be exposed to cyclic compressive or tensile loads (**Figure 14**). However, inclusions in the neck region would reduce the swan neck's design load-bearing capacity, consequently

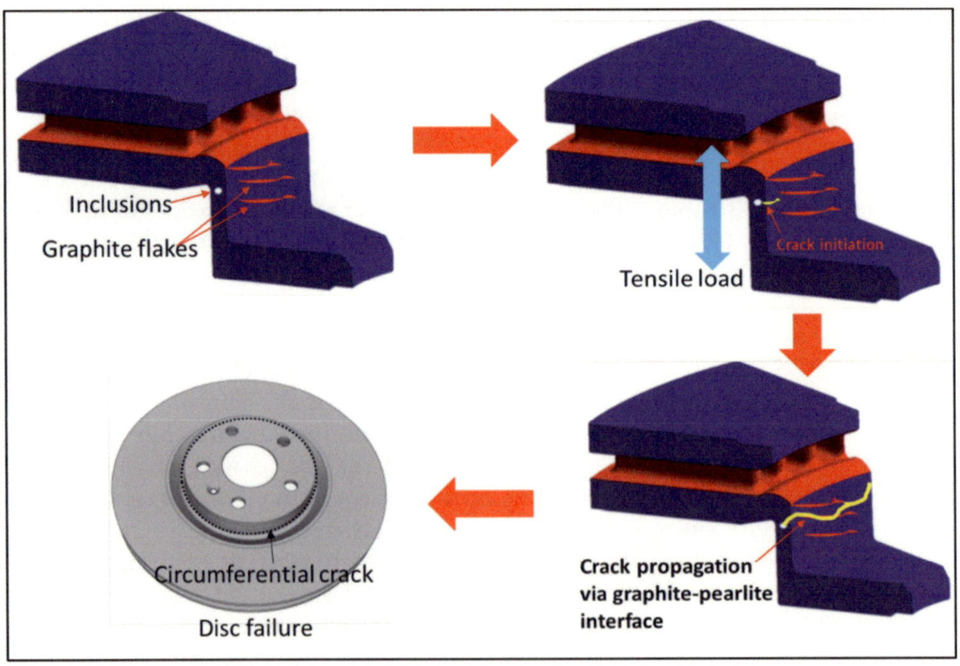

Figure 14.
A schematic illustration of the mode of failure of the rotor.

subjecting this region to higher stress during service. Due to the high thermal expansion potential of the inclusion particles, they would disintegrate into smaller particles [11, 35] and produce cavities that generate microcracks around the inclusion particles (**Figure 9**). Also, during the application, the brake rotor continuously experiences thermal (heating and cooling) and mechanical loadings that cause local concentrations of strain, induce residual stresses and initiate microcracks because of the thermal expansion mismatch between the inclusion particles and the pearlitic matrix. These microcracks propagate further through the interface between the graphite-pearlitic matrix (**Figure 7b**) and ultimately cause the rotor's failure.

After understanding and establishing the failure root cause(s), it is essential to formulate corrective or preventive measures/actions to forestall future recurrence. Judging from the various inspections performed on the failed rotor, the outcomes indicated that the inclusions (sand and slag materials) embedded in the rotor's swan neck during the casting process were responsible for its premature failure. This observation indicated that the failure root cause originated from the rotor manufacturing, implying that the preventive recommendations should be focused on the casting process to ensure quality castings. Therefore, the following strategies should be strictly implemented and practiced by the metal casters for casting quality enhancement and to avert nonmetallic inclusions during the following casting operations.

i. Mold design and preparation: Good gating design and adequate compaction of the molding sand during mold preparation are the initial steps to ensure quality casting. However, the gating system and the mold cavity should be free of sharp corners to minimize erosion wear. When assembling the mold, any loose sand particles around the down sprue or at the bottom of the sprue should be eliminated to prevent sand entrapment during mold filling.

ii. Melt preparation and pouring: Melt cleanliness is paramount to sound castings. Metal casters should implement a comprehensive approach to preventing slag particles from getting entry into the mold cavity by using suitable purifying additives (fluxing and floating agents) that protect liquid iron from further oxidation during melting and allow the slag to flow on top of the molten iron for easy removal before the pouring operation. A ceramic screen should be integrated into the gating system to separate slag material from entering the mold during melt pouring. Besides, a bottom-poured ladle should be used to eliminate slag entrapment. Melt pouring should be done to minimize turbulent flow during mold filling.

iii. Quality control management: Metal casting is a complex process that requires standard quality assurance management. A poorly managed quality control system in the foundry industry would impair the quality of castings. Therefore, metal casters should rigorously inspect the quality of all the casting processes and castings to diminish products' imperfections. An in-line nondestructive evaluation (NDE) solution (radiographic and ultrasonic inspections) should be developed and implemented in the foundries for the early discovery of cast imperfections such as nonmetallic inclusions, gas porosity, cracks, shrinkage porosity, and other casting irregularities to ensure a 100% quality assurance of their products. Continuous training of foundry workers regarding quality inspection in all facets of casting operations should be implemented, and negligent attitudes among foundry workers should be discouraged.

5. Conclusions

This chapter highlights the impact of nonmetallic inclusions on the quality and performance of a cast product and how this can ultimately cause premature component failure. The practical application of root cause analysis methods in resolving failure problems is discussed using a failed automotive brake rotor as a case study. The investigation concluded that the automotive brake rotor failed prematurely due to the influence of nonmetallic inclusions in the swan neck of the rotor. The inclusions were concentrated at approximately 1.620 ± 0.237 mm away from the disc's inner radial perimeter, reducing the design thickness of the rotor's swan neck. A cluster of inclusions in the rotor's neck region provided stress concentration and crack initiation sites. Multiple microcracks were initiated within and around the oxide inclusions and propagated through the interface between the flaky graphite and the pearlitic matrix. Examination of the rotor's hardness, chemical composition and microstructure suggested that the failed disc had been produced from the appropriate gray cast iron material and met the specification requirements; and that there were no deficiencies or abnormalities found with the disc material. The evaluation revealed that nonmetallic inclusions were the primary cause of the failure, and preventive recommendations were provided to forestall future recurrence and ensure high-quality castings.

Acknowledgements

The component investigated in this publication was provided by the Quality Control Unit, Automotive Components Floby.

Conflict of interest

The author declares no 'conflict of interest' regarding the work published in this paper.

Notes/thanks/other declarations

Thank you.

Author details

Samuel A. Awe
Research and Development Department, Automotive Components Floby, Floby, Sweden

*Address all correspondence to: Samuel.awe@acfloby.com

References

[1] Failure Analysis by Thermal Analysis Application Examples. Thermal Analysis Applications Booklet. Switzerland: Mettler-Toledo GmbH, AnalyticalCH-8606 Nänikon; n.d

[2] Bolton W, Higgins RA. Causes of failure. In: Materials for Engineers and Technicians. Seventh ed. Abingdon, Oxon; New York, NY: Routledge; 2021. pp. 367-390. DOI: 10.1201/9781003082446-26

[3] Alam MR, Hassan SF, Amin MA, Arif-Uz-Zaman K, Karim MA. Failure analysis of a Mobile crane: A case study. Journal of Failure Analysis and Prevention. 2018;**18**:545-553. DOI: 10.1007/s11668-018-0437-1

[4] Mukhopadhyay G. Construction failures due to improper materials, manufacturing, and design. In: Handbook of Materials Failure Analysis. Oxford, United Kingdom: Butterworth-Heinemann; 2018. pp. 59-81. DOI: 10.1016/B978-0-08-101928-3.00004-5

[5] Aadnøy BS, Looyeh R. Principal and Deviatoric stresses and strains. Petroleum Rock Mechanics: Drilling Operations and Well Design. 2011:27-39. DOI: 10.1016/B978-0-12-385546-6.00003-6

[6] Chen W. Modeling and prediction of stress corrosion cracking of pipeline steels. In: El-Sherik AM, editor. Woodhead Publishing Series in Energy, Trends in Oil and Gas Corrosion Research and Technologies. Woodhead Publishing; 2017. pp. 707-748. DOI: 10.1016/B978-0-08-101105-8.00030-9. ISBN: 9780081011058

[7] Horynová M, Juliš M, Čelko L, Švejcar J. Failure analysis of induction hardened injector body. Engineering Failure Analysis. 2015;**56**:538-544. DOI: 10.1016/j.engfailanal.2014.10.017

[8] Papadopoulou S, Vazdirvanidis A, Toulfatzis A, Rikos A, Pantazopoulos G. Failure investigation of products and components in metal forming industry: Root cause analysis and process-based approach. Journal of Failure Analysis and Prevention. 2020;**20**:106-114. DOI: 10.1007/s11668-020-00801-4

[9] Dondapati S, Trivedi M, Dondapati RS, Chandra D. Investigation on the mechanical stresses in a muffler mounting bracket using root cause failure analysis (RCFA), finite element analysis and experimental validation. Engineering Failure Analysis. 2017;**81**:145-154. DOI: 10.1016/j.engfailanal.2017.08.010

[10] Ghosh D, Roy H, Subramanian C. Metallurgical failure investigation of premature failed platen Water Wall tube in a thermal power plant boiler. Journal of Failure Analysis and Prevention. 2021;**21**:733-737. DOI: 10.1007/S11668-021-01137-3/TABLES/4

[11] Awe SA. Premature failure of an automobile brake disc: Effect of nonmetallic inclusions. Engineering Failure Analysis. 2022;**137**:106263. DOI: 10.1016/j.engfailanal.2022.106263

[12] Zhou S, Li Z, Jiang L, Wang X, Xu P, Ma Y, et al. An investigation into the role of nonmetallic inclusions in cleavage fracture of medium carbon pearlitic steels for high-speed railway wheel. Engineering Failure Analysis. 2022;**131**:105860. DOI: 10.1016/j.engfailanal.2021.105860

[13] Collignon M, Cristol AL, Dufrénoy P, Desplanques Y, Balloy D. Failure of truck

brake discs: A coupled numerical-experimental approach to identifying critical thermomechanical loadings. Tribology International. 2013;**59**:114-120. DOI: 10.1016/j.triboint.2012.01.001

[14] Li B, Liang X. Failure analysis of escalator step induced by the bolt surface defects. In: Haeri H, editor. Materials in Environmental Engineering: Proceedings of the 4th Annual International Conference on Materials Science and Environmental Engineering. Berlin, Boston: De Gruyter; 2017. pp. 265-280. DOI: 10.1515/9783110303568-026

[15] Awe SA, Thomas A. The Prospects of Lightweight SICAlight Discs in the Emerging Disc Brake Requirements. Dresden, Germany: Proceedings of the Eurobrake 2021; 17-21 May 2021. pp. 1-6

[16] Reif K. Fundamentals of Automotive and Engine Technology: Standard Drives, Hybrid Drives, Brakes, Safety Systems. Germany: Bosch Prof. Springer Vieweg; 2014

[17] Awe SA. Developing material requirements for automotive brake disc. Modern Concepts in Material Science Science. 2019;**2**:1-4. DOI: MCMS.MS.ID.000531

[18] Aranke O, Algenaid W, Awe S, Joshi S. Coatings for automotive gray cast iron brake discs: A review. Coatings. 2019;**9**(9):552. DOI: 10.3390/COATINGS9090552

[19] Bon DG, Ferreira MH, Wladimir W, Filho B, Luiz GW. Fracture micromechanisms evaluation of high-strength cast irons under thermomechanical fatigue conditions. International Journal of Metalcasting. 2020;**14**:696-705. DOI: 10.1007/s40962-019-00399-w

[20] Stefanescu DM. Casting Defects. In: Doru M, editor. Cast Iron Science and Technology. Vol. 1A. USA: ASM Handbook, ASM International; 2017. pp. 355-372. DOI:10.31399/asm.hb.v01a.a0006338

[21] Trojan PK. Inclusion-Forming Reactions. In: Viswanathan S, Apelian D, Donahue RJ, DasGupta B, Gywn M, Jorstad JL, et al., editors. Casting. Vol. 15. ASM Handbook, ASM International; 2008. pp. 74-83. DOI: 10.31399/asm.hb.v15.a0005193

[22] Jamil M, Khan AM, Hegab H, Sarfraz S, Sharma N, Mia M, et al. Internal cracks and non-metallic inclusions as root causes of casting failure in sugar mill roller shafts. Materials. 2019;**12**(15):2474. DOI: 10.3390/ma12152474

[23] Brown C. Casting Failure Analysis Techniques and Case Studies. In: Viswanathan S, Apelian D, Donahue RJ, DasGupta B, Gywn M, Jorstad JL, et al., editors. Casting. Vol. 15. ASM International; 2008. pp. 1183-1191. DOI: 10.31399/asm.hb.v15.a0005343

[24] Aliya D. The failure analysis process—An overview. Journal of Failure Analysis and Prevention. 2022;**22**:42-57. DOI: 10.1007/S11668-021-01328-Y/TABLES/6

[25] Shipley RJ, Miller BA, Parrington RJ. Introduction to failure analysis and prevention. Journal of Failure Analysis and Prevention. 2022;**22**:9-41. DOI: 10.1007/S11668-021-01324-2/FIGURES/43

[26] EASE. Root Cause Analysis: How to Use a Fishbone Diagram. San Clemente, CA, USA: EASE, Inc.; n.d. Available from: https://www.ease.io/root-cause-analysis-how-to-use-a-fishbone-diagram/ [Accessed: August 2, 2022]

[27] Peter K. How to Tell if Rotor Is Bad (15 Bad Rotor Symptoms) - Off-Roading

Pro n.d. Available from: https://offroadingpro.com/how-to-tell-if-rotor-is-bad/ [Accessed: August 5, 2022]

[28] The European Standard: Microstructure of Cast Irons – Part 1: Graphite Classification by Visual Analysis (ISO 945-1:2008). Geneva, Switzerland: European Committee For Standardization; 2011

[29] Mohammadnejad A, Bahrami A, Goli M, Nia HD, Taheri P. Wear induced failure of automotive disc brakes-a case study. Materials. 2019;**12**:1-10. DOI: 10.3390/ma1224214

[30] Wang Y, Karasev A, Park JH, Jönsson PG. Nonmetallic inclusions in different ferroalloys and their effect on the steel quality: A review. Metallurgical and Materials Transactions B. 2021;**52**:2892-2925. DOI: 10.1007/S11663-021-02259-7

[31] Regordosa A, Llorca-Isern N. Chemical and structural characterization of slag compounds formed in the melting processes produces spheroidal graphite cast irons. International Journal of Metalcasting. 2016;**10**:421-434. DOI: 10.1007/s40962-016-0025-7

[32] Jonczy I. Diversification of phase composition of metallurgical wastes after the production of cast iron and cast steel. Archives of Metallurgy and Materials. 2014;**59**:481-485. DOI: 10.2478/amm-2014-0079

[33] McEvily AJ. Defects. Metal Failures: Mechanisms, Analysis, Prevention, Hoboken, NJ, USA: John Wiley & Sons, Inc.; 2013, p. 316-35. DOI: 10.1002/9781118671023.ch12.

[34] Balan KP. Metallurgical Failure Analysis: Techniques and Case Studies. Amsterdam, Netherlands: Elsevier; 2018

[35] Zerbst U, Madia M, Klinger C, Bettge D. Defects as a root cause of fatigue failure of metallic components. II: Nonmetallic inclusions. Engineering Failure Analysis. 2019;**98**:228-239. DOI: 10.1016/j.engfailanal.2019.01.054

Chapter 11

Creep Failure of 25Cr-35Ni Centrifugally Cast Reformer Tube

Kanhirodan Ravindranath, Abdulmuhsen Akbar, Bader Al-Wakaa and Zak Abdallah

Abstract

Cast 25Cr-35Ni alloys are extensively being used in the petrochemical and petroleum refining industries for high-temperature applications. A typical application of such alloys in the industry is in the manufacture of cast catalyst reformer tubes for the production of hydrogen. The cast 25Cr-35Ni catalyst reformer tubes possess the required mechanical properties, creep resistance, oxidation resistance, and high-temperature stability. Though reformer tubes are designed to give a service life of over 100,000 hours at temperatures beyond 900°C, there are incidents of failure due to creep damage, which is the predominant failure mechanism in reformer tubes. The paper discusses an investigation conducted on the premature failure of a 25Cr-35Ni reformer tube. The investigation involved microstructural assessments and the evaluation of mechanical properties. The microstructure and mechanical properties of the service-exposed reformer tube were also compared with a new tube. The investigation revealed that the failure of the tube was due to creep embrittlement. The creep embrittlement was due to the microstructural degradations that occurred as a result of overheating. Adherence to the design and operational parameters is critical in mitigating creep embrittlement failures.

Keywords: creep embrittlement, cast reformer alloy, microstructure, tensile test, overheating

1. Introduction

Creep embrittlement is the predominant failure mechanism in cast reformer tubes in the fertilizer, petrochemical, and petroleum refining industries [1–7]. Reformer tubes are used in the process industry for the production of hydrogen by reaction between natural gas and steam in the presence of a catalyst. The reforming reaction is highly endothermic. Reformer tubes are very critical components being exposed to severe conditions of temperature and process for a long time. The tubes typically operate at temperatures over 900°C for prolonged time durations. Failure of reformer tubes in the process industry mostly results in plant shutdowns and thus results in economic loss due to lost production, in addition to replacement costs. Considerable efforts have been made in alloy development to mitigate creep embrittlement and to extend the service life of reformer tubes [8–11]. The reformer alloys are not controlled by any international specifications and are mostly proprietary. The alloys contain high contents of nickel and chromium along with minor alloying additions of elements such as niobium, titanium, silicon, etc. The alloying elements provide the required oxidation and carburization resistance along with excellent creep resistance. Nickel additionally provides a stable austenitic matrix and microstructural stability. Nickel also retards the precipitation of intermetallic phases in high-performance alloys [3, 12, 13]. The microstructure of as-cast alloys is the austenitic dendritic type with inter-dendritic carbides. In service, when exposed to high temperatures, microstructural changes occur in reformer alloys with precipitation of secondary carbides along with intermetallic secondary phases [9, 10, 14–25]. The precipitation of secondary phases that occurs during high-temperature service drastically affects the mechanical properties of the reformer tubes [7, 8, 18, 21]. The high-temperature exposure for prolonged durations can also lead to the initiation of creep voids in the material. Eventually, the creep voids will link together to form cracks and tube rupture.

The new generation reformer alloys of type 25Cr-35Ni modified with minor alloying additions typically perform well under reforming conditions at temperatures well beyond 900°C [3, 21]. The minor alloying elements added to the reformer alloys such as titanium, niobium, zirconium, tungsten, etc., provide a fine dispersion of carbides that are stable at temperatures well in excess of 900°C [8–10, 16, 21]. The elements also promote the fragmentation of the as-cast microstructure and partial replacement of chromium-rich carbides by more stable alloy carbides [9]. The distribution and geometry of carbides play an important role in imparting creep resistance to the reformer alloys [6]. Optimally distributed fine carbides act to restrict the movement of dislocations, thus enhancing the creep resistance [3]. The primary carbide network typically consists of chromium-rich $Cr_{23}C_6$ and niobium-rich carbides, which get enriched with other alloying elements during high-temperature service [3, 22]. Continued exposure of the tubes to high temperatures can lead to coarsening and coalescence of the precipitated secondary carbides [3, 7, 26]. The reformer tubes were designed for a life of 100,000 hours at the operating temperature. Although reformer tubes have a service life of 12–15 years, premature failure of tubes is often encountered mainly because of microstructural degradations due to overheating during service. Overheating of tubes even for short durations can lead to precipitation and coalescence of carbides in 25Cr-35Ni reformer tubes, in turn leading to premature failure predominantly by creep embrittlement [27]. There are reports of premature failure of reformer tubes due to overheating and the resultant creep [1, 3–7, 26, 27]. HP40Nb microalloy grade reformer tube which was operating at 880°C suffered creep damage and failure in 2 years of service when the tubes experienced temperature excursion up to 1150°C for a short period of 5–10 minutes [1]. Similarly, the HP40Nb reformer tube failed in 7 years of service when the tubes operating at 870°C were exposed to temperatures higher than the design for short durations. The tensile strength and elongation of the tube specimens exposed to 1000°C were significantly lower due to the microstructural degradation [7]. The overheating of the tubes accelerates the dissolution of secondary carbides and coarsening of primary carbides at the inter-dendritic boundaries [1, 5, 7], leading to the initiation of creep cavities. The failure initiates with the nucleation of cavities and their evolution into fissures and microcracks and final rupture [1, 5].

The 25Cr-35Ni reformer alloys are termed HP alloys [28, 29]. The alloys contain about 0.40% carbon. The high carbon content in the alloy provides the required high-temperature strength and resistance to creep by forming carbides with the alloying elements. The carbide precipitation along the grain boundaries restricts grain boundary sliding, while finely dispersed carbides within the grains provide strength, both together impart high-temperature strength and creep resistance [3].

Reformer tubes are manufactured through the centrifugal casting process [28]. The centrifugal force during the casting process produces a hollow cylinder product with minimal wall thickness variations. Directional solidification of the tube starting from the outer surface in contact with the mold leads to a high-quality sound cast metal, free of inclusions and cavities. The reformer tubes are made by pouring the molten alloy with controlled chemistry into a refractory-lined rotating die. The charge is pre-melted in an electric induction furnace. The die rotates at high speeds, which facilitates uniform metal thickness on the inside surface of the die. The molten metal quickly solidifies starting from the die-metal interface. Due to the cooling effect of the die, metal solidifies quickly forming columnar grains, extending from the outside of the tube toward the inner surface. As the solidification of molten alloy progresses, the rate of solidification slows down resulting in the formation of equiaxed grains near the inner tube surface. The Proper control of parameters is critical to ensure the manufacture of reformer tubes with good quality. The large rotational forces drive the lighter suspended particles, such as non-metallic phases and gas bubbles to the inner liquid phase before the complete solidification of the tube. The tubes are thus supplied with the inner surface machined to remove the layer containing defects. The tubes are typically manufactured with a rough external surface. The rough external surface gives a better radiation heat transfer. After the casting process, the tubes are subjected to inspection and testing to ensure that they are supplied defect free. The inspection includes visual inspection, liquid penetrant, and eddy current inspection, while the tests conducted are to ensure adherence to chemical composition and mechanical properties.

Despite continuous efforts in alloy development and creep life predictions, there are incidents of premature failure of reformer tubes in the fertilizer, petrochemical, and petroleum refining industries [1, 3–7, 26, 30, 31]. The paper discusses an investigation conducted on the premature failure of a 25Cr-35Ni reformer tube. The failed reformer tube was in hydrogen production service for 8 years at a temperature of 880°C. Since the tubes were designed considering the service life of 100,000 hours, the failure of the tube in 8 years is considered a premature failure, and conducted an investigation. The investigation involved microstructural studies and the assessment of mechanical properties. The properties of the service-exposed tubes were compared with that of a new alloy.

2. Experimental

The chemical analyses of the service-exposed tube and the new tube were conducted by optical emission spectrometry (OES). The service-exposed reformer tube section was subjected to visual examination to understand the macro features of failure. The features of the tube section were recorded using a digital camera. After the visual examination, specimens were cut from a location near the location of rupture for microscopic assessments and hardness measurements. The microscopic assessments were aimed at identifying the presence of cavities and cracks characteristics of creep damage [1–3], while the hardness measurements were made to understand the effect of secondary phase precipitation on the hardness [1–3] of the alloy. The specimens were mounted and metallographically prepared for the examination of the cross-sections. The mounting of the specimens was done in phenolic powder using a Buehler Simplimet 3000 automatic mounting press. A Buehler Automet 250 with Ecomet 250 power head was used for the grinding and polishing of the mounted specimens.

The metallographic examinations were made in as-polished and etched conditions. The etching of the metallographically prepared specimens was made using 10% oxalic acid electrolytically at an applied voltage of 6 V, as reported elsewhere [3]. The optical microscopic examination was carried out using a Zeiss Axio Observer Z1m microscope to study the microstructure and to observe the cavities. The specimens prepared from the service-exposed tube section were subjected to scanning electron microscopic (SEM) examinations using TESCAN TS 5135 Vega to further examine the cavities and microstructure. Chemical characterization of the precipitated phases was carried out using Oxford Xmax 20 energy dispersive spectrometer (EDS) attached to the SEM. Microhardness measurements of the samples were carried out on the reformer tubes at a 200 g load with a dwell time of 5 s, while hardness measurements were made on the tube samples at a load of 5 kg and a dwell time of 5 s. Tensile tests were performed on the service-exposed and new tubes at room temperature as per ASTM A370 [32].

The secondary phases precipitated in the service-exposed reformer tube were extracted using electrochemical extraction as reported elsewhere for the extraction of secondary phases from the service-exposed stainless steel grade 347 heater tube [33] for the chemical characterization. The composition of the electrolyte was 5 g oxalic acid and 200 ml hydrochloric acid made up to 1000 ml in distilled water. A sample measuring about 10 cm × 1 cm × 1 cm was cut from the reformer tube and prepared to 220 grit finish by grinding using a water-cooled abrasive paper and grinding machine. Electrochemical extraction was carried out in a 250-ml beaker. The reformer tube sample served as the anode, and a stainless steel piece served as the cathode. The cell voltage was controlled at 1.5 V. The extraction process continued till a large section of the test specimen was dissolved. The secondary phase particles settled at the bottom of the beaker were collected and washed with distilled water. After repeated washing, the extracted secondary phase particles were separated by a centrifuge and dried in an oven. The collected secondary phase particles were characterized using a Panalytical X'pert X-ray diffractometer (XRD).

3. Results and discussion

The chemical composition of the service-exposed and new reformer tube samples are given in **Table 1**. The analyses meet the requirements of HP40-Nb reformer heater tubes for chemical composition.

A photograph of the service-exposed tube sample with a closer view of the cracks is shown in **Figure 1**. The sample showed cracks opened up in the longitudinal direction. The crack edges were thick-lipped and appeared brittle in nature. The tube section exhibited a limited degree of expansion but did not show any localized bulging. The macroscopic features of the failed tube section were similar to the reformer tubes that failed due to overheating and creep damage [1, 26]. The general appearance of the tube surface was black in color.

Tube material	C	Cr	Ni	Fe	Si	Mn	Nb
Service-exposed	0.41	24.1	35.2	36.8	1.14	1.12	0.85
New tube	0.46	23.8	33.8	38.7	1.22	0.89	0.74

Table 1.
The chemical composition of the reformer tube samples.

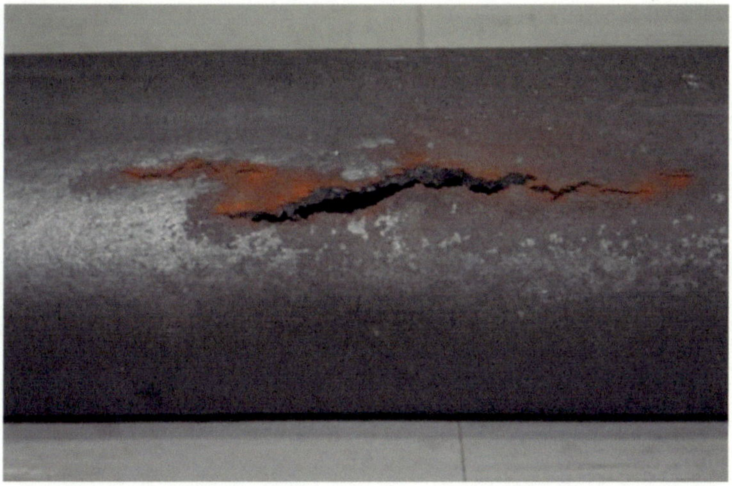

Figure 1.
Photograph of the service-exposed reformer tube sample.

Optical micrographs of the service-exposed tube and the new tube are shown in **Figures 2–5**. The micrographs of the service-exposed tube in the as-polished condition showed a dendritic network of primary carbides with numerous voids (**Figure 2a**). The voids were predominantly within the primary carbide network. The aligned voids were also linked at some locations and formed fissures (**Figure 2b**). The micrograph of the new tube also showed a dendritic network of primary carbides, but less continuous in comparison to that observed in the service-exposed tube (**Figure 3**). Voids or other imperfections were absent in the new tube sample. The micrographs of the service-exposed tube in the etched condition are shown in **Figure 4**. The formation of cracks by the linkage of aligned voids is evident in the micrographs. The primary carbides precipitated at the interdendritic boundaries appeared to have been coarsened to form a continuous network of carbides. In addition to the primary carbide network, the micrographs also showed numerous finely distributed secondary carbides within the grains. The secondary carbides were also found agglomerated at some locations. Similar observations on the presence of voids, linkage of aligned voids, precipitation of secondary carbides, and coalescence of

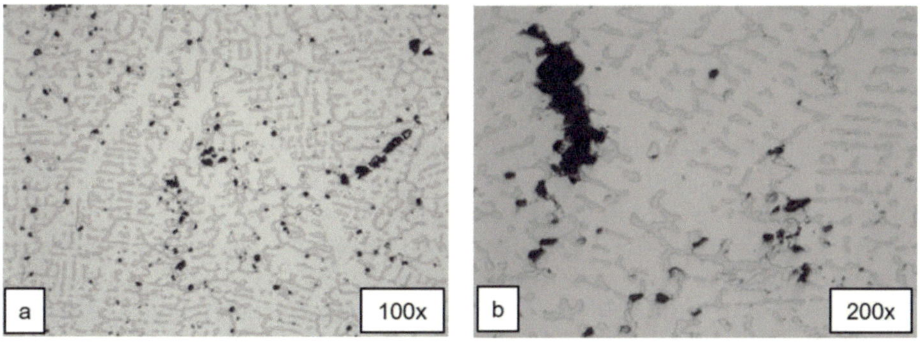

Figure 2.
Optical micrographs of the service-exposed reformer tube in the as-polished condition.

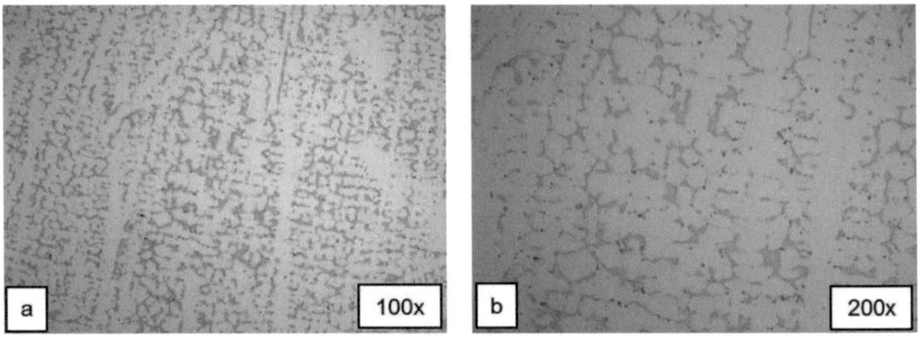

Figure 3.
Optical micrographs of the new reformer tube in the as-polished condition.

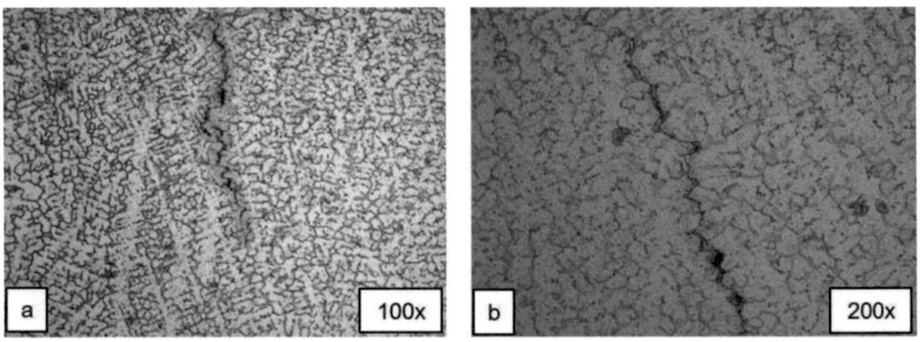

Figure 4.
Optical micrographs of the service-exposed reformer tube in the etched condition.

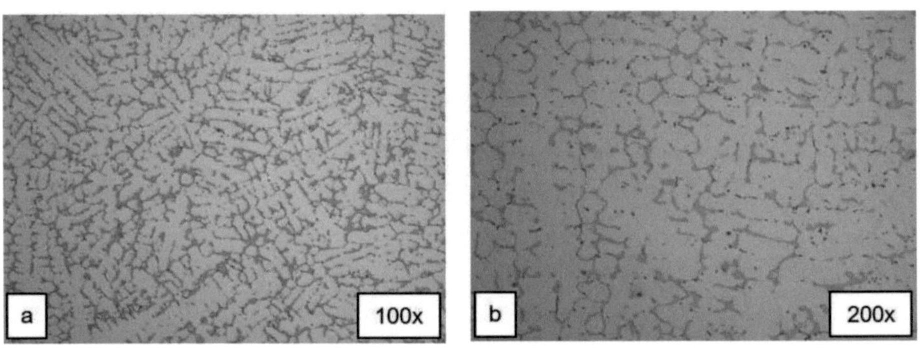

Figure 5.
Optical micrographs of the new reformer tube in the etched condition.

carbides have been reported in the literature [1–3, 9, 14, 17, 34, 35]. The micrographs of the new tube sample in the etched condition showed the primary carbide network, while the secondary carbides were absent (**Figure 5**).

Figures 6 and 7 show the SEM micrographs of the service-exposed tube sample. The presence of voids and the formation of fissures by the linkage of aligned voids are evident in the micrographs. The voids are predominantly initiated within the dendritic primary carbide network, at the interface between the carbide precipitate

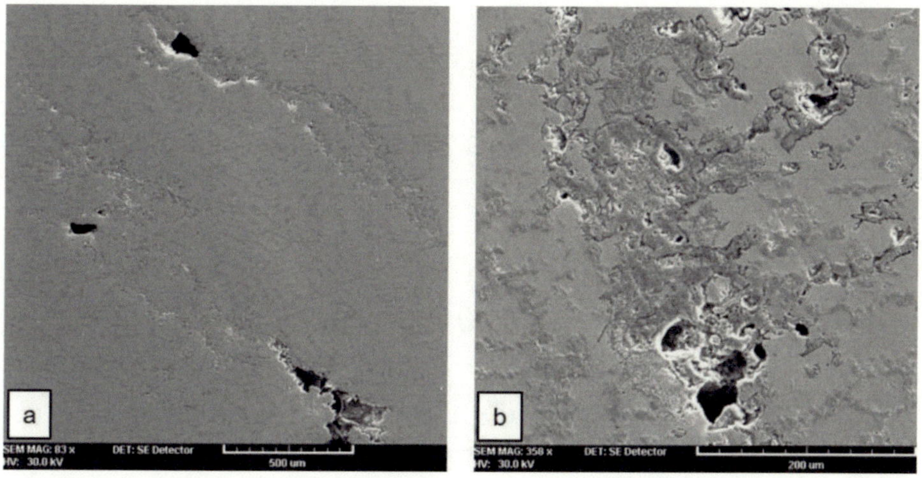

Figure 6.
SEM micrographs of the service-exposed reformer tube sample.

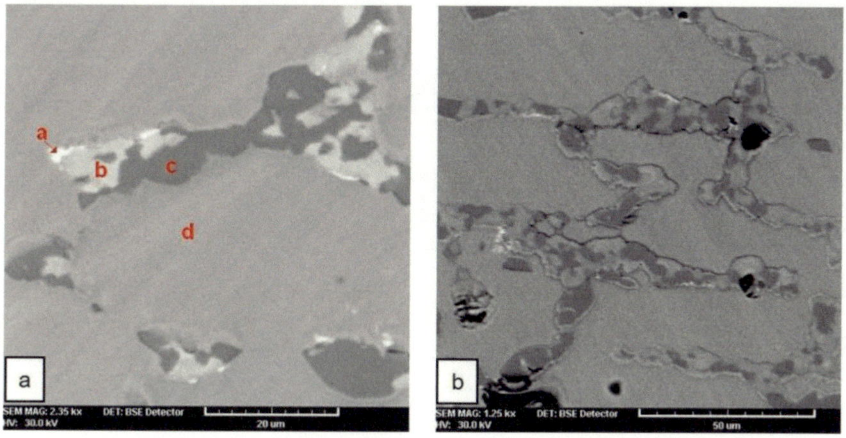

Figure 7.
*SEM micrographs of the service-exposed reformer tube showing carbide precipitates and voids within the carbide precipitate. The EDS spectra obtained for the precipitates marked in **Figure 7a** are shown in **Figure 8**.*

and the matrix. The micrographs further showed that the dendritic network was comprised of different precipitates as indicated by different contrasts in the backscattered image (**Figure 7a**). The initiation of voids and formation of cracks by the linkage of voids within the carbide network has been reported in reformer tubes that suffered creep failure [1–4]. The creep voids typically initiate at the primary carbide-matrix interface.

The EDS spectra obtained for the precipitates with different contrasts as seen in **Figure 7a** are shown in **Figure 8**. The spectra indicate the enrichment of different elements in the precipitate. The brightest precipitates are enriched in niobium, with relatively minor amounts of other alloying elements such as chromium, nickel, and iron (**Figure 8a**). The precipitate is most likely niobium-rich carbide. The light gray precipitate is enriched in chromium with some amounts of nickel, niobium, silicon, and iron (**Figure 8b**). G-phase is rich in niobium and silicon along with chromium. The darker precipitate is chromium-rich (**Figure 8c**) along with iron and nickel.

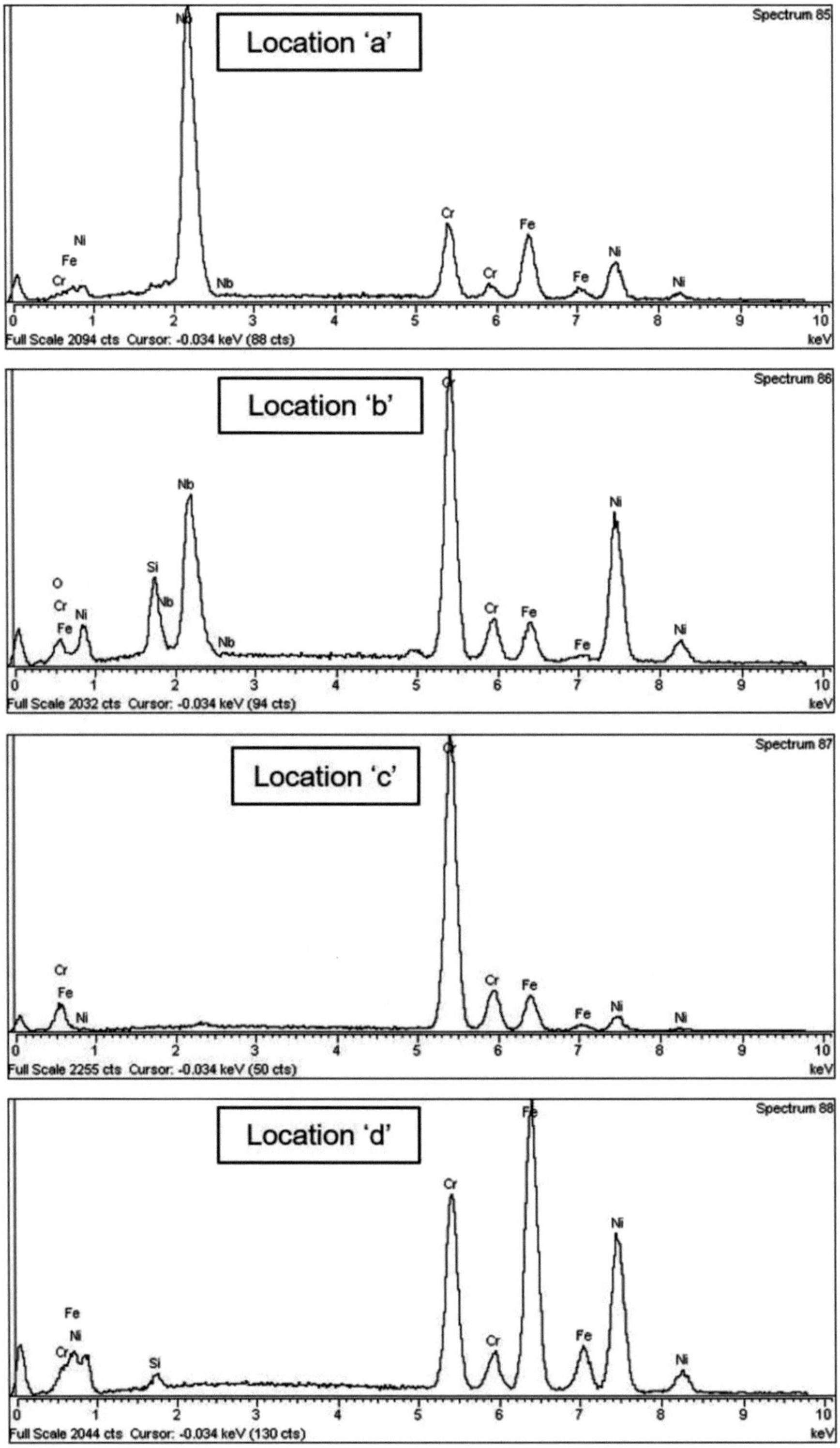

Figure 8.
*EDS spectra obtained for the precipitates observed in the service-exposed reformer tube sample at locations marked in **Figure 7a**.*

These precipitates are likely to be chromium-rich carbides. The spectrum obtained for the matrix shows peaks corresponding to iron, chromium, and nickel, the alloying elements present in the tube material (**Figure 8d**). Chemical characterization of phases precipitated in service-exposed reformer tubes has been the subject of several studies and the studies have characterized the precipitated phases [7, 9, 17, 22–25]. Niobium-rich carbide, niobium-silicon- rich G-phase, and chromium carbide and typically identified in the service-exposed reformer tube samples. The present results are also well in agreement with the reported studies. The precipitation of such phases is a precursor to the initiation of creep voids and creep embrittlement of reformer alloys that experienced overheating [7, 10, 14, 15].

Figure 9 shows the XRD spectrum obtained for the phases extracted from the service-exposed reformer tube sample. The phases that precipitated in the service-exposed tube sample are chromium carbides ($M_{23}C_6$ and M_7C_3), niobium carbide, and austenite. The presence of the phases is evident in the EDS spectra of the carbides also. The presence of these carbides in the service-exposed tube sample is well in agreement with the XRD data reported in the literature [36–38]. The precipitation and coarsening of carbides occur in reformer alloys during long-term exposure at elevated temperatures. Such precipitated phases significantly affect the properties of reformer alloys, due to the chemical nature of the precipitated phases and the crystallographic mismatch [6, 10].

The measured hardness values of the service-exposed and new reformer tube samples are given in **Table 2**. The hardness values indicate the higher hardness of the respective phases in the service-exposed tube sample in comparison to the new reformer tube sample. The higher hardness observed for the respective phases in the service-exposed reformer tube sample is due to the precipitation and subsequent coarsening of the secondary carbides that occurred due to exposure to high temperatures. The phases that precipitate in the alloy due to high-temperature exposure are hard and thus leading to an increase in the hardness.

The tensile properties of the service-exposed and the new reformer tube samples are given in **Table 3**. The service-exposed reformer tube sample showed a relatively higher yield strength and tensile strength in comparison to the new tube sample. The service-exposed tube possessed an elongation of only 3% compared to 12% observed for the new reformer tube sample. The microstructural changes that occurred in the alloy during exposure to high temperatures i.e., precipitation of secondary carbides

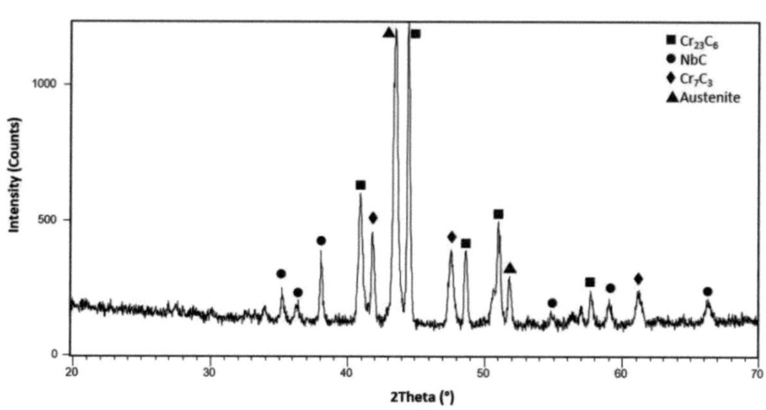

Figure 9.
XRD spectrum of the secondary phases extracted from the service-exposed reformer tube sample.

Tube material	Microhardness (HV)		Hardness (HV)
	Matrix	Dendritic carbide	
Service-exposed	197	219	213
New tube	164	206	198

Table 2.
The hardness of the reformer tube samples.

Tube material	Yield strength (MPa)	Tensile strength (MPa)	Elongation (%)
Service-exposed	398	481	3
New tube	368	456	12

Table 3.
The tensile properties of the reformer tube samples.

and coalescence of carbides, induced brittleness in the alloy, thus resulting in a significant drop in the ductility of the alloy. The secondary phases that precipitated during exposure to the high temperature are brittle and hard. The hardness measurements also indicated an increase in the hardness of the service-exposed tube sample.

The microstructure of the service-exposed tube contained finely distributed secondary carbides, in addition to the primary dendritic carbide network. The primary carbide network was more continuous in the service-exposed tube. The precipitated secondary carbides have coalesced at some locations. The observed microstructure of the service-exposed reformer tube is typical of centrifugally cast reformer tubes exposed to high temperatures [1–3]. The optical and electron microscopic studies revealed the presence of voids and micro-cracks in the service-exposed sample. The cracks have been formed by the linkage of voids and followed the interdendritic zones. The observed voids and micro-cracks are typical of creep damage [1–3]. The presence of creep voids and fissures in the service-exposed tube indicates that the failure of the service-exposed reformer tube was due to creep embrittlement.

The fine and uniformly distributed secondary carbides provide the required high-temperature strength, while carbide precipitation at grain boundaries strengthens grain boundaries [3]. High-temperature strength and creep resistance are important properties required for reformer alloys. As the temperature of exposure of reformer tubes increases, the carbides coalesce. At higher temperatures, the coalescence of carbides becomes reversible. It has been reported that $M_{23}C_6$ carbides are stable up to 1250°C in some high-temperature alloys [3, 39]. However, the coalescence of fine carbides due to exposure to high temperatures negatively impacts the creep resistance and strength. The creep voids predominantly initiate within the primary carbide network and eventually, the initiated creep voids link to form fissures and cracks. The type of precipitated carbides and the mismatch between the precipitated carbides and the matrix plays an important role in the initiation of creep voids in reformer tubes [6, 8–10]. The present results indicate that carbide precipitates coalesced, which affected the creep property of the alloy. The reason for the coalescence of carbides is the exposure of the alloy to higher temperatures. The presence of voids, formation of cracks by the linkage of aligned creep voids, and coalescence of carbide, as in the failed reformer tube are the typical features of reformer tubes failed due to overheating and the resultant creep damage [1–4]. The premature failure of the alloy due to overheating and creep damage points to the need to adhere to the design and operational conditions to avoid the reoccurrence of such failures. The failure investigation also points to the need to further enhance the capabilities of material research to develop high-temperature alloys to withstand much higher temperatures so that accidental overheating is not leading to catastrophic failures in critical services.

4. Conclusions

The investigation conducted on the service-exposed reformer tube sample revealed the presence of voids and fissures along the interdendritic boundaries. The voids were initiated predominantly within the dendritic primary carbide network. The carbides were found precipitated along the interdendritic boundaries and also inside the grains in the service-exposed tube, while in the new tube, the carbides were precipitated along the interdendritic areas only. The carbides precipitated in the service-exposed tubes were $M_{23}C_6$, M_7C_3, and NbC. The precipitated secondary carbides were also found coalesced. The microstructural change as a result of the high-temperature exposure marginally increased the strength, while dropping the ductility. The investigation conducted on the service-exposed reformer tube indicated that the failure of the tube was due to creep embrittlement. The microstructural changes of the service-exposed reformer tube and creep embrittlement in a short duration further indicate that the reformer tube was subjected to overheating, which resulted in creep embrittlement. Adherence to the design and operational parameters is critical in mitigating the premature failure of reformer tubes.

Author details

Kanhirodan Ravindranath[1*], Abdulmuhsen Akbar[1], Bader Al-Wakaa[1] and Zak Abdallah[2]

1 Petroleum Research Center, Kuwait Institute for Scientific Research, Safat, Kuwait

2 Steels and Metals Institute, College of Engineering, Swansea University, UK

*Address all correspondence to: kravi@kisr.edu.kw

References

[1] Swaminathan J, Gogulath K, Gunjan M, Roy P, Ghosh R. Failure analysis and remaining life assessment of service exposed primary reformer heater tubes. Engineering Failure Analysis. 2008;**15**:311-331

[2] Swaminathan J, Prasad P, Gunjan MK, Gugloth K, Roy PK, Singh R, et al. Mechanical strength and microstructural observations for remaining life assessment of service exposed 24Ni-24Cr-1.5Nb cast austenitic steel reformer tubes. Engineering Failure Analysis. 2008;**15**:723-735

[3] Rampat K, Maharaj C. Creep embrittlement in aged HP-Mod alloy reformer tubes. Engineering Failure Analysis. 2019;**100**:147-165

[4] Whittaker M, Wilshire B, Brear J. Creep fracture of the centrifugally-cast superaustenitic steels, HK40 and HP40. Materials Science & Engineering A. 2013;**580**:391-396

[5] Guglielmino E, Pino R, Seervetto C, Sili A. Creep damage of high alloyed reformer tubes. In: Handbook of Materials Failure Analysis with Case Studies from the Chemicals, Concrete and Power Industries. Oxford: Butterworth Heinemann; 2016. pp. 69-91. DOI: 10.1016/B978-0-08-100116-5.00004-1

[6] Wahab AA, Kral MV. 3D analysis of creep voids in hydrogen reformer tubes. Materials Science and Engineering A. 2005;**412**:222-229

[7] Han Z, Xie G, Cao L, Wang L, Sun G. Material degradation and embrittlement evaluation of ethylene cracking furnace tubes after long term service. Engineering Failure Analysis. 2019;**97**:568-578

[8] Yan J, Gao Y, Yang F, Yao C, Ye Z, Yi D, et al. Effect of tungsten on the microstructure evolution and mechanical properties of yttrium modified HP40Nb alloy. Materials Science and Engineering A. 2011;**529**:361-369

[9] Almeida LH, Ribeiro AF, May IL. Microstructural characterization of modified 25Cr-35Ni centrifugally cast steel furnace tubes. Materials Characterization. 2003;**49**:219-229

[10] Attarian M, Taheri AK. Microstructural evolution in creep aged of directionally solidified heat resistant HP-Nb steel alloyed with tungsten and nitrogen. Materials Science and Engineering A. 2016;**659**:104-118

[11] Mohri T, Takemura K, Shibasaki T. Application of advanced material for catalyst tubes for steam reformers. In: 37th Annual Safety in Ammonia Plants and Related Facilities Symposium, Annual Meeting. Vol. 33. San Antonio: AIChE; 1992. pp. 86-100

[12] Practical Guidelines for the Fabrication of Duplex Stainless Steels. London, UK: International Molybdenum Association; 2014

[13] Lo KH, Shek CH, Lai JK. Recent developments in stainless steels. Materials Science and Engineering R. 2009;**65**:39-104

[14] Abbasi M, Park I, Ro Y, Nam J, Ji Y, Kim J, et al. Microstructural evaluation of welded fresh-to-aged reformer tubes used in hydrogen production plants. Engineering Failure Analysis. 2018;**92**:368-377

[15] Abbasi M, Park I, Ro Y, Ji Y, Ayer R, Shim J. G-phase formation in

twenty-years aged heat-resistant cast austenitic steel reformer tube. Materials Characterization. 2019;**148**:297-306

[16] Schutze M, Hasselbacher B, Hoffman JJ. Development of an atlas of microstructures for the latest heat-resistant cat alloys, NACE. Corrosion. 2007: Paper No. 07425

[17] Shi S, Lippold JC. Microstructure evolution during service exposure of two cast, heat-resisting stainless steels – HP-Nb modified and 20-32Nb. Materials Characterization. 2008;**59**:1029-1040

[18] Liu CJ, Chen Y. Variations of the microstructure and mechanical properties of HP40Nb hydrogen reformer tube with time at elevated temperature. Materials and Design. 2011;**32**:2507-2512

[19] Rodriguez J, Haro S, Velasco A, Colas R. A metallographic study of aging in a cast heat-resisting alloy. Materials Characterization. 2000;**45**:25-32

[20] Tawancy HM, Ul-Hamid A, Mohammed AI, Abbas NM. Effect of materials selection and design on the performance of an engineering product—An example from petrochemical industry. Materials and Design. 2007;**28**:686-703

[21] Bonaccorsi L, Guglielmino E, Pino R, Seervetto C, Sili A. Damage analysis in Fe-Cr-Ni centrifugally cast alloys tubes for reforming furnaces. Engineering Failure Analysis. 2014;**36**:65-74

[22] Ray AK, Kumar S, Krishna G, Gunjan M, Goswami B, Bose SC. Microstructural studies and remnant life assessment of eleven years service exposed reformer tube. Materials Science and Engineering A. 2011;**259**:102-112

[23] Mostafaei M, Shamanian M, Purmohamad H, Amini M, Saatchi. Microstructural degradation of two cast heat resistant reformer tubes after long term service exposure. Engineering Failure Analysis. 2011;**18**:164-171

[24] Kondrat'ev SY, Anastasiadi GP, Ptashnik AV, Petrov SN. Evolution of microstructure and phase composition of a subsurface of cast HP-type alloy during a long term high-temperature aging. Materials Characterization. 2019;**150**:166-173

[25] Laigo CF, Le Gall R, Tancret F, Furtado J. SEM, EDS, EPMA-EDS and EBD characterization of carbides in HP type heat resistant alloys. Materials Characterization. 2008;**59**:1580-1586

[26] Ray AK, Sinha SK, Tiwari YN, Swaminathan J, Das G, Chaudhuri S, et al. Analysis of failed reformer tubes. Engineering Failure Analysis. 2003;**10**:351-362

[27] Perez IU, Junior LN, Bueno LO, de Almeida LH, da Silveira TF. Short duration overheating in a steam reformer: consequences to the catalyst tubes. Journal of Failure Analysis and Prevention. 2013;**13**:779-786

[28] Webb GM, Taylor WK. Reformer tubes: Not a commodity. Ammonia Technical Manual. 2006:125-134

[29] Tillack DJ, Guthrie JE. Wrought and Cast Stainless Steels and Nickel Alloys for Use in the Refining and Petrochemical Industries. Publication 10071. Toronto: Nickel Institute; 2002

[30] Zhao Y, Gong J, Yong J, Wang X, Shen L, Li Q. Creep behaviours of Cr25Ni35Nb and Cr35Ni45Nb alloys by modified theta method. Materials Science and Engineering A. 2016;**649**:1-8

[31] Xiang S, Chen X, Fan Z, Chen T, Lian X. A deep learning-aided prediction

approach for creep rupture time of Fe-Cr-Ni heat-resistant alloys by integrating textural and visual features. Journal of Materials Research and Technology. 2022;**18**:268-281

[32] ASTM A370. Standard Test Methods and Definitions for Mechanical Testing of Steel Products. West Conshohocken, Pennsylvania: ASTM International; 2021

[33] Ravindranath K, Tanoli N, Al-Wakaa B. Effect of long-term service exposure on the localized corrosion and stress corrosion cracking susceptibility of Type 347 stainless steel. Corrosion. 2018;**74**:350-361

[34] Wang WZ, Xuan FZ, Wang ZD, Wang B, Liu CJ. Effect of overheating temperature on the microstructure and creep behavior of HP40Nb alloy. Materials and Design. 2011;**32**:4010-4016

[35] Dessolier T, McAuliffe T, Hamer WJ, Hermse CGM, Britton TB. Effect of high temperature service on the complex through-wall microstructure of centrifugally cast HP40 reformer tube. Materials Characterization. 2021;**177**:111070

[36] Ghatak A, Robi PS. A comparative study of constitutive equations for the creep deformation of HP40Nb micro-alloyed steel. Materials Science and Engineering A. 2015;**648**:418-427

[37] Voicu R, Andrieu A, Poquillon D, Furtado J, Loacaze J. Microstructure evolution of HP40-Nb alloys during aging under air at 1000 °C. Materials Characterization. 2009;**60**:1020-1027

[38] Soares GDA, de Almeida LH, da Silveira TL, May IL. Niobium additions in HP heat-resistant cast stainless steels. Materials Characterization. 1992;**29**:387-396

[39] Pankiw RI, Voke DP, Muralidharan G, Evans ND, Stevens CO, Liu KC, et al. Precipitation and its effect on the design of cast heat resistant alloys. In: Paper No. 07424, CORROSION 2007. Houston: NACE International; 2007

Chapter 12

Fatigue Behavior of Reinforced Welded Hand-Holes in Aluminum Light Poles with a Change in Detail Geometry

Cameron R. Rusnak and Craig C. Menzemer

Abstract

Welded aluminum light poles often contain hand-holes. These hand-holes are used to give access for electrical wiring installation and maintenance purposes. Wind load may cause light poles to be loaded in a cyclic manner. This cyclic loading can cause localized fatigue cracking around the hand-hole. Fatigue failure around hand-holes has been observed in the field, but studies surrounding the resistance of the hand-holes are few and far between. This study included four-point bending fatigue tests on welded aluminum poles containing hand-holes. Eight welded aluminum specimens, each with two hand-holes, were tested in fatigue. These 16 details were loaded at the same stress range. Each specimen had a slightly different geometry or treatment applied to the hand hole. These different details mimicked traditional reinforced hand holes, similar to those evaluated in previous studies. Changes in the treatment and/or geometry included milling the inside of hole, milling the inside of the hole as well as the cast insert prior to welding, and milling the cast insert itself prior to welding. Among the 16 details tested, 15 failed as a result of fatigue cracking. It was found that specimen failure would originated in the throat of the fillet weld and then proceeded to propagate into the reinforcement ring/casting. A finite element analysis was used in addition to the experimental study.

Keywords: fatigue test, static test, welded aluminum hand-hole details, design S-N curve, high cycle fatigue

1. Introduction

The illumination of outdoor recreational areas, roadways, sidewalks, and parking lots is of the utmost importance at night. To illuminate these areas, different forms of poles are used to support an overhead light fixture. Typically, the best choice of material for these poles aluminum due to its corrosion resistance, lightweight, ease of joining, ease of handling and a high strength to weight ratio. Wind loads are often the prominent force that is applied to poles and can cause localized fatigue cracking at different areas interest [1, 2].

Modern fatigue design utilizes a lower bound S-N curve that is typically established from full-scale test data. Specimens usually contain some sort of stress concentration around a point or detail of interest. Stress concentrations usually occur around connections, cutouts, keyways, copes as well as other locations [3, 4]. One possible way to improve fatigue life would be to reduce the impact of these stress concentrations by minimizing abrupt changes in cross-section. This may be done by providing "smooth" transitions between parts. The fatigue behavior of electrical access hand-holes in welded aluminum light poles is largely unknown. The majority of existing data that has been developed was collected from full-scale welded steel poles [5]. In this study, points of interests where these stress concentrations occur are between the pole itself and the welded hand-holes.

NCHRP report number 176 contains results from experiments conducted at Lehigh University on both unreinforced and reinforced hand-holes in welded steel structures. During this experiment, 13 of the specimens contained some form of hand-hole. Different geometries were evaluated during this study. Over the course of the experiments, none of the hand-holes cracked. To provide an estimate of the stress concentration around the hand-holes, finite element models were created. AASHTO Category E' was recommended [6].

Observations have confirmed the existence of fatigue cracking associated with different hand-holes in the field. NCHRP report number 469 describes fatigue cracks found on welded steel structures near and around the hand-holes in several states. These states included New York, California, New Mexico and Minnesota [7]. In Iowa, there was a failure in a high-mast welded pole that was found to contain cracking around the hand-hole. This failure prompted further investigation into other poles and towers, in which multiple contained some form of fatigue cracking. Fatigue cracking was found in welded aluminum light poles mounted on the Mullica River Bridge after a violent storm in 2011 [8]. **Figure 1** depicts a fatigue failure associated with a hand-hole on an aluminum light pole that occurred in the field.

Figure 1.
Fatigue crack in welded aluminum light pole hand-hole.

Twenty light poles were tested in bending fatigue at the University of Akron. Static tests were conducted in addition to the fatigue tests in an attempt to gain a better insight into how the strain is distributed around the handhole. The results from this study concluded that the hand-hole fatigue test data fell above category D and E design S-N curves [9]. In addition, the University of Akron performed a study on the effect of a change in the diameter of the specimens, from 10 in to 8 in. It was found that this change had a small, but slightly beneficial effect on fatigue life [9].

In comparison to the previous research conducted on the fatigue behavior of welded aluminum light poles, the novelty of this study comes from examination of changes to the welded hand-hole detail that have not been explored. Previous studies [9–11] utilized an identical weld detail but with different diameter specimens. The specimens included typical 10in diameter as well as 8in diameter poles loaded in four point bending. The change in the detail geometry could yield a better fatigue life that could be used in the field. During this study, eight light pole specimens, each with a different hand-hole detail treatment or geometry, were tested in fatigue. Finite element models were developed to improve the understanding of the stress field around the hand-hole.

2. Experiments

2.1 Pole geometry and material properties

Eight aluminum specimens were tested under cyclic loading to examine the behavior of the modified reinforced hand-hole details (**Figure 2**). Each of the specimens consisted of a 25.4 cm (10 in) diameter extruded aluminum alloy tube with a 0.635 cm (¼ in) thick wall. Each of these 6063 aluminum alloy tubes had two hand-holes with reinforcement welded in place using a GMAW (Gas Metal Arc Welding) process with 4043 filler (**Figure 2**) [3, 12]. Each specimen was 3.66 m (144 in) in length, with the hand-holes paced 1.37 m (54 in) in from each end respectively. Support rollers for the specimens were inserted 15.2 cm (6 in) from either end.

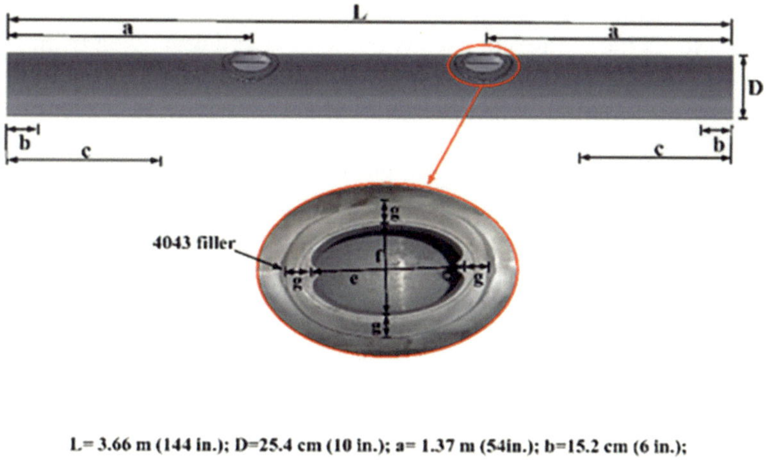

Figure 2.
The geometry of a welded aluminum hand-hole detail in four-point fatigue testing.

Part Name	Alloy	Tensile Yield Strength	Ultimate Tensile Strength
Tube	6063-T6	31 ksi (213.7 Mpa)	35 ksi (241.3 Mpa)
Fillet Welding	4043	N/A	29.2 ksi (201.3 Mpa)
hand-hole	A356-T6	20 ksi (137.9 Mpa)	30 ksi (206.8 Mpa)

Table 1.
Mechanical properties of the aluminum hand-hole tubes.

Table 1 summarizes the minimum mechanical properties of the welded aluminum hand-hole specimens [3, 12].

2.2 Geometry changes to reinforcement

Each of the eight poles contained two hand holes and each sample a different hand-hole treatment or modified geometry. The first set of two consisted of the same welded hand-hole detail used in the field. These hand-holes measured 150 mm (6 in) in the longitudinal direction and 100 mm (4 in) in the transverse direction. The other groups of poles had holes that were milled around the inside before welding, utilized reinforcement castings that were milled the outside before welding or a combination of the two. In the milling of both the hole and cast insert, only about a 6.25 mm (¼ in) of material was removed. As the hand-holes are cut with a plasma process, removal of material around the perimeter of the hole is intended to eliminate any hot – short cracking.

2.3 Fatigue tests

The fatigue setup that was used in the lab may be seen in **Figure 3**. A control system with a 245 KN (55 kip) MTS servo-hydraulic actuator was utilized to apply load to the specimens. A structural load frame capable of support 1335 KN (300 kips) was used to mount the actuator. The specimens were loaded by a spreader beam with supports that were rollers machined to fit the profile of the tube.

Strain gages were used to monitor strains in the specimens. They were mounted around the hand-holes using adhesive. Tests were conducted in load control. The typical location of the strain gages on the specimen may be seen in **Figure 4**. Strain gage placement was the same as the older study conducted and was taken from [9]. Strain gage resistance was 350 ohms and were 3.175 mm (⅛ in) in. Strains were recorded every two hours, intermittently for 10 second using a Micro-Measurements System 8000 data acquisition device.

Specimens had their hand-hole openings facing “downward” during testing. This was to ensure that they were in tension. Failure occurred when the maximum target load on the specimen could not be supported. Not being able to support the target load is an indication that cracking occurred. When a displacement over 10% of the target was realized, the test was shut down. This 10% displacement was set in order to ensure that the specimens would not fail catastrophically, and the damaged detail could be repaired. This repair was in the form of a moment clamp placed over the failed hand-hole. There was a single case where this moment clamp was unable to repair the specimen to continue testing.

Eight poles, each with two hand-holes, were tested at a nominal stress range of 5.4 ksi (37.4 Mpa). Each of these poles were loaded between 227 kg (500 lbs) and 3402 kg (7500 lbs) to provide a direct comparison between all tests. Strain gages were placed

Figure 3.
Fatigue test set-up of welded aluminum hand-hole details.

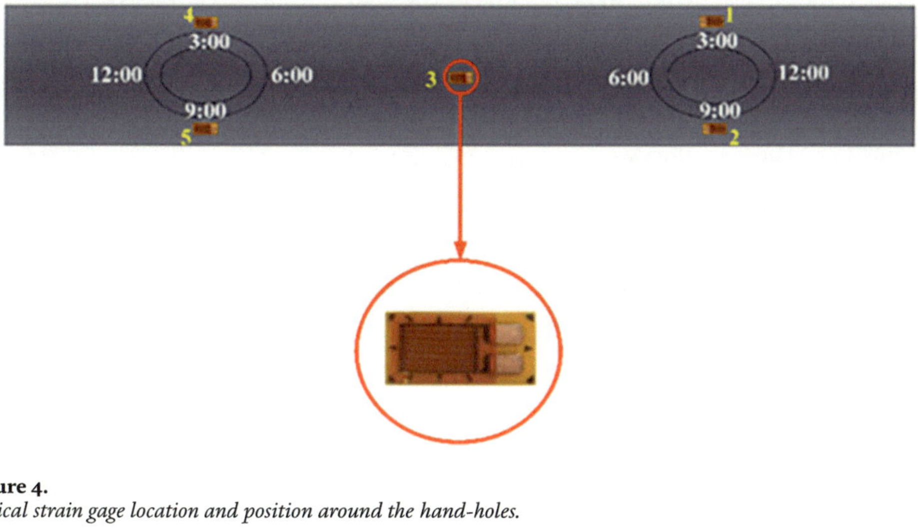

Figure 4.
Typical strain gage location and position around the hand-holes.

as shown in **Figure 5** and installed on the surface of the tube adjacent to the fillet weld that joins the cast hand-hole to the tube. Strain gage placement was the same as the older study and was taken from [9]. A strain gage was placed at the 3 and 9 o'clock positions respectively, with the addition of a gage in the middle of the specimen. The center of the gage was placed within 2 ~ 3 times the tube thickness away from the welding throat. The strain gages were connected to the data acquisition system to measure the nominal stress ranges.

All of the specimens were cycled at 2 Hz and tested around the clock. Visual inspections of the hand-holes were conducted several times daily. Of the 16 hand-hole details tested, 15 failed.

Figure 5.
The position of strain gages installed around a hand-hole.

The data shown in **Figure 6** reveals that changes to the weld detail do not have a drastic effect on fatigue life. In fact, some of the changes appeared to have a negative effect on the fatigue life. This was most prevalent in the milling of the cast insert. The only test that appeared to have any positive effect was when the hole itself was milled. In this case, the fatigue life was increased slightly. **Figure 7** shows the compilation of previous data compared to the change in details [11].

During testing, cracks were first observed in the throat of the fillet weld that joins the tube to the reinforcement ring. These cracks were first observed along the minor axis of the hand-hole at either the 3:00 or 9:00 position in Ref. to a clock. A sample of one of these cracks may be seen in **Figure 8**. These cracks would normally start at the weld root and propagate through the throat of the weld and then progress around the handhole. **Figure 9** shows an example of the crack surface after failure. Note that this

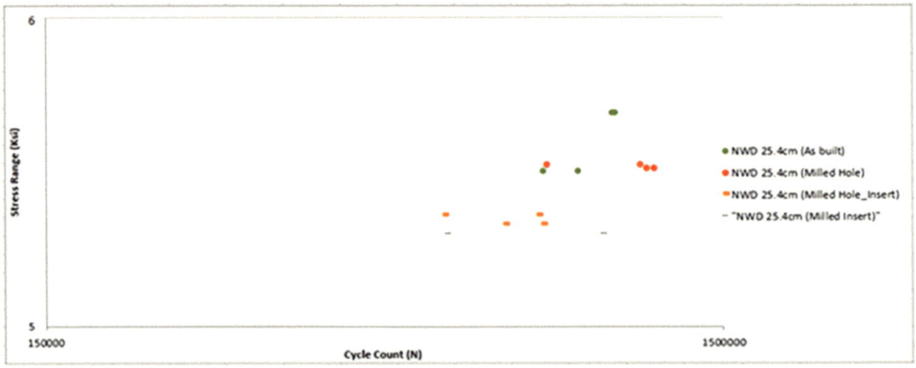

Figure 6.
Fatigue test results of new weld hand-hole details.

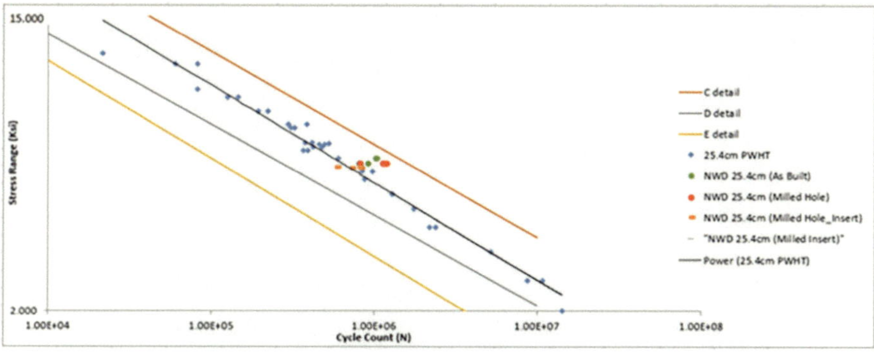

Figure 7.
New weld detail vs. old 10in data.

Figure 8.
Fatigue crack through weld throat between reinforcement and tube.

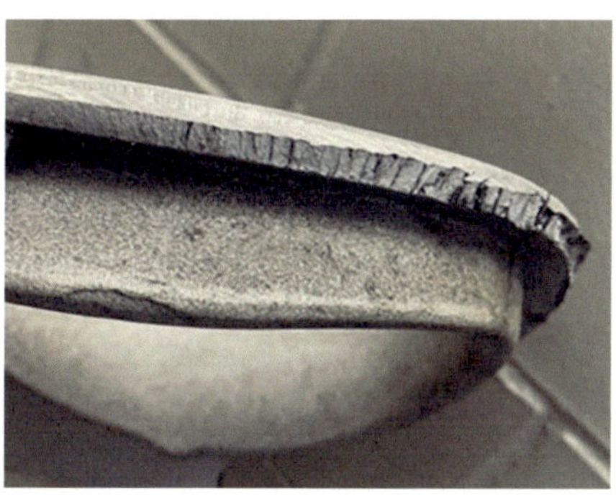

Figure 9.
Jagged crack edge.

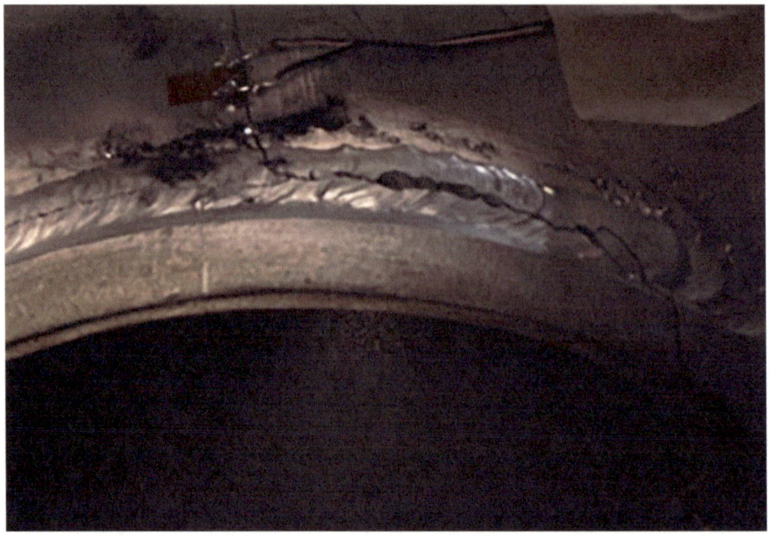

Figure 10.
Typical final failure of hand-hole details.

is similar to that of a previous study conducted at the University of Akron [10]. The edges of the jagged surface appear to coincide with the weld bead ripples.

Final failure typically occurred as the crack progressed through the cast insert near either the 11:00 or 5:00 position (**Figure 10**). In these cases, the fatigue cracks spent most of the cyclic life while small. In every case there was a definite transition from the time of the development and propagation of visible fatigue cracks within the fillet weld throat to the final failure. In most cases, cracks would develop and progress around the hand-hole. At this point, they would appear to stop, and only after a significant number of load cycles were applied, would failure occur.

3. Finite element modeling

A finite element (FE) model was created for the four-point bending specimen in an attempt to understand the stress distribution adjacent to the hand-holes. This model was constructed to be identical to the weld detail and geometry of the specimens. The model also included an initial imperfection in the detail along the line where cracking would first be observed. The purpose was to determine how a crack through the throat of the fillet weld would affect the stress distribution. Local stresses were mesh dependent for this study. Finer mesh sizes often increase the stresses local to important geometric details, whereas a coarse mesh often results in a reduction in local stress. The model as a whole consisted of 633,324 nodes with 346,747 elements. Each consisted of a mix of both hexahedral and tetrahedral element types. **Figure 11** depicts the mesh and model as a whole.

Stress "hot spots" indicate where fatigue cracking is likely to develop. **Figure 12** shows a longitudinal stress map along the "Z" axis. Under bending, longitudinal stress developed and the distribution, sense, and magnitude are affected by the presence of the hand-hole detail. This is shown in **Figure 12** with the shade of blue being a relatively low stress while elevated stress (**Figure 11**) is green. Hot spots appear alongside most of the hand-hole, maximum values between the 10:00 and 2:00 positions,

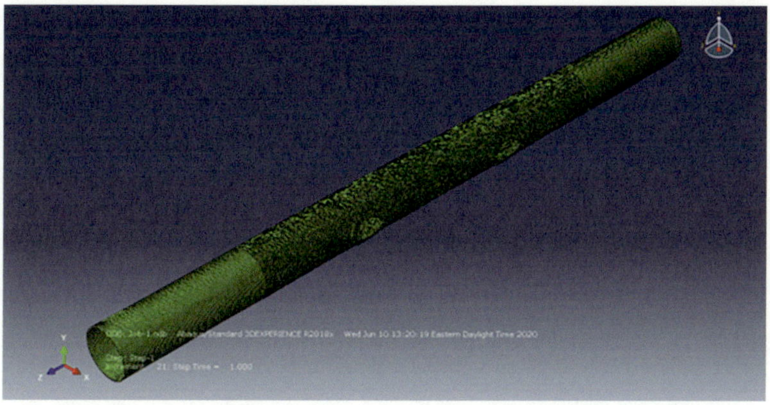

Figure 11.
Overall model.

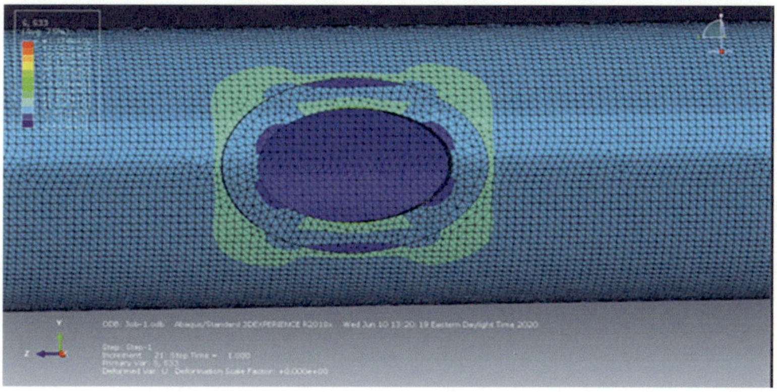

Figure 12.
Contours of longitudinal stresses around the hand-holes.

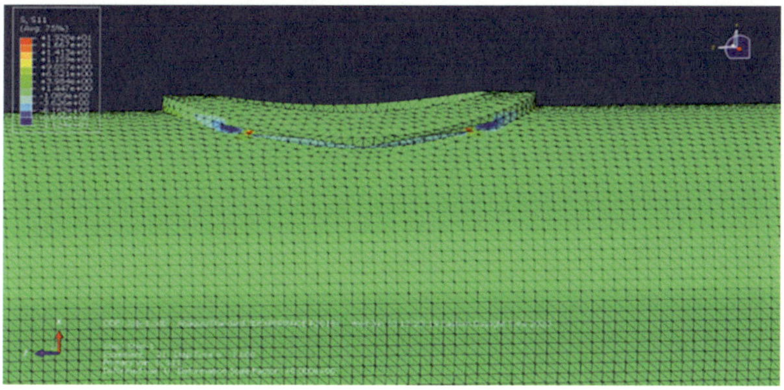

Figure 13.
Stress concentration X direction.

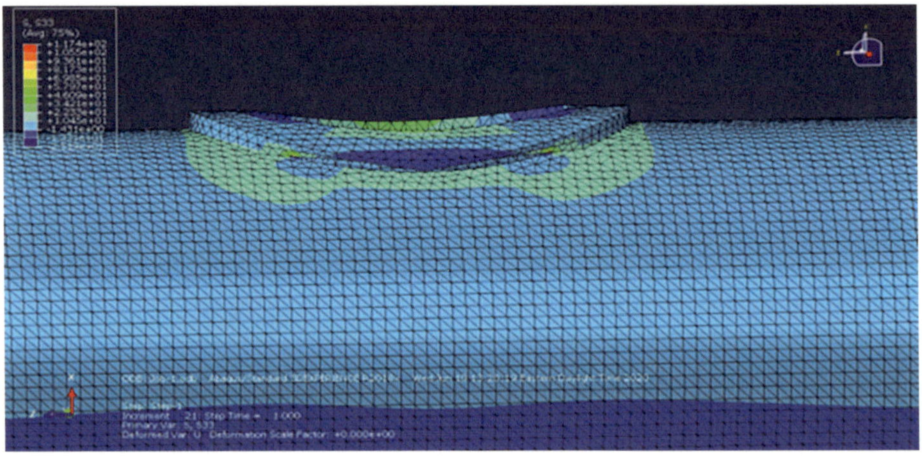

Figure 14.
Stress concentration Z direction.

as well as between the 5:00 and 7:00, indicated by a lighter shade of blue. Elevated stresses are apparent on the inside edge of the reinforcement. Stresses are larger near the junction of the inside and outside legs of the reinforcement [9]. This makes sense because, as the tube bends, the cast reinforcement attempts to elongate on the tension side. This stretches the reinforcement insert and results in transverse bending.

Figures 13 and **14** show how a crack within the detail effects the stress hot spots. In **Figure 13**, to be precise, the stress is along the "X" axis and shows that the highest concentration of stress is located at the "end" of the crack with a red color. **Figure 14** shows this exact same phenomenon along the "Z" axis.

4. Conclusions

Fatigue tests conducted on aluminum light pole samples containing a variety of welded hand-hole details revealed that there were no significant increases in the fatigue life with changes in the treatment of the hand holes or changes in the detail geometry. A total of 8 tests were conducted resulting in 15 data points, all subjected to nearly the same stress range and loading rate. Fatigue test results indicated that while there was some deviation in the results when plotted by itself, the changes made did not have a drastic effect on the fatigue life of the specimens. The finite element models show how an initial crack effects the stress concertation around the handhole. From the results, it can be seen how the stress "pools" at the location where that initial cracks were observed. This was consistent with the results in the lab.

Acknowledgements

The authors would like to convey their gratitude to HAPCO Company for manufacturing the samples for experiments and providing funding for the study.

Author details

Cameron R. Rusnak* and Craig C. Menzemer
Department of Civil Engineering, The University of Akron, Auburn Science and Engineering Center (ASEC 210), Akron, OH, United States

*Address all correspondence to: crr44@zips.uakron.edu

References

[1] Murthy MVV, Rao KP, Rao AK. On stresses around an arbitrarily oriented crack in a cylindrical shell. International Journal of Solids and Structures. 1974;**1974**(10):1243-1269

[2] Durelli AJ, Parks VJ, Feng HC. Stresses around an elliptical hole in a finite plate subjected to axial loading. Journal of Applied Mechanics. 1966;**1966**(33):192-195

[3] Aluminum Design Manual. Specification for Aluminum Structures. Arlington, VA, USA: The Aluminum Association; 2010

[4] Fisher JW, Kulak GL, Smith IFC. A Fatigue Primer for Structural Engineers. National Steel Bridge Alliance: AISC; 1998

[5] AASHTO (American Association of State Highway and Transportation Officials). Standard Specifications for Structural Supports for Highway Signs, Luminaires and Traffic Signals (LRFDLTS-1). Washington, DC, USA; 2015

[6] Roy S, Park YC, Sause R, Fisher JW, Kaufmann EJ. Cost-Effective Connection Details for Highway Sign, Luminaire, and Traffic Signal Structures. NCHRP 10-70 Web-Only Doc. 176. Washington, DC, USA: Transportation Research Board; 2011

[7] Dexter RJ, Ricker NJ. Fatigue-Resistant Design of Cantilevered Signal, Sign, and Light Supports, NCHRP Rep 469. Minneapolis, MN, USA: Univ of Minnesota; 2002

[8] Menzemer C. Examination of several Mullica river bridge light poles, corresponding to J. Bowman, Hapco, USA. 2012

[9] Daneshkhah AR, Schlatter CR, Rusnak CR, Menzemer CC. 2019 Fatigue behavior of reinforced welded hand-holes in aluminum light poles

[10] Rusnak CR. Fatigue Behavior in Reinforced Electrical Access Holes in Aluminum Light Pole Support Structures. Master's thesis. Akron, OH, USA: the University of Akron; 2019

[11] Schlatter CR. Fatigue Behavior of the Reinforced Electrical Access Hole in Aluminum Light Pole Structures. Master's thesis. Akron, OH, USA: the University of Akron; 2017

[12] Hilty E, Menzemer C, Manigandan K, Srivatsan T. Influence of welding and heat treatment on microstructure, properties and fracture behavior of a wrought aluminum alloy. Emergency Mater Research. 2014;**2014**(3):230-242

Chapter 13

Failure Analysis of Steam Generator Tubes

Gonzalo Diego and Susana Merino

Abstract

Steam generator fabricated of alloy 600 tubes has been replaced in the last decades by newer versions built with alloy 690 or alloy 800 tubes. At first, it was thought that both alloys were resistant to SCC mechanisms but after several years of operation, some NPPs developed denting processes and IGSCC, especially in alloy 800. Corrosion products accumulated in the crevice area between tube-sheet and secondary tube surface of SG increase the stresses and produce a chemical more aggressive to the environment. These two points are the main factor to occur "denting" and increase the possibility of ODSCC can be developed. This chapter describes the failure analysis of an alloy 800 SG tube with only nine years of operation. Microstructural characterization of sludge was carried out after the cleaning by mechanical tube sheet water lancing. SEM/EDX, surface analysis by Auger, and XPS spectroscopy were used to identify chemical species over the OD surface and the fracture surface of some open cracks. "Denting" mechanism is related to flow accelerated corrosion and oxidant particles introduced into secondary side of SG. Therefore, it is convenient to prevent the impurities intake into SG and perform a chemical cleaning process to avoid the hard sludge formation.

Keywords: steam generator (SG), denting, intergranular stress corrosion cracking (IGSCC), alloy 800, steam generator maintenance

1. Introduction

Nuclear power plants steam generators (NPPs SGs) are normally vertical cylindrical vessels made up of inverted U-tubes and with steam-water separators at the top of the device. Boiling occurs on the shell side named the secondary side while on the primary side, inside of the tubes, the coolant never reaches the state of steam (310–330°C and 15 MPa). Most NPPs PWR have between two and four SGs.

Actually, most steam generator tubes are fabricated with thermally treated Alloy 690 even though is still supplied with replacement Alloy 800NG tubing. In this manner, progressively replacing Alloy 600 MA has been performed due to many failures caused by stress corrosion cracking (SCC) in this alloy. **Figure 1** shows a Fe-Cr-Ni ternary diagram where is shown the composition of the alloys used in nuclear power plants.

Alloy 690 is a solid substitutional alloy consisting of approximately 60% nickel, 30% chromium, and 10% iron. The concentration limits of these elements as well as

IntechOpen

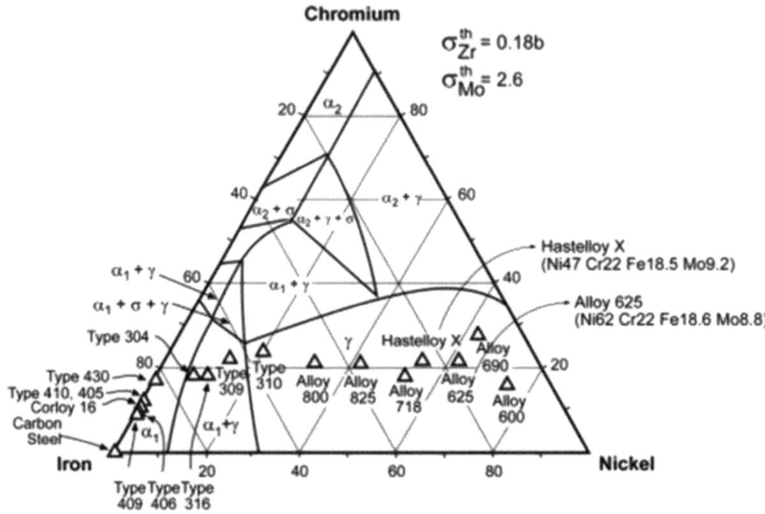

Figure 1.
Fe-Cr-Ni ternary diagram for 400°C for NPPs SGs [1].

Standard	Designation
ASTM/ASME	B163/SB163
UNS	N06690
DIN	17,459
ISO	NiCr29Fe9
AFNOR	NC30Fe

Table 1.
Designations for alloy 690.

the rest of the constituent elements present slight differences depending on the manufacturer or institution that sets the specification. **Table 1** presents various designations and standards for this alloy. The chemical composition of alloy 690 is listed in **Table 2**.

The control of carbon within the limits established according to **Table 2** is essential to obtain the desired corrosion resistance and mechanical properties. In general, a maximum limit is established, although in some specifications a minimum limit is indicated [2]. Carbon range between 0.015 and 0.025% is the optimum to obtain a microstructure with inter and intragranular carbide precipitation. The lower carbon limit (0.015%) is necessary to ensure enough carbon is available for correct carbide precipitation. The upper limit (0.025%) is specified to ensure that most of the carbides formed can be dissolved during mill-annealing heat treatment [3]. Limits have also been specified for sulfur and phosphorus since these elements could segregate at grain boundaries and have a detrimental effect on corrosion resistance. In addition, sulfur decreases the ability of these alloys to be hot worked, so it is necessary to keep its level at low values.

On the other hand, Alloy 800 is an alloy closer to titanium stabilized stainless steel. This alloy has been used for industrial applications with slight differences in its composition and with different thermal treatments to ensure optimal behavior in the

Chemical Requirements (ALLOY 690)	
Nickel	**remainder**
Chromium	27.0–31.0
Iron	7.0–11.0
Manganese	0.50 max.
Carbon	0.05 max.
Copper	0.50 max
Silicon	0.50 max
Sulfur	0.015 max.

Table 2.
Composition limits, %. ASTM designation B163 or ASME SB163: "Standard Specification for Seamless Nickel and Nickel Alloy Condenser and Heat-Exchanger Tubes."

operating environment. In the nuclear industry, the composition and heat treatment of alloy 800 has been optimized, establishing a clear difference between the standard alloy and its modified version named modified alloy 800, **Table 3**.

The compositional variations introduced in the alloy 800 have important repercussions in terms of the corrosion resistance of this material (**Table 4**). For example, the reduction of the carbon content to 0.03% and the stabilization ratio Ti/C>12 and Ti/C +N>8 have been established to ensure good resistance to sensitization and, therefore, resistance to intergranular corrosion in conditions representative of acid-sulfate chemistry. In addition, the total amount of aluminum and titanium is important. Al+Ti contents above 0.55%, produce the precipitation of γ'phase (Ni_3(Al, Ti)) when the material is subjected to thermal treatments between 500 and 700°C. This gamma phase and the $M_{23}C_6$ type carbides produce an increase in creep resistance with a clear decrease in ductility [4, 5]. Increasing the minimum chromium level to 20% results in good pitting corrosion resistance. On the other hand, increasing the minimum nickel content to 32% is intended to improve resistance to transgranular stress corrosion cracking.

Most of the failures that occur in NPPs SGs tubes involve various mechanisms, including:

- High cycle of fatigue in the tube support plate (TSP)
- Primary water stress corrosion cracking (PWSCC)
- Outer diameter stress corrosion cracking (ODSCC)

Standard	Designation
ASTM/ASME	B163/SB163
UNS	N08800
EN	2.4642 - NiCr29Fe
ISO	6207
AFNOR	Z5 NC 35–20

Table 3.
Designations for alloy 800.

Chemical Requirements (alloy 800)	
Iron	**remainder**
Niquel	30.0–35.0
Chromium	19.0–23.0
Manganese	1.50 max.
Carbon	0.10 max.
Copper	0.75 max
Silicon	1.0 max
Sulfur	0.015 max.
Aluminum	0.15–0.60
Titanium	0.15–0.60

Table 4.
Composition limits, %. ASTM designation B163 or ASME SB163: "Standard Specification for Seamless Nickel and Nickel Alloy Condenser and Heat-Exchanger Tubes."

- Tube support denting (TSD)
- Fretting
- Others: pitting, wastage.

Most nuclear power plants operate under AVT (all-volatile treatment) conditions on the secondary side of the steam generators. Common volatile conditioning agents are ammonia, amines, and hydrazine (or hydrazine substitutes). With AVT, the pH of the feed water ranges from 8.8 to 9.8.

During normal operation of the SGs, nonvolatile compounds rise to a concentration level higher than the feed water. For this reason, the SGs are periodically purged using continuous blowdown. Although this operation should be sufficient to avoid corrosion damage, the permanent ingress of corrosion products produces deposits accumulation on the tube-sheet increasing the risk of the initiation corrosion process. Physical and chemical changes may occur depending on the impurities level in the deposits or the sludge composition, such as solubility product, precipitation of compounds, hydrolysis, etc [6]. Impurity sources can be treatment plant quality, cooling water (condenser leaks), repairs and impurities in conditioning chemicals, etc. The insulating effect of a deposit of corrosion products can cause overheating of the tube metal and subsequent failure. Because they are generally porous, corrosion product deposits can also provide sites for boiler water concentration and thus the potential for caustic attack. In general, the concentration of species due to the presence of crevices and sludge accumulation gives rise to acidic or basic environments far away from normal secondary chemical conditions.

For this reason, many nuclear plants periodically perform sludge removal by tube-sheet lancing (SL) and inner bundle lancing (IBL) on the steam generator's secondary side. Lancing uses high-pressure jets to mechanically remove sludge from the tube-sheet surface, and in this way, impurities concentration is eliminated and denting formation is avoided.

2. Experimental procedure

The failure analysis was performed on the steam generator tube in different areas with circumferential cracks previously detected using Eddy current and UT technique. Three areas were characterized on the surface of the tube, called: the free tube, transition zone, and expanded zone (**Figure 2**). Using the stereo microscope, the area of interest was identified and isolated by a silicon carbide disc cutter. The cutting procedure was dry, without refrigeration, so as not to disturb the composition of the existing deposits on the external surface.

The characterization of the deposits was performed by EDX, XPS, and XRD. These deposits or sludge frequently accumulate in the area of the tube sheet.

The characterization of the fracture surface was performed by longitudinally cutting the tube into two equal half parts. One of them was tested for fatigue to obtain the fracture surfaces (**Figure 3**).

The fracture surface was examined on SEM to quantify the length and depth of existing cracks. The deposits found on the fracture surface of the two cracks were analyzed by EDX and by Auger spectroscopy, five zones of each crack from OD, (point 1) to inside the crack (point 5) (**Figure 4a**).

Metallographic characterization of the axial sections was carried out on the other half of the tube piece as shown in **Figure 4b**.

In addition, the microstructural characterization of the hard sludge steam generator has been carried out. The samples to be examined were collected during the last cycles and appeared in form of deposits or sludge and accumulate in the area of the tube sheet. All samples were weighed and dried at 60° for 48 hours and then classified by sieve size into three types: dust (D), small size (SS), and large size (LS). Microstructural and physical characterization of each individual particle was also performed using SEM and EDX analysis, stereomicroscope, and microhardness.

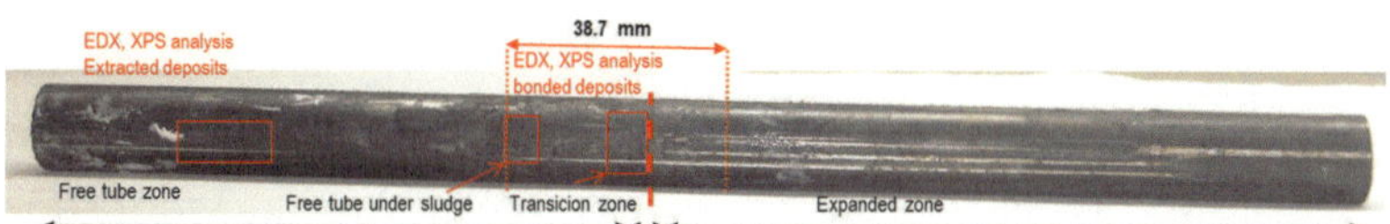

Figure 2.
As-reception segment tube. showing deposits analysis areas.

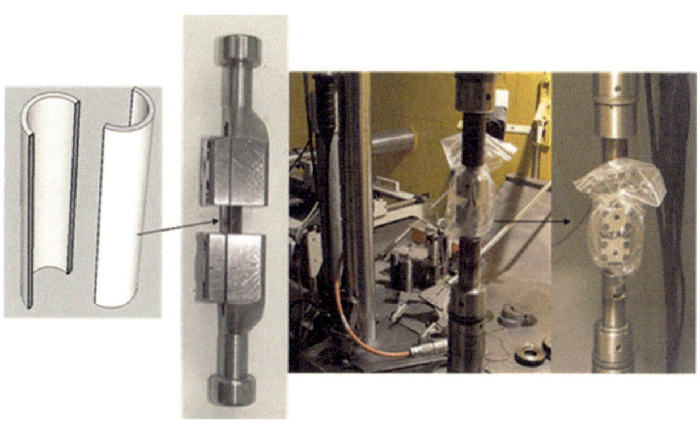

Figure 3.
Devices used to obtain the fracture surface.

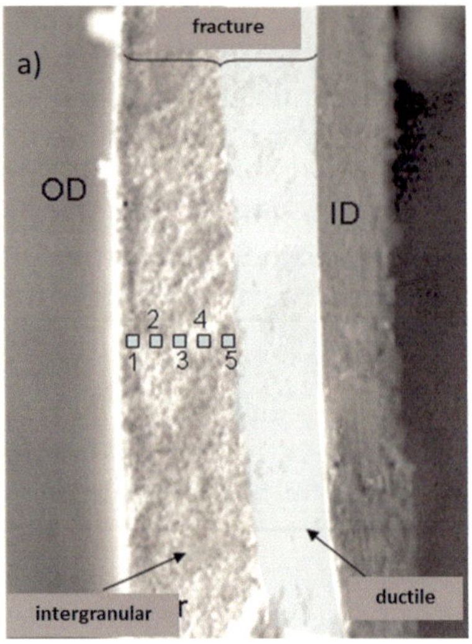

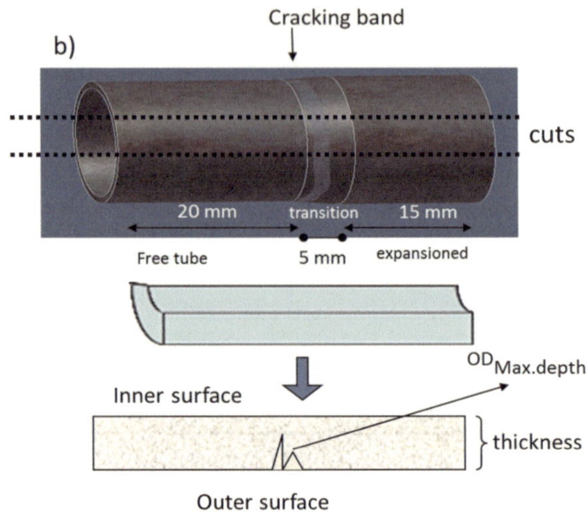

Figure 4.
a) Points analyzed on the fracture surface by Auger spectroscopy. b) Metallographic specimens from the other half of the tube in the area of interest.

3. Results

3.1 Visual inspection

Circumferential cracks and denting were detected by Eddy's current technique before the extraction in the expanded transition zone. Three different areas were identified for subsequent tests: free tube, transition zone, and expanded zone (**Figure 5a**). Zero is also marked for recording observations collected along the tube perimeter.

Multiple circumferential small cracks separated by ductile ligaments in an area of 5 mm width are shown in **Figure 5b**. These cracks were located at 3 mm from the last contact point between the tube and the tube sheet. Damage as "wastage," pitting or other has not been detected. Different color is observed among these areas. The expanded region shows a red coloration (tube sheet contact) while the transition zone presents a dark gray color because this area has been covered by sludge.

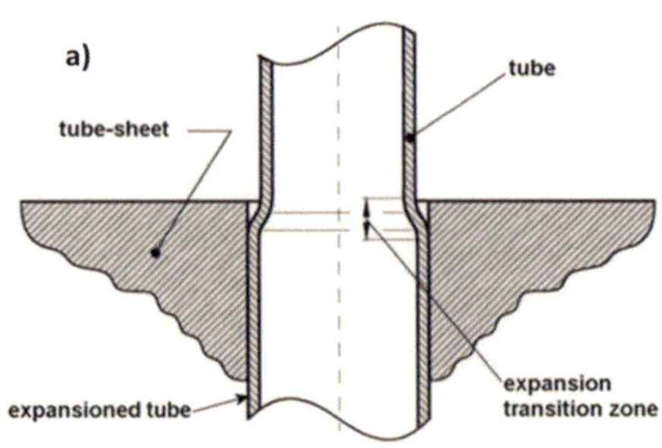

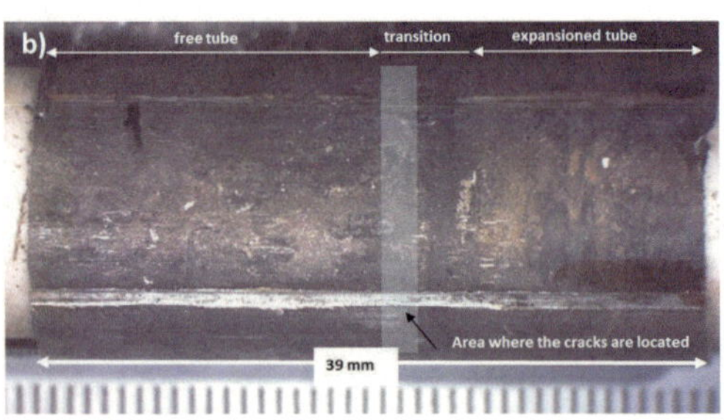

Figure 5.
a) As-reception segment tube extracted in the area of the sketch. b) Image analysis to detect the expansion end of the tube by stereoscopy microscopy.

3.2 Deposits characterization on free tube under sludge.

Table 5 shows a comparative analysis of the deposits removed from the free tube surface. EDX results of sludge deposits, located in the expansion transition and extracted in the free tube, indicate a process of impurities concentration on the crevice area formed between the tube sheet and the tube. Significant concentrations of Si, Mg, Ca, S, Cu, Zn, Na, and Cl were detected. This process of impurity concentration is not as evident in the top of the sludge area where EDX analysis indicates that the detected impurities are present only in part of the analyses. Moreover, Pb has been found in low concentrations.

XPS and XRD results of these deposits showed the existence of metallic Cu and FeS2, which correspond to reducing environment of the typical operation secondary side in SG. Likewise, reduced sulfur species, such as sulfides or sulfites, were detected in the expansion transition zone and the area under sludge. Moreover, XPS has identified the presence of metallic Fe and Cr in the free tube under the sludge.

Other compounds identified are a mixture of silicates, mainly of Al, and Mg. Na is sometimes associated with Cl and S, although this association is qualitative by EDX analysis. XPS analysis has identified the presence of Ca (as Ca chlorides) and a mixture of metallic oxides with K and Mg. S is detected in both oxidized and reduced forms, like sulfates and sulfides.

A comparative XPS analysis of deposits at both free tube and expansion transition zone is summarized in **Table 6**. Fe is detected as magnetite and mixed spinels. Metallic Fe is present in the free tube deposits of the surface and is not detected in the transition area. Al-silicates with Mg, Na, CaCl2, and S, like sulfates and sometimes as sulfites, are detected in both areas.

3.3 Cracking characterization on OD side

Figure 6a shows the cracks detected on the OD side of the tube. Cracks have a circumferential orientation with corrosion deposits and they were detected in an area around the tube located above the last point of contact between the tube and tube sheet. The highest density of the cracks was located at 5 mm of this last point. **Figure 6b** shows the intergranular morphology of the cracks at high magnification and the aspect of the corrosion deposits on the tube surface. EDX analyses show important amounts of Na, Si, S, and Cl in these areas. Also, the presence of Cu and traces of Pb and Zn were also detected in five analyses.

Elements	EDX	XPS	DRX
Fe	M: 41% atomic	M:Magnetite and spinels. Hidróxide and Fe metallic.	M: Spinels and sulfide form
Mg, Al y Si	M: 29% atomic	M: Al Silicates	—
Na,K,Mg, Ca	M: 25% atomic (all elements)	X: CaCl2, only detected	X: Ca, K y Mg as mixed metallic oxides.
Cr, Ni*,Mn, Ti	X: Cr*, Ni*	X: Spinels, mixed oxides	X: Spinels, mixed oxides
Cu, Zn	X	X: Cu met. and óxide Cu+.	X: Cu+, as (KCuO2)
P, S y Cl	X	X: Cl-	X: Sulfides and sulfates

**Possibility of these elements belongs to base material. M: Majority phase or element; X: Minority phase or element.*

Table 5.
Overview of EDX, XPS and XRD analyses on deposits removed from the free surface of the tube.

Element	Free tube under sludge	Transiction zone
Fe	Fe2O3/ Fe3O4, FeO, metallic Fe	Fe_3O_4, FeO, $FeSO_4$
Ni	NiO (probably)	**
Cr	Cr_2O_3, spinels of Fe, Cr	**
Ca	$CaCl_2$	CaCl2, silicates
Cl	chlorides	chlorides
S	sulfides *	Sulfates, sulfites
Si	Silicates, SiO2	Silicates, SiO_2
Al	Al2O3, Al(OH)3, silicates	Al_2O_3,silicates
Mg	Silicates, oxide	Silicatos, oxido

** sometimes as sulfites*
*** minority phase (mixed oxides)*

Table 6.
XPS analysis on free tube under sludge and transition area.

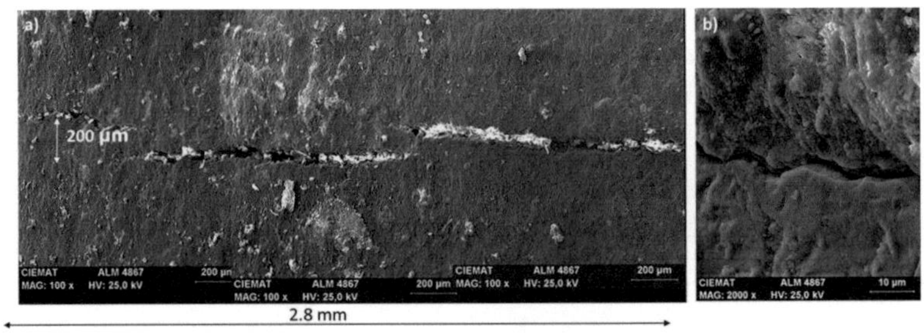

Figure 6.
a) Circumferential cracking on OD transition zone. b) Detail of the intergranular morphology of the cracks.

3.4 Fracture Surface characterization

The tube segment with the cracks was cut into axial two halves. In one of them, a fatigue process for obtaining the fracture surfaces was performed. This fracture was examined by SEM in order to measure the length and depth of the open cracks. Two fractures were identified and these were formed by small cracking linked between them by ductile ligaments. Dimensions of both cracks are summarized in **Table** 7.

Morphology of the fracture is shown in **Figure** 7. Crack A presents an intergranular morphology with several ductile ligaments. Maximum depth of the intergranular crack was 937 mm, 82% of the tube thickness. Crack B, also of intergranular

Crack measurements	Crack A	Crack B
Length	10 mm	6.5 mm
Average depth from OD	637 µm (56%)	492 µm (43%)
Max. depth	937 (82%)	633 (55%)

Table 7.
Dimensions of the fracture surface of the two cracks.

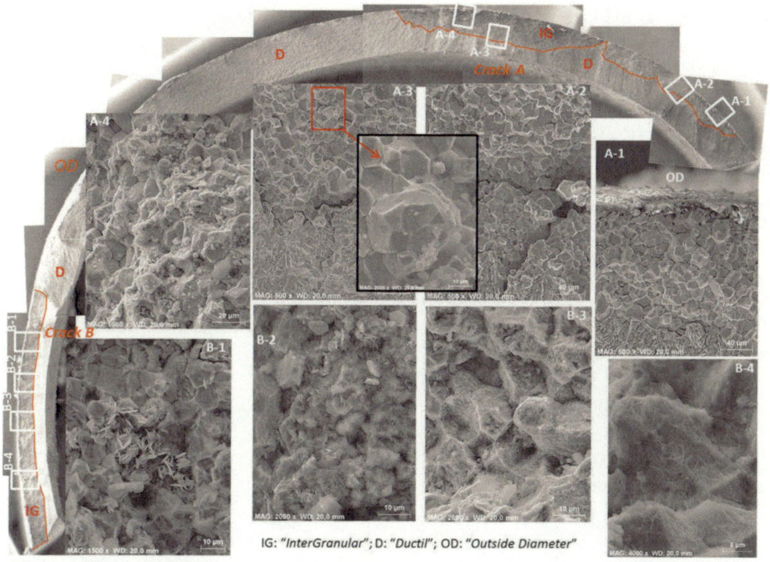

Figure 7.
Fracture surface with details of the intergranular crack and deposits morphology (A and B).

morphology, is not complete, due to the initial cut to obtain the two halves. Maximum depth of this crack B is 55% of the tube thickness.

Intergranular morphology of the two fractures showed some different aspects between them (**Figure 7**). Fracture surface of crack B was covered by a continuous gelatinous film, while the fracture surface of crack A showed corrosion deposits only in the initial area, near the outer surface, while the remainder of the fracture was completely clean.

Significant differences in the EDX analyses of the two areas (A and B) of the fracture were not observed. Crack B has a higher sulfur average concentration than crack A. Occasional sulfur concentrations can reach 29.1% at crack B. Silicon concentrations are similar in both cracks while the chlorine concentration is higher at crack A. It should be noted that Na was not detected in any of the cracks. On the other hand, Zn is more frequently found at crack A than at crack B where only traces of this element were present. Small concentrations of barium have also been detected in the two fractures.

Metallographic characterization of selected cracks by stereoscopy microscopy was prepared in the thickness plane of the other half segment tube. **Figure 8** shows the morphology of the crack in the metallographic samples. The maximum depth detected

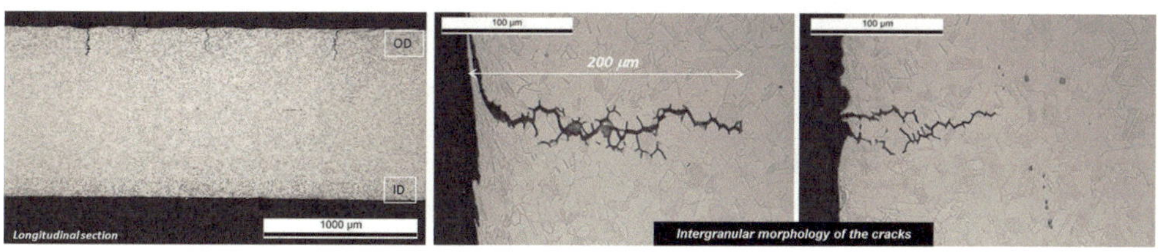

Figure 8.
Intergranular cracks in metallographic samples.

was 210 μm, being in all cases of intergranular morphology. In one of them, it can be observed that the crack initiation is parallel to OD surface, possibly due to the existence of a hardened surface (shot peening process). In other areas, the cracks propagated with normal orientation to OD surface.

Figure 9a shows the sulfur distribution inside the cracks by x-ray mapping. In some cases, S seems to be associated with Ni or Cu and not with oxygen. It is very possible that these elements are in sulfide form. A Fe and Ni (Cr depleted) rich layer

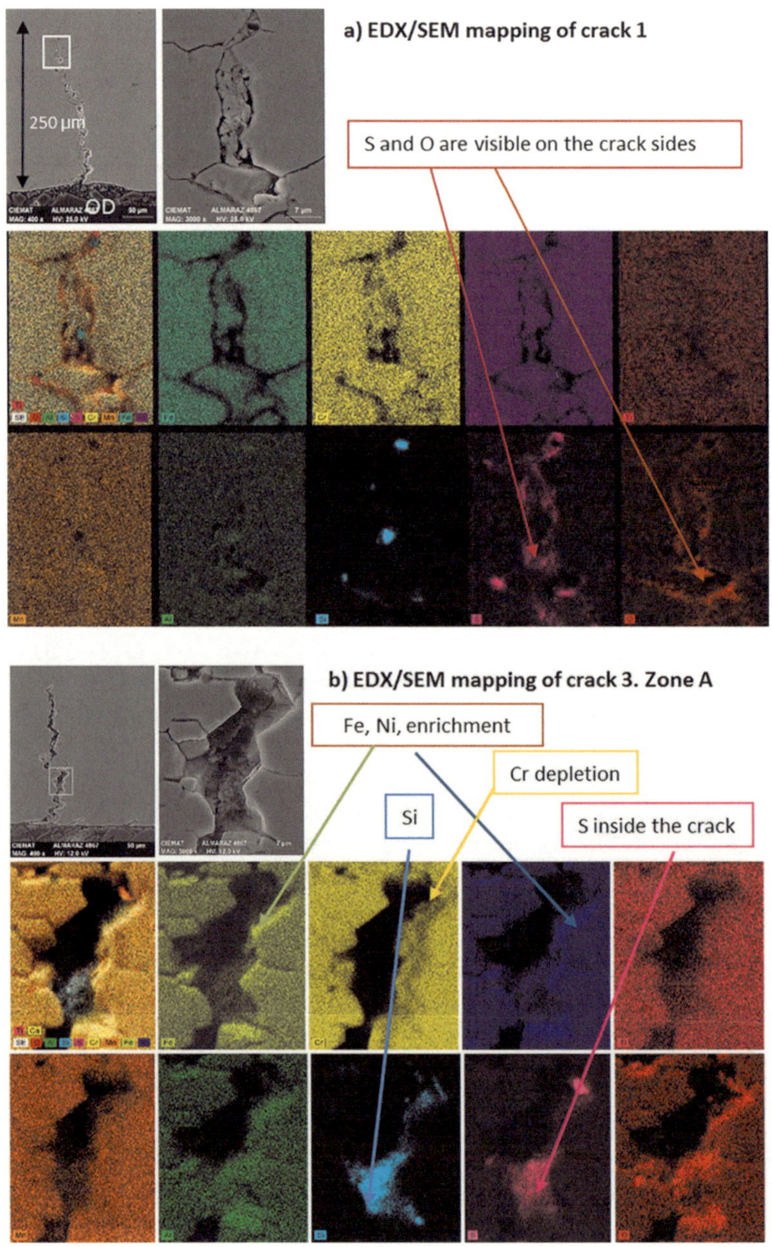

Figure 9.
a) EDX/SEM mapping of crack 1. b) EDX/SEM mapping of crack 3.

can be observed in **Figure 9b**. Besides, Si and oxygen have been detected in very internal areas of the crack. Also, the presence of S is remarkable in this area.

Deposits located on the fracture surface of cracks A and B were analyzed by Auger spectroscopy. Analyses were performed in five areas of each crack from OD to the tip of the crack. Sputtering process was performed after prior analysis in these five areas in order to observe the concentration depth profile from fracture surface to base metal. **Figure 10** shows the profiles corresponding to point 1 (near OD surface) and point 5 (bottom or tip of the crack). **Table 8** summarizes the results of these analyses where it is indicated (in nm and with arrows) the enrichment and depletion of some elements at different depths.

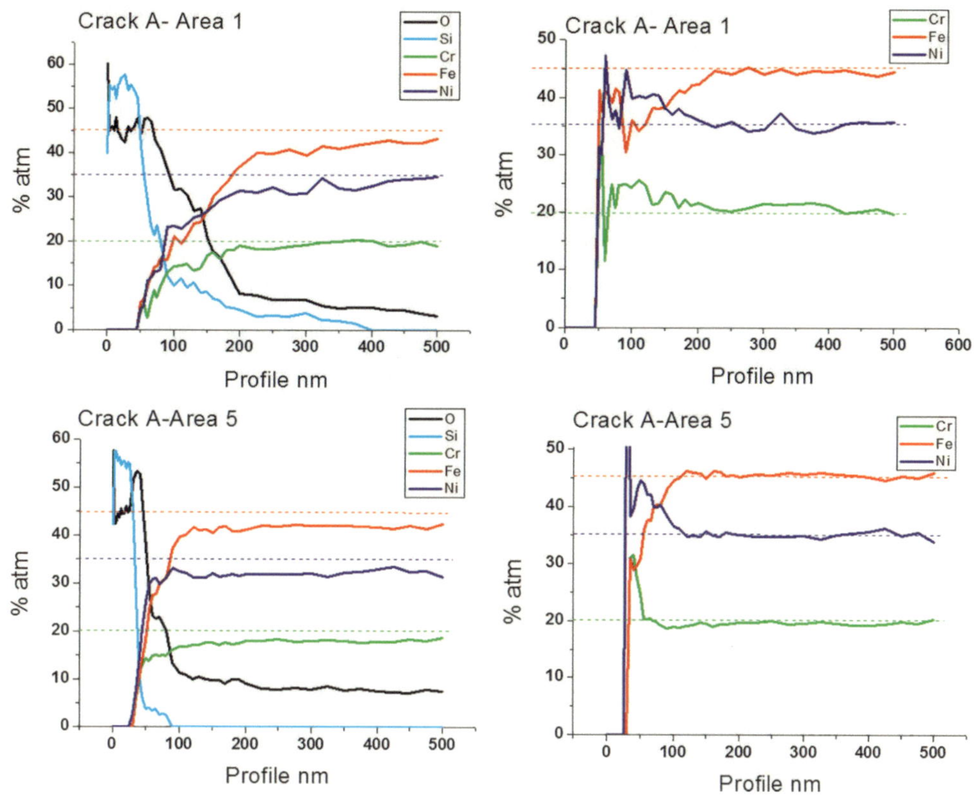

Figure 10.
Auger profile concentration for O, Si, Cr, Fe, and Ni and Cr, Fe, and Ni. Areas 1 and 5 of crack A.

Area	Si	O	Cr	Fe	Ni
1	375	375	225↑	225↓	225↑
2	800	800	100↑650↓	250↓-700↑	600↑
3	225	225	100↑	250↓	225↑
4	110	150	50↑	100↓	150↑
5	100	110	70↑	100↓	110↑

Table 8.
Profile concentration summary by auger spectroscopy on the five areas of crack a. up arrow (↑) indicates enrichment and down arrow (↓) depletion, relative to base material concentration.

Concentration profiles at crack A (area 1), show a thicker silicate layer around 400 nm. This point is the closest to OD. In the other points, far from OD (area 5), the silicate layer is thinner (<100 nm). Ni enrichment is always detected at the five points of the analyses. Fe is always depleted in all cases, even though in the first layers this depletion is less pronounced. Cr is enriched in the first nm and after it is stabilized with a concentration corresponding to the base material or slightly depleted.

3.5 Sludge characterization

After sub-samples separation, M (magnetite layered), C (collars), and F (flakes) pieces were identified visually to characterize macroscopic features such as size, shape, and color. **Figure 11** shows macroscopic photographs of deposit sub-samples. Typical magnetite pieces were present as small "gravel," generally on the order of a few millimeters in length. The colors of these pieces were grey with areas of orange tone and showed geometries and curvature compatible with the triangular shape of the space between tubes in SG tube sheet (**Figure 12**).

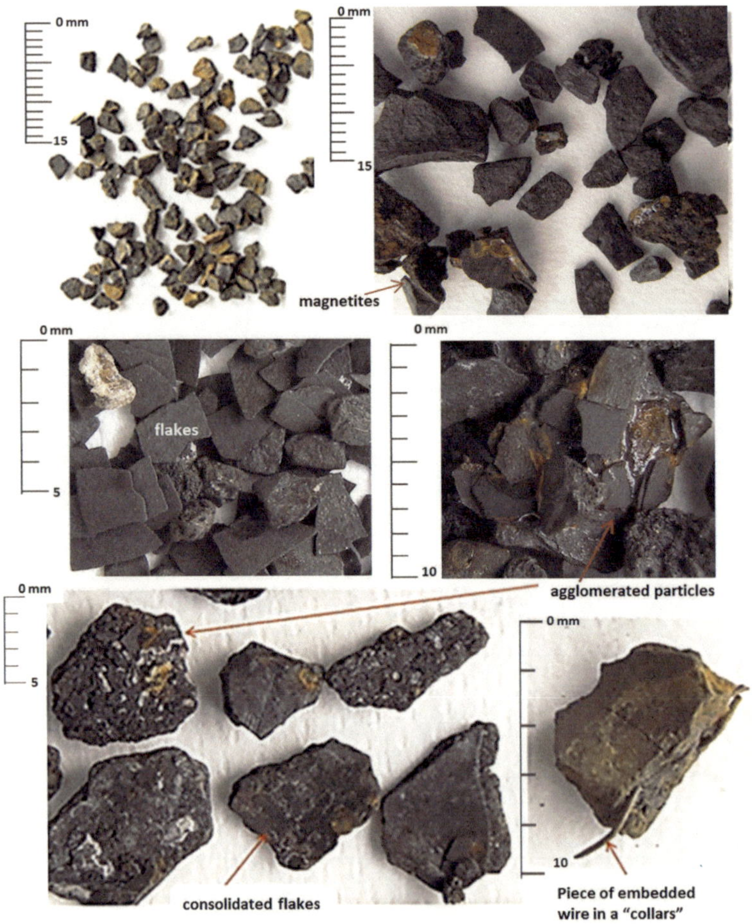

Figure 11.
Different morphology of deposits collected from steam generator.

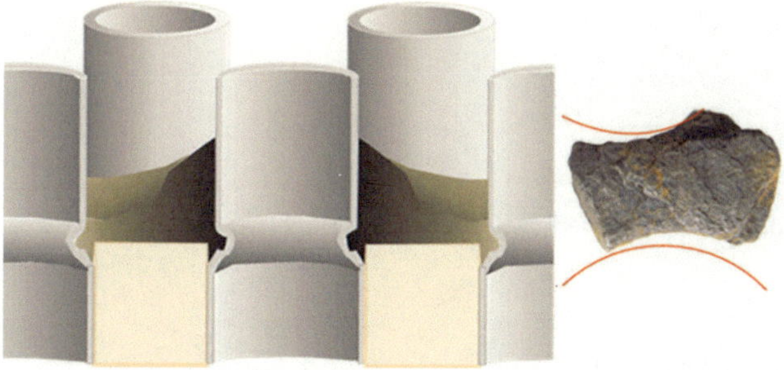

Figure 12.
Individual particle from tube-sheet SG. Dimensions fit the space between the tubes of tube sheet.

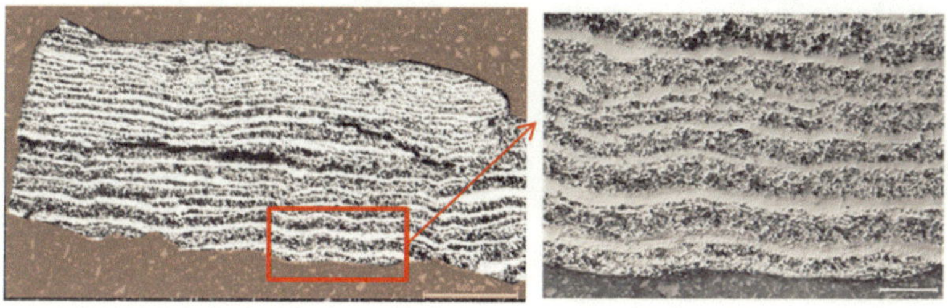

Figure 13.
Cross-section of magnetite layered particle. Optical and SEM images.

Magnetite layered pieces of several mm of thickness presented very high hardness values (876 Hv maximum). On the other hand, "flakes" present moderate hardness values (533 Hv maximum). "Collars" showed hardness slightly lower and heterogeneous values.

Most of the magnetite deposits are part of the oxide layer formed on the low carbon steel alloy and its composition is $Fe3O4$. **Figure 13** shows the morphology in the cross-section of magnetite pieces by optical microscopy. The structure consists of alternate layers with greater or lesser porosity. In the layers with higher porosity is easy to distinguish magnetite crystals with loss of cohesion. SEM/EDX measurements were performed on the cross-section of the magnetite layered in order to quantify the concentration of the local elements at the different layers.

Figure 14 shows the elemental distribution on the magnetite layers with enriched bands in Mn. Dot analysis shows only Fe and O in the compact layers while in the porous layers exists an enrichment of Mn, Ni, Si, and other elements in a lower concentration (**Figure 15**).

The total of these elements forming the enriched bands is only 2%, the rest is Fe and O. Additional analyses performed detected an increase in S and Ni with isolated precipitates of Cu (**Figure 16**).

As discussed at the start of this section, "collars" pieces were detected Spanish NPPs (**Figure 17**). Cross-sections of these samples are formed by a number of metallic inclusions embedded in a matrix of Al and Si species (**Figures 18** and **19**). These last elements were enriched preferably near the interface with SG tube surface. Si/Al ratio is higher when the silicates are the binding agents.

Regarding "flakes" or tube scales is only observed in some NPPs before CC treatment. These particles are not detected with subsequent FFA additions and after chemical cleaning treatment.

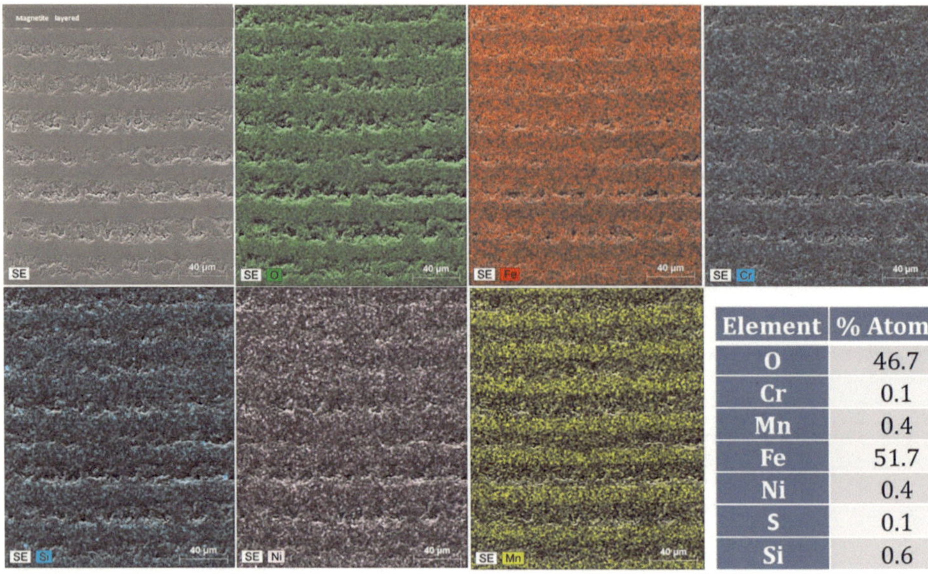

Element	% Atomic
O	46.7
Cr	0.1
Mn	0.4
Fe	51.7
Ni	0.4
S	0.1
Si	0.6

Figure 14.
Mapping of elements distribution in the cross-section of magnetite layered.

% atm.	EDX-A	EDX-B
O	49,6	48,8
Cr	0,1	0,0
Mn	1,1	0,0
Fe	48,2	51,2
Ni	0,4	0,0
Mo	0,2	0,0
Si	0,4	0,0

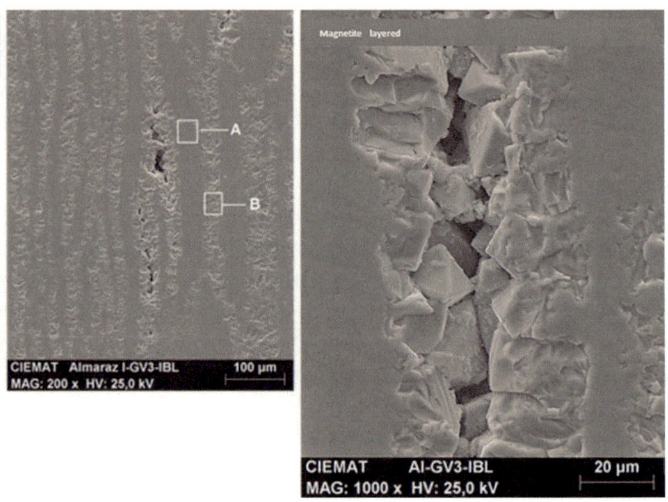

Figure 15.
Dot EDX analysis in the cross-section of magnetite layered sample.

Normally, flakes are formed by an inner layer in contact with the tubing surface and an outer layer with high porosity (**Figure 20**).

4. Failure analysis. Discussion

Alloy 800NG steam generator tubes are a composition modified austenitic stainless steel for use as steam generator tubes in PWRs. The main modifications concern the carbon content, C<0.03% by weight, and the Ti/C>12 ratios to avoid sensitization of

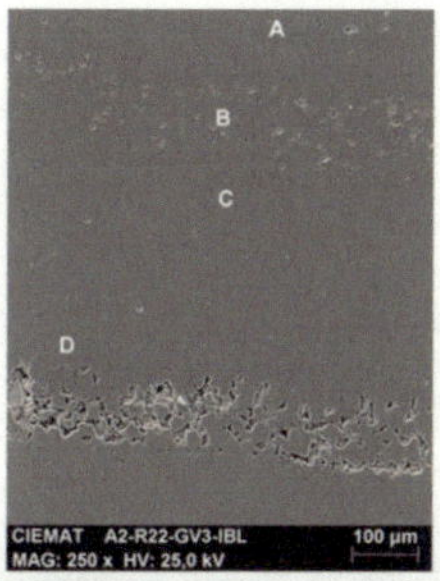

(% atom.)	EDX-A	EDX-B	EDX-C	EDX-D
O	24,1	50	43,8	40,1
Si	0	0	0,4	0,3
S	14,5	0	0	10,1
Cr	0	0	0,1	0,4
Mn	0,1	0	0,8	1
Fe	32,5	50	54,4	39
Ni	0,4	0	0,4	9,1
Cu	28,4	0	0	0

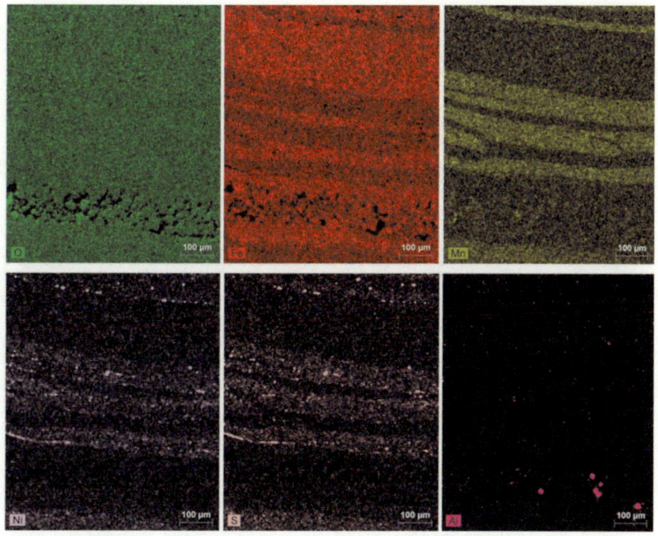

Figure 16.
Mapping of elements distribution in the cross-section of Magnetite layered.

Figure 17.
Morphology of "collars" removed from NPPs.

the material. The final stage during the tube manufacturing process is a shot-peening treatment of its external surface in order to prevent stress corrosion cracking mechanisms. The shot-peening process creates an outer layer with high plastic deformation of the surface that prevents crack propagation under SCC conditions. Circumferential cracks in the expansion transition were detected before extraction of the analyzed tube by Eddy's current inspection besides "denting" degradation and a height of sludge, about 46mm.

Destructive examination of the expansion transition zone has revealed a 5 mm wide band with multiple circumferential cracks separated by small ductile ligaments. These cracks were located 3 mm from the last point of contact between the tube and the tube sheet. No other type of damage such as "wastage," pitting or others has been detected.

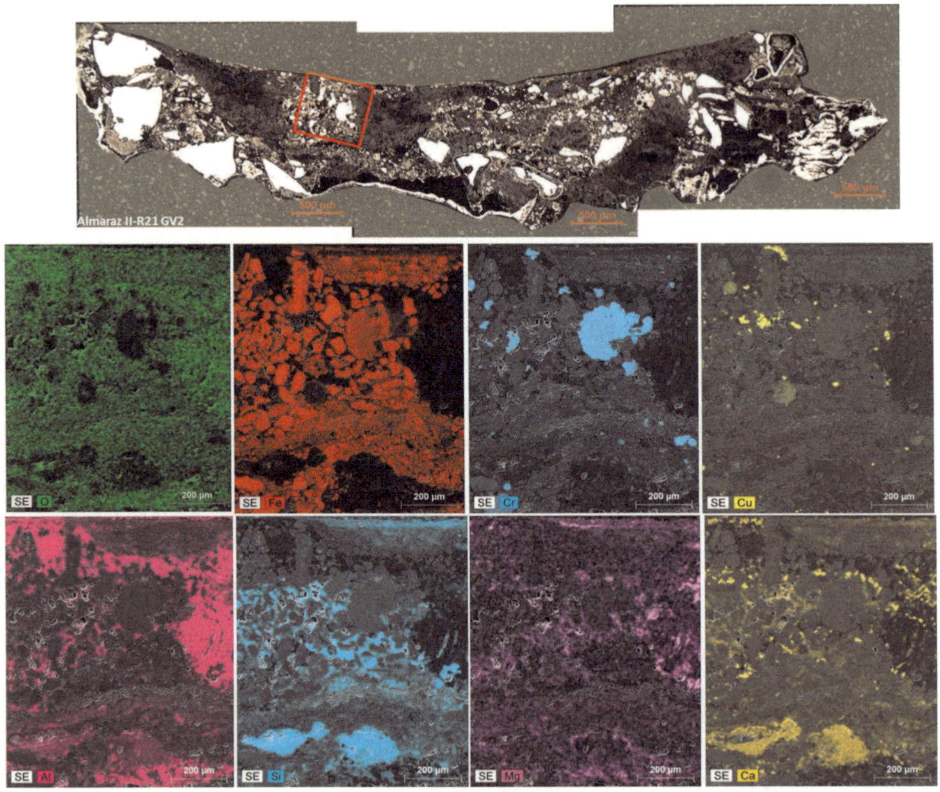

Figure 18.
Cross-section collar. Spatial distribution of chemical elements by SEM-EDX.

Two cracks have been characterized during the fractographic examination both of intergranular morphology, which at first glance presents different aspects. Crack B's fracture surface was completely covered by a viscous morphology film while crack A was covered with deposits in the initial area, near the external surface of the tube, the rest of the fracture was apparently clean.

OD-initiated corrosion of circumferential cracks in SG tubes has been described by different authors [7]. Destructive examination performed by EDF [8] found circumferential and longitudinal cracks in forty tubes examined. Circumferential cracks were located below the top of the tube sheet, while axial cracks were detected above the tube sheet under a sludge layer. The existence of "denting" was not explicitly mentioned.

Extraction of deposits from the expansion transition zone and free tube, next higher area, were compared by EDX analysis. These results indicated a process of impurities concentration in the gap between the tube sheet and the tube. Si, Mg, Ca, S, Cu, Zn, Na, and Cl were detected in this area but these concentrations were lower in the upper part of the sludge zone. It should be noted that traces of Pb have been found in some of the EDX analyses carried out on the initial area of only one of the analyzed cracks. Some authors suggest that alloy 800 could be more susceptible to SCC in alkaline solutions with lead (Pb) [9].

XPS and DRX of the extracted deposits revealed the existence of metallic Cu and FeS_2, typical water chemistry of secondary side SG. Likewise, reduced sulfur species such as sulfides or sulfites have been detected in the expansion transition zone and

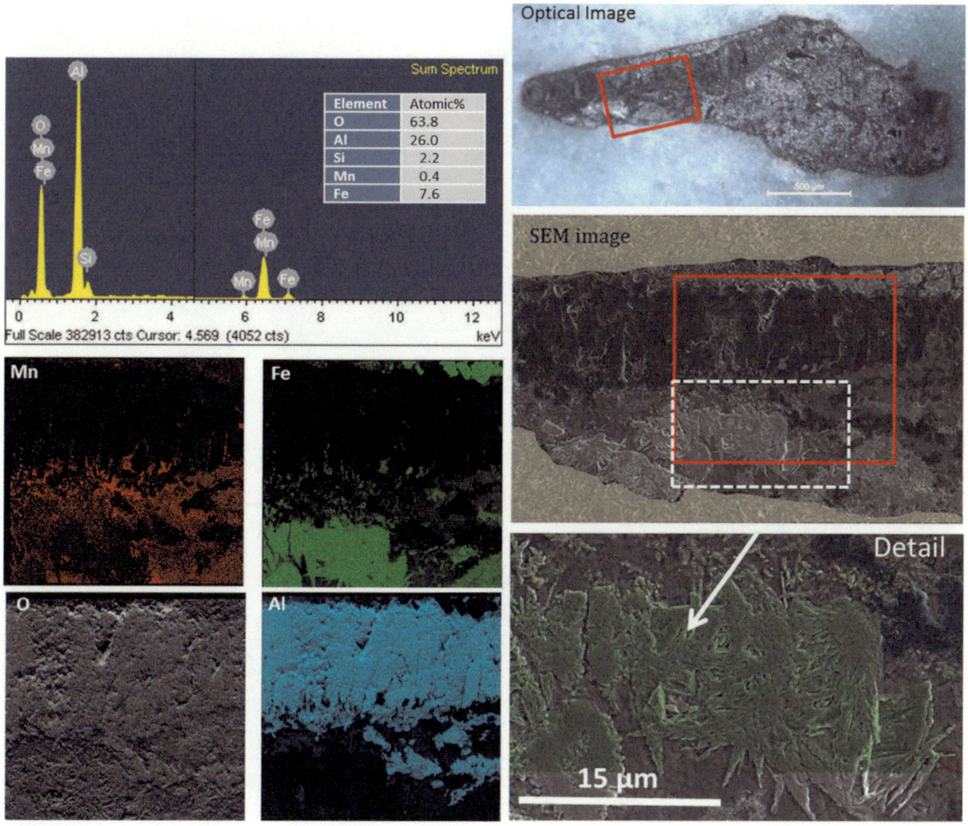

Figure 19.
Cross-section collar. Spatial distribution of chemical elements by SEM-EDX.

under the sludge. On the other hand, XPS analyses detected the presence of Fe and Cr metallic in the free tube deposits under the sludge. In addition, there is a need to clarify the species reduction that could be produced due to ionic sputtering carried out for the analysis. However, metallic Fe has also been identified in the existing deposits in the expansion transition zone, in which ion sputtering was not carried out.

EDX analyses performed on the fracture surface of the A and B cracks confirmed the presence of impurities inside the cracks. Crack B fracture surface was completely covered by a film rich in silicates with discrete particles very rich in sulfur, besides the impurities mentioned above. However, the analyses performed in crack A revealed only deposits in the initial area of the fracture, which did not differ from those analyzed in crack B, except that they had a lower concentration of sulfur. The presence of impurities inside the cracks is also confirmed by the EDX analyses carried out inside the cracks prepared for metallographic characterization.

Silicon compounds, like silica or silicates, were continuously detected on different areas of the external surface tube and over fracture surfaces of the cracks. Many scientific papers describe the silicate's role influence in the SCC susceptibility through intergranular mechanism on the SG secondary side. Failure analyses in tubes of French plants described Al-silicate deposits over a chromium-rich brittle layer associated with corrosion damage. The presence of Al-silicates confirms that cracks formation has not been produced in strong alkaline media where this silicate is not stable

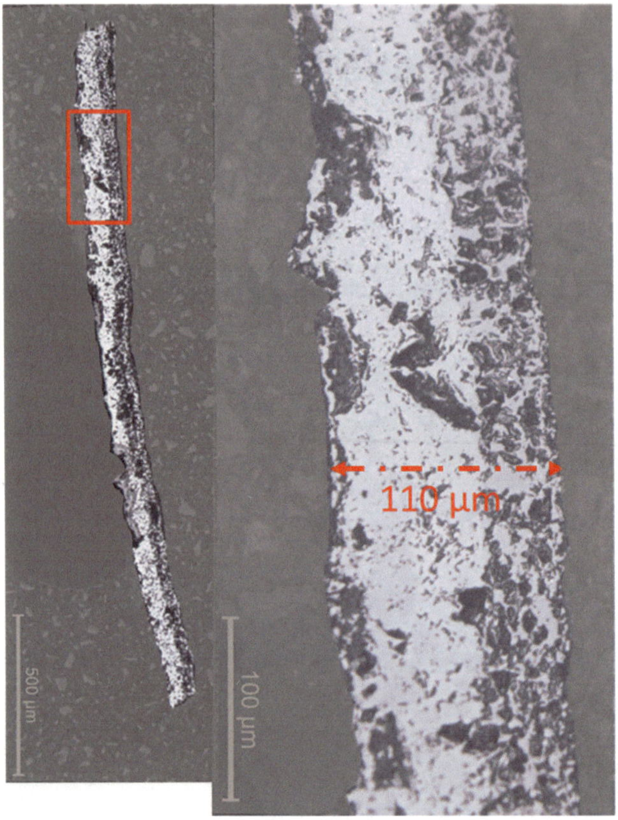

Figure 20.
Cross-section of "flake."

(pH>10 at 300°C). EPRI (Rev.7 of the secondary guides) notes that when Si/Al ratio decreases the response of alloy 600 to IGA/SCC is worse [10]. In addition, a threshold value ~2 of Si/Al ratio could produce an IGA/IGSCC increase for alloy 600 MA.

In the case of alloy 800NG, deposits formation of silicates depends on microstructure. Tubes 800 NG are manufactured of Fe-base alloy. Oxide layer of this alloy expose to secondary water of PWR present a strong affinity for silica [11] and iron, even nickel can perform a cation exchange reaction with this element. Other elements such as Ca, Mg, and other alkalines can form mixed silicates producing local acid in the environments and contributing to increasing corrosion mechanisms due to the formation of protons. This mechanism could perform under the deposits in contact with an external surface of the tube due to the inhibitory capacity of silica as is demonstrated in the results of plants.

However, the silicate's role is unknown in the SCC of Incoloy 800. In fact, laboratory results show that the formation of the Al-silicate layers depends on the microstructure. These layers are very thin when they are formed on rich-chromium film, thicker on rich-nickel film, and 20–30 times thicker on rich-Fe film. Iron is the major element in Incoloy 800 it has a strong affinity for silica. In this way, iron can migrate with the Al-silicate by cation exchange reaction. Similar behavior can be attributed to nickel but this mechanism is not observed for the chromium. Meanwhile, calcium is fixed on the surface of silicates. This type of reaction can produce local acid in the environment under the deposits and contribute to material dissolution [8].

Furthermore, silica is considered an element with inhibitory capacity. This remark is based on the results of plants where lower levels of IGA/SCC were produced when silica above 40 ppb was detected during the blowdown [5].

On the other hand, many experiments have been carried out related to the effect of sulfur in different alloys for steam generator tubes in the middle of the secondary with the aim of carrying out comparative tests between alloys 600, 690, and 800. Most of the tests in the laboratory cannot be extrapolated to plant conditions, but they provide a study of the resistance of these alloys in a certain medium with sulfur species, such as sulfates in an acid medium. For example, alloy 800 was susceptible to IGSCC in a sulfate acid environment (pH = 4, at room temperature). This susceptibility was increased when the pH was lower. On the other hand, the alloy 600 produced wastage in an acid sulfate environment as was demonstrated in tests performed at CIEMAT labs in "model boiler test" [12].

The behavior of alloy 800 NG secondary water chemistry of PWR with sulfates in a neutral or slightly alkaline environment has been studied by Bouvier of EDF [13]. The dilution used was intended to be representative of existing environments in flow restricted areas and consisted of deionized water with additions of sulfuric acid and caustic soda to achieve pH between 5 and 9.5 at 320°C and sulfate concentrations of 100ppm, 5000ppm (0.05M), and 57000ppm (0.6M). Alloy 800 exhibited IGSCC / IGA in the 0.05 and 0.6 M sulfate solutions at pH=5, while no damage was observed at pH 6. Cracking was only observed in slightly polarized specimens (+100mV/Ecorr), contrary to what was observed in caustic media [14], where copper significantly increases the aggressiveness of the solution for alloy 800. Results obtained by Westinghouse also indicate that the presence of 0.6M sulfate concentrations has no accelerating effect on the stress corrosion cracking of alloy 800 at pH 6 to 8 at 320°C, while it has a significant effect at pH 5 [15]. In summary, one would only expect a significant contribution of sulfur species to the cracking detected in an acidic pH medium, which if present should have been identified by the presence of chromium-rich oxide layers in the corroded zones or some type of damage, as dissolution of material in typical of acidic media [16].

The results of the EDX analyses and Auger analyses carried out on the fracture surfaces of cracks A and B allow us to infer the medium in which the cracks have been formed. For this purpose, the concentration profiles obtained on cracks A and B with those obtained on samples tested in the laboratory were compared (**Figure 21**). In acidic media, a first layer rich in iron and nickel is observed, which is identified as a layer deposited on the specimen during cooling, and below it a layer with enrichment in chromium and depletion in nickel and iron is visible. In caustic media, an oxide

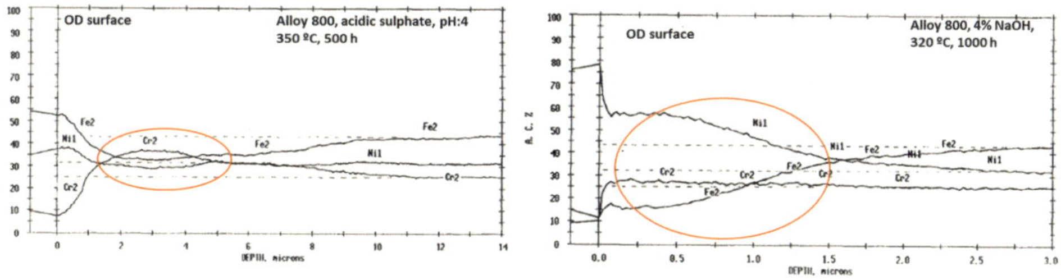

Figure 21.
Auger analyses. Concentration profile of deposits formed on the external surface of a C-ring specimen tested in acid sulfates at pH =4 at room temperature and in caustic media.

layer with clear nickel enrichment, slight chromium enrichment in its outermost part, and iron impoverishment is observed.

Figure 22 shows the concentration profiles obtained on the crack fracture surface formed in an Incoloy 800 tube tested in a "model boiler" in the laboratory, in acid sulfate media, with pH=4, at room temperature.

If we compare these results with the concentration profiles obtained at the bottom of cracks A and B (**Figure 21**), we observe that the main difference between these curves and those of the laboratory sample is the existence of nickel enrichment in the latter, which, as shown in **Figure 23**, is the typical trace left by alkaline environments.

Therefore, we can conclude that the cracks have been formed in a neutral or slightly alkaline environment. Alkaline environments with pH> 10 can be discarded due to the presence of silicates. Sulfur presence would suggest acidic environments, although this evidence is not fully conclusive in all performed analyses. However,

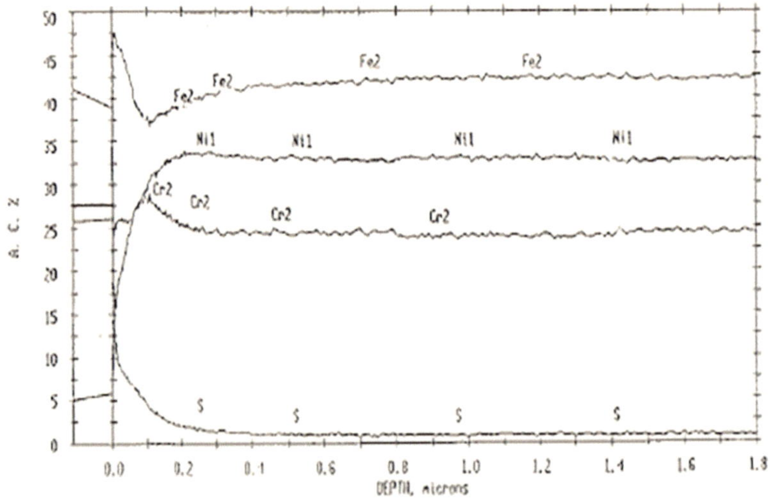

Figure 22.
Auger analyses. Concentration profiles of the oxides at the bottom of a crack in an Incoloy 800 tube were tested in a model boiler in acid sulfate media.

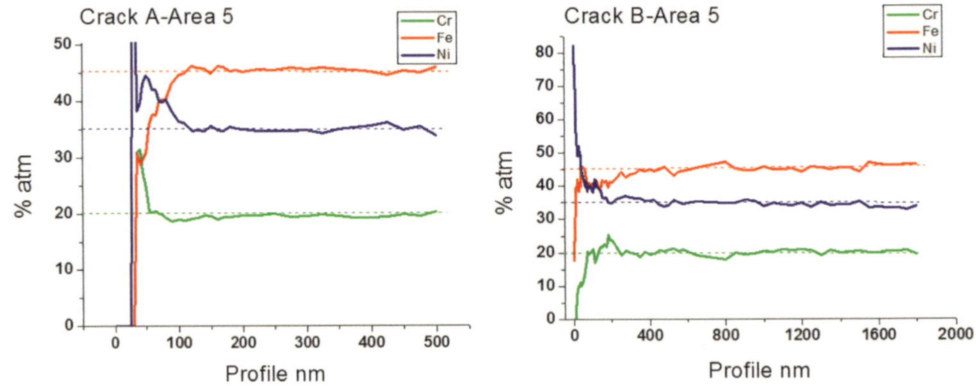

Figure 23.
Concentration profiles obtained in the segment of the tube on the fracture surface of cracks A and B in the area near the bottom of the crack.

since at pH =5, sulfates considerably increase the susceptibility of alloy 800 to IGSCC, although in significant concentrations, we cannot rule out the widespread presence of sulfur in the deposits and oxide layers on the external surface and inside the cracks did not enhance the stress corrosion cracking process to some extent. In addition, it must be considered that, as a consequence of the expansion tube and shot peening process, the susceptibility to SCC in the expansion transition zone would have been higher than the tube-free zone contributing to an increase in the denting mechanism [17].

Regarding sludge characterization of particles collected on the tube-sheet as a result of secondary water chemical used, it should be noted that a root or contributing cause for the degradation mode by denting is produced by magnetite deposition over the tube-sheet surface (known as steam generator fouling). Flow accelerated corrosion (FAC) is the main source of iron from carbon steels and low alloy steels, which causes deposition in the secondary cycle. FAC is an electrolytic process where a mass transfer is produced between oxide/fluid interfaces. The most dominant variables are temperature, fluid velocity, fluid pH, the water amine, oxygen content, steam quality, void fraction of the fluid, piping geometry, and the pipe material composition [18]. Deposition process can be approached as a chemical and physical process where crystallization mechanisms and adhesion of solid particles to the surface are carried out. Soluble iron is deposited on the surfaces when the solubility of the iron decreases. It has been suggested that soluble iron can bind magnetite particles (consolidation) and reduce their re-entrainment [19, 20].

Magnetite solubility is very strongly dependent on both pH and temperature (**Figure 24**). For 300°C there is a minimum in solubility (about 5·10–9 M) found at about pH_{300}=7. However, the solubility increases one order of magnitude with the change of one unit of pH at 300ºC in both directions [21]. As a rule, the magnetite solubility presents a minimum value for each temperature and pH, diminishing for lower temperatures and higher pH. This is why efforts to avoid the deposition of magnetite by ETA injection before shutdown and, thus, reduce the magnetite solubility and FAC. Obviously, if the amount of iron in solution is high the magnetite deposition will occur during the shutdown, especially in the cold legs. Magnetite layer is formed by a competitive process between deposition and solubility due to the small variations around this minimum of the solubility.

A protective layer of magnetite can increase during normal operation of the plants, by these previous arguments, up to a certain thickness from which magnetite layer spallation is produced. This would explain the structure found in samples named magnetite layered where alternate layers are formed during the oxidation process of low carbon steels (tube sheet). A first oxide layer is initially composed mainly of iron and oxygen and other alloying elements that depending on their different diffusion coefficients will form enriched layers in the oxide microstructure. The thickness of this first oxide layer can increase during operation up to a limit depending on several factors: temperature, environment, and time. This process will lead to a rise in the stress between a metal surface and oxide layer. When the stress in the scale increases to the limit accommodated by elastic strain, it will deform or fracture. The spallation of the scale removes its protective function allowing direct access of the environment to the metal beneath and leads to a rapid increase in oxidation forming a second oxide layer. During the oxidation, the stresses can grow due to the different specific volumes of the oxide and the volume of the metal being consumed in the reaction, which can also be influenced by specimen geometry. Moreover, thermal stresses caused by differential thermal expansion or contraction in the oxide and the substrate during temperature change can also lift the oxide layer. FFA additions can increase the

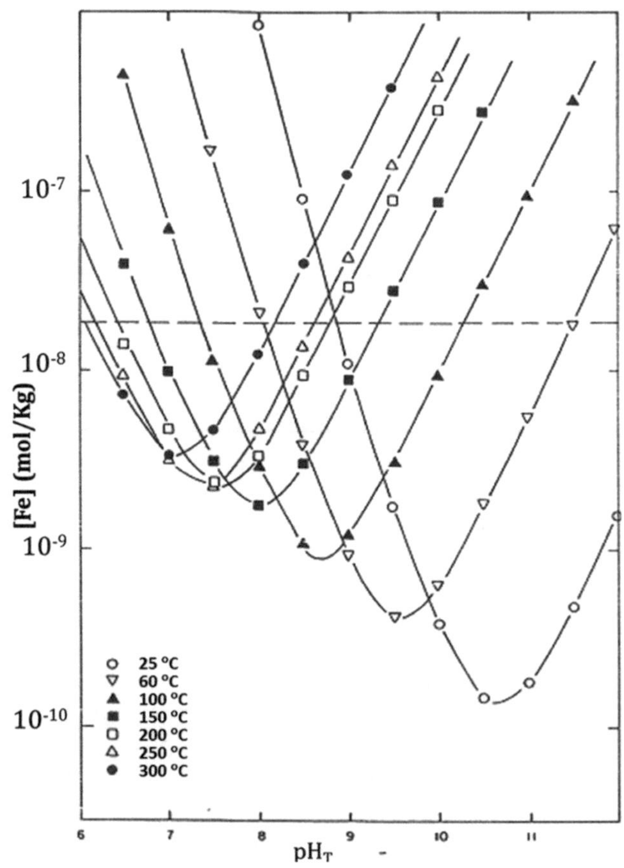

Figure 24.
Calculated solubilities of magnetite as a function of pH over the temperature range 25°C–300°C. p{H2} = 1 atm [20].

spallation process due to hydrophobic behavior over the low carbon steel surface. This is one of the main reasons why the magnetite layered samples represent the highest fraction of the deposits collected.

Low Si/Al ratios have been associated with an increased risk of degradation of Alloy 600 MA tubing but no evidence for Alloy 800 tubing because this alloy is much more resistant to SCC. However, binding agents may promote the sludge consolidation, particularly in the "collars," increasing the concentration of minor species as potential impurities for the tubing degradation [10]. On the other hand, metallic inclusions embedded in the collar matrix have been attributed to denting process when they are oxidized, although not all inclusions detected were associated with alloys with a low resistance to corrosion. Some metallic inclusions came through stainless steel or stellites. Moreover, scales formed during operation on the external tubing surface are composed of a silicate inner layer in contact with the tube surface and a magnetite outer layer with high porosity. In this way, magnetite is always present in the composition of the different types of collected samples in the sludge and its consolidation.

As is seen in the previous figures, the formation of the "collars" depends mainly on the silicon and aluminum concentration. These elements are initially in form of colloids or aggregates of colloids and they may be oxides, oxyhydroxides, or mixed. The size of the colloids depends mainly on the environmental conditions: pH and temperature. Colloids with a large surface area produce a strong interaction with surface oxides, such as tube-sheet or SG tubing. This interaction is subjected to two main types of forces: "van der Waals" and "electrostatic repulsion" due to the presence of their surface charges. There are several theories that exist to explain the interaction between colloids and the coolant ingredients, or otherwise, the sum of the van der Waals and electrostatic interaction potentials between particle pairs. This mechanism forms the basis of the well-known Derjaguin–Landau–Verwey–Overbeek (DLVO) theory for colloid stability [22, 23]. The final result of this mechanism is a consolidation process of the deposits over the tube sheet that leads to the formation of hard sludge or "collars," which are responsible mainly for the mechanism of denting.

Since denting needs a sludge accumulation plus contaminants and ODSCC is closely related to that, some strategies of NPP are focused on: a) Reducing the contaminants intake into SG; b) Decreasing the entry of erosion-corrosion products into SGs by means of FFA injection before shutdown and ETA injection during normal operation; c) Removing the accumulated sludge, especially hard sludge from the SG's during outages by lancing and inner bundle lancing.

On the other hand, chemical cleaning process is used to remove magnetite from the gtube scale and hard tube-sheet deposits accumulated on the tube-sheet surface. However, a high concentration of silicon and aluminum was found, due to the higher temperatures at the interface of the collar and tube or tube sheet precluded the removal of the hard sludge with chemical cleaning at the dented areas in the hot leg sludge piles. Moreover, dissolution rates vary with the surface area of the deposit available to the solvent and, therefore, tube-sheet and crevice deposits will dissolve slowly if the surface area deposits are low [24].

5. Conclusion

Failure analysis of an SG tube was performed in order to know the degradation mechanisms that occurred during the operation. Samples of sludge piles were collected during sludge lancing cleanings to understand the nature of these deposits. Several types of samples were detected: Magnetite layered, "collars" associated with Si/Al ratio, and tube scale "flakes." Sample morphology was associated with a hard sludge, for example the "collars," related to impurities intake into SG, like silicon, aluminum, and iron. Iron is ingressing in the secondary side of SG due to oxidant conditions and maintenance work and during refueling. Carbon steels and low alloy steels are the components with a greater contribution of iron at a secondary cycle by flow accelerated corrosion (FAC) process. Specific chemical conditions produced in the tube-sheet of SG on the secondary side lead to the precipitation of soluble species and solid particles performing denting mechanism by hard sludge, which is the prelude of SCC. Due to these facts, multiple intergranular cracks were formed in the OD located at 3 mm above the last contact point between the tube and the TS, practically in the whole perimeter of the tube. The fracture surface of these cracks was partially covered by a silicon-rich layer and sulfur. EDX and Auger analyses point out Ni enrichment with slight Cr enrichment and Fe depletion. These results are compatible with neutral or moderately caustic environments evidenced by comparative analyses performed for Alloy 800 tested in caustic and acid sulfate environments.

6. Recommendations

Measures to prevent the impurities intake into SG and performing a chemical cleaning process are actions that should be taken into account to avoid sludge formation.

Acknowledgements

The authors would like to acknowledge Spanish NNPs for providing the study materials.

Conflict of interest

The authors declare no conflict of interest.

Author details

Gonzalo Diego* and Susana Merino
CIEMAT, Madrid, Spain

*Address all correspondence to: g.diego@ciemat.es

References

[1] Pugh JW, Nisbet JD. A study of the iron-chromium-nickel ternary system. JOM. 1950;**2**:268-276

[2] Feldman H, Wolfe R. Advanced nuclear technology: Alloy 690 steam generator tubing specification sourcebook. 2016. EPRI report: 3002009412. Available from: https://www.epri.com/research/products/000000003002009412

[3] Hickling J, King C. Materials Reliability Program Resistance to PWSCC of alloys 690, 52 and 152 in Pressurized Water Reactors (MRP-11). 2004. pp 2.1-2.14, EPRI report: 1009801. Available from: https://www.epri.com/research/products/000000000001009801

[4] Tavassoli AA, Colombe G. Mechanical and microstructural properties of alloy 800. Metallurgical Materials Transaction A. 1978;**9**:1203

[5] Dehmolaei R, Shamanian M, Kermanpur A. Microstructural changes and mechanical properties of Incoloy 800 after 15 years service. Materials Characterization. 2009;**60**:246-250

[6] Riznic J. Steam Generators for Nuclear Power Plants. UK, Woodhead Publishing: Elsevier Ltd; 2017

[7] Karwoski KJ. Circumferential cracking of steam generator tubes. United States: N. p., 1997. DOI: 10.2172/477675

[8] Boursier JM, Dupin M, Gosset P, Rouillon Y. Secondary Side Corrosion of French PWR Steam Generator Tubing: Contribution of Surface Analyses to the Understanding of the Degradation Process. Ninth International Symposium on Environmental Degradation of Materials in Nuclear Power System Water Reactors. TMS. 1999. pp. 555-565

[9] Capell B, Wolfe R, Lumsden J, Eaker R. PbSCC of alloy 800NG steam generator tubing in alkaline environments. In: Proceeding of: 17th International Conference on Environmental Degradation of Materials in Nuclear Power Systems – Water Reactors. Ottawa, Canada; 2015

[10] Feldman H. Steam Generator Management Program: Steam Generator Deposits Characterization for Steam Generator Tube Degradation prediction and Management. 2008. EPRI report: 1018249. Available from: https://www.epri.com/research/products/000000000001018249

[11] Sala B, Organista M, Henry K, Erre R, Gelpi A, Cattant F, et al. Laboratory Study of Corrosion of Steam Generator Tubes: Preliminary Results. United States: NACE International; 1995

[12] Gómez-Briceño D, García MS, Hernández F. Effect of Secondary Cycle Sulphuric Acidic Injection on Steam Generator Tubes. Fontevraud 3th International Symposium. Avignon, France: French Nuclear Energy Society; 1994. p 565.

[13] Bouvier O, Prieux B, Vaillant F, Bouchacourt M, Lemaire P. Níckel alloy stress corrosion cracking in neutral and lightly alkaline sulfate environments. In: Proceedings of the Ninth International Symposium on Environmental Degradation Materials in NPP. California: TMS; 1999. pp. 695-702

[14] Castaño ML, Gomez-Briceño D, Esteban F. Steam generator replacement:

Inconel 690TT and Incoloy 800 mod as an alternative to Inconel 600. In: Procedings of Eurocorr. Barcelona. pp. 447-457

[15] Staehle R, Properties of passive films on high nickel alloys as affected by low valence sulfur. Workshop on Effects of Lead and Sulfur on the Performance of Secondary Side Tubing of Steam generators in PWRs. Argonne National Laboratory, 1012780; 2005. p. 34

[16] Laire C, Platbrood G, Stubbe J. Characterization of the secondary side deposits of pulled steam generador tubes. In: Procedings of Seventh International Symposium on Environmental Degradation Materials in NPP. Breckenridge, Colorado; 1995. pp. 387-397

[17] Wolfe R, Feldman H. Steam Generator Management Program: PWR Steam Generator Top-of-Tubesheet Denting. EPRI 3002002197. 2014. pp. 3-15. Available from: https://www.epri.com/research/products/000000003002002197

[18] NEA. Flow Accelerated Corrosion (FAC) of Carbon Steel & Low Alloy Steel Piping in Commercial Nuclear Power Plants. NEA/CSNI/R. 2014;**6**:2015

[19] Vepsäläinen M, Saario T. Magnetite Dissolution and Deposition in NPP Secondary Circuit. Espoo: VTT Technical Research Centre; 2010 VTT-R-09735-10. Available from: https://www.vttresearch.com/sites/default/files/julkaisut/muut/2010/VTT-R-09735-10.pdf

[20] Hermansson H-P. The stability of magnetite and its significance as a passivating film in the repository environment. SKI Report. 2004;**07** Available from: https://inis.iaea.org/collection/NCLCollectionStore/_Public/42/022/42022527.pdf?r=1

[21] Macdonald DD, Shierman GR, Butler P. The thermodynamics of metal-water systems at elevated temperatures. In: Part 2: The Iron Water System. AECL-4137. Atomic Energy of Canada Ltd; 1972 Available from: https://inis.iaea.org/collection/NCLCollectionStore/_Public/04/065/4065211.pdf?r=1

[22] Derjaguin BV, Landau L. Theory of the stability of strongly charged lyophobic sols and of the adhesion of strongly charged particles in solution of electrolytes. Progress in Surface Science. 1993;**43**(1–4):30-59

[23] Verwey EJW, Overbeek J. Theory of the stability of lyophobic colloid. Journal of Physical Chemistry. 1947;**52**(3):631-636

[24] Thomas RC. Utility Experience with Steam Generator Chemical Cleaning. 1994 EPRI TR-104553. Available from: https://www.epri.com/research/products/TR-104553

Chapter 14

Perspective Chapter: Bridge Deterioration and Failures

Kenneth C. Crawford

Abstract

The purpose of this chapter is to present research on bridges failures, the conditions that produce structural failures in bridges, and to better understand the processes that impact and degrade the performance and service life of bridges. While rare, bridge failures can have devastating consequences with loss of life. Bridge deterioration and the risk of failure is a continuing challenge for transportation infrastructure owners. Bridge deterioration is caused by many factors, to include increased service loads, deicing chemicals, and aggressive environmental conditions. By looking at basic bridge types and their components, the focus of research in this chapter examines seven bridges failures and the conditions that led up to their failures. The research in this chapter produced two significant findings. The causes of bridge failures can be traced back to human error in the life cycle of the bridge. The second result is insufficient systematic analysis of bridge failures, their causes, and how to prevent them. While analyzing the causes of bridge deterioration and the processes that contribute to bridge failures the goal of this chapter is to provide a better understanding how bridges deteriorate to minimize failures in the future and to build better bridges. Bridge failures are preventable.

Keywords: bridges, deterioration, failure, design, construction, inspection

1. Introduction

One of the challenges for infrastructure owners of large national networks of highway and railroad bridges is maintaining the performance of the bridges over their designed service life without the risk of failure. But bridges do fail for a number of reasons. A bridge failure is most often the result of some failure in the life cycle of a bridge from concept through design, construction and maintenance during its service life. The direct failure of bridges is the result a number of factors to include structural over loading, material failures, poor designs, faulty construction, inadequate oversite and control of bridge construction, inadequate inspections, delayed maintenance, earthquakes, and flooding.

In research work on the frequency and causes of bridge failures K. Wardhana, et al., studied over 500 failures of bridge structures in the United States between 1989 and 2000 with the age of the failed bridges ranging from 1 year (during construction) to 157 years, having an average age of 52 years. They state "The most frequent causes of bridge failures were attributed to floods and collisions. Flood and scour, with the major flood disaster in 1993, contributed to the frequency peak of bridge failures (almost 53% of all failures). Bridge overload and lateral impact forces from trucks, barges/ships, and trains constitute 20% of the total bridge failures. Other frequent principal causes are design, detailing, construction, material, and maintenance. Comparison made among three periods of similar studies (1977–1981, 1982–1988, and 1989–2000) revealed almost similar trends, with most failures occurring during the bridge's service life. Also, human-induced external events occurred frequently in all three periods, but were most dominant in the first and third periods [1]. While Wardhana studied three periods of bridge failures in the United States the statistics are representative of bridge failures around the world in all years.

This chapter examines bridge failures and structural deterioration for the purpose of understanding the elements that cause deterioration in bridge structures and the factors that lead to the structural failures. There are a number of elements that cause the material in bridge structural components to degrade and deteriorate over time, in both steel and RC (reinforced concrete) bridges. The two dominate elements causing deterioration are deicing chemicals (chlorides) and vehicle over loading. These two elements affect all types of bridges and their components. In extreme cases extensive deterioration potentially leads to structural failure. Of the bridge failures presented in this chapter each failure has one or more of these factors which contributed to the bridge collapsing.

2. Background

The means and methods for bridge failure analysis used in this research paper involved a literature survey of papers that studied and analyzed the causes of bridge failures. Their papers address the types of bridge failures, their root causes, and events and processes that led up to their failures. In most all cases the cause of failure can be traced back to human error.

To provide a basis for the examining the six bridge failures presented in in this chapter a description of the seven bridge types is provided together with an outline of the basic components of a bridge structure, to include foundation, substructure, and superstructure. It is worthy to note that in the service life of a bridge deterioration can occur in any of these components to a degree that can increase the potential for a bridge structural failure. Deterioration is a process changing an object, in our case bridges, to a lower quality state in which a degraded condition can lead to structural failure. A steel bridge with structural beams in a rusted condition is an example of advanced deterioration.

For this chapter bridge failure analysis is the process to understand why structures, components, systems, methods, and processes fail. Failure analysis is a science to analyze how structural systems fail by determining the sources and causes of failure. The bridge failure analysis process includes structural analysis which incorporates the fields of mechanics, dynamics and failure theories. From a theoretical perspective, the primary goal of structural analysis is the computation of deformations resulting from internal and external forces and stresses. In practice, structural analysis reveals the structural performance of the engineering design and ensures the soundness of the structural integrity in design. Using an array of methods, to include nondestructive testing, the failure analysis process collects data on failed components for examination and analysis to determine the cause of failure. The objective of bridge failure analysis is to develop corrective actions and better structural designs with improved reliability to prevent bridge failure over its service life.

The description of a bridge failure is presented in a paper by G. Zhang, et al.

"Bridge failure, is generally associated with serious economic and life losses is dedefined as the incapacity of a constructed bridge or its components to perform as specified in the design and construction requirements. Principal causes can be divided into internal causes and external causes or natural factors and human factors. Design error, construction mistakes, hydraulic, collision, and overload are the top 5 leading causes of bridge failures, resulting in more than 70% of the bridge failures. Causes of bridge failures are closely related to regional economy, structural type, type of use, material type, and service age. The failure rate is high for steel bridges, which is inseparable from excessive emphasis on structural strength and the lack of consideration on structural stability and fatigue in early years." [2].

The paper from Zhang makes an important point. The issue involves understanding the rate of fatigue in structural elements in both steel and reinforced concrete bridges and the impact on bridge performance relative to the weakest point(s) in the bridge structure. Not identifying through inspection and testing the weakest points in a bridge increases the risk of failure.

A description of four highway and two pedestrian bridge failures are presented in this paper. The failure analysis process used for each bridge looks at the bridge type and structure, its failure mode, the events leading up to the failure, and the factors contributing to its failure. Of all the factors contributing to a bridge failure the one common factor in all the failures is the role human errors played in the chain of events leading up to the bridge collapse. Stakeholders, designers, bridge engineers, inspectors, and maintainers are responsible for preventing bridge failures. Bridge failures are preventable.

3. Bridges

Seven basic types of bridges represent a majority of the highway, railroad, and pedestrian bridges in North America, and around the world. These include truss, tied-arch, suspension, arch beam, cable-stayed, and cantilever. Each type of bridge fulfills a specific requirement for a specific location. The distribution of stresses, both horizontal and vertical, determines the type of bridge structure required. It is interesting to note many bridge designs have been in practice for a number of centuries without major changes. For example the arch bridge was perfected by the Romans over 2000 years ago.

3.1 Arch bridge

Arch bridges are the oldest bridges in existence, with the Arkadiko bridge in existence today with many still in use. The basic design of the arch bridge allows the arch span to transmit lateral pressure to supporting abutments on a solid Greece, from the thirteenth century BC, still in service. Over 900 roman arch are in foundation. The simple design of the arch bridge with stone under compression allows for an extremely stable bridge, which explains why older arch bridges are still in existence and in use today, **Figures 1** and **2**.

An example of older arch bridges still in service is provided in a paper by D. Trajber et al. [3], in which they examine the condition of the bridges and assess the degree and rate of deterioration from anthropogenic and environmental factors, with the goal of providing accurate condition assessments and establishing necessary maintenance. In their paper they state "Historical masonry arch bridges still form an important part of Croatian transportation network. There are approximately 680

Figure 1.
Roman Alcantura Arch Bridge in Spain Built 103–106 AD. CC BY-SA 3.0.

Figure 2.
Stari Most Arch Bridge Built 1557 Mostar, Bosnia Hersogevina. CC BY-SA 3.0.

masonry arch bridges and culverts currently being used for railways and roadways. Many of these bridges are relatively old (more than hundred years in most cases) but still in usage. Increasing vehicle load and speeds as well as deterioration due to anthropogenic and environmental influence have highlighted the need for reliable assessment of their service condition and regular maintenance. The aim of this study is to provide a review of existing masonry arch bridges in Croatia. Firstly, a historical review of bridges is given showing the time period in which they were built, indicating the materials and design principles used for their construction. Next, bridge typologies are presented as well as their detailed analysis of geometric characteristics for brickwork bridges. Finally, a short review of damages and their impact on serviceability of bridges is given. This review presents masonry arch bridges in Croatia and the need for reliable method of assessing their service condition in order to provide proper maintenance, repairing and retrofitting." [3].

3.2 Beam bridge

One of the simplest types of bridge, **Figure 3**, is the beam bridge with abutments supporting two or more beams over relative short spans. The two main beams can have cross beams to add strength and stability. The beam bridge does not transfer load stress as in an arch bridge. Many small and medium beam bridges on main and secondary roads are beam bridges.

3.3 Cable-stayed bridge

A cable stayed bridge, **Figure 4**, consisting of cables connected to load-bearing tower pylons and the deck below are used to span large distances. By connecting the cables to the pylons a fan like pattern is created. In effect the cable-stayed bridge is a statically indeterminate continuous girder bridge where the dead and live load internal forces are smaller than a girder bridge. With their structural members in tension the cable-stayed bridge makes more efficient use of materials.

In a paper by J. Radic et al. [4] the authors point out a great number of parameters are required for the shaping of cable-stay bridges, where principal requirements for the shaping of beams, stay cables and pylons are explained. Cable-stayed bridge design analysis must take into account strong interactions between principal load-bearing structural elements. As an example the principal properties of the Jarun bridge cable-stayed bridge in Croatia are explained in the light of guidelines for the shaping of cable stay bridges, and an accent is placed on specific features of this bridge [4].

Figure 3.
Example of a steel beam bridge. CC BY-SA 3.0.

Figure 4.
Cable-stayed bridge Rio Antirrio Bridge in Greece. Opened 2004 length 2.8 km. CC BY-SA 3.0.

3.4 Cantilever bridge

The cantilever bridge, **Figure 5**, made from structural steel or pre-stressed concrete, using simple trusses and beams, connects two cantilever arms in a suspended span center piece with no direct support underneath. Horizontal beams and diagonal bracing support the bridge load with no vertical bracing. The first cantilever bridge in 1866 was the Hassfurt Bridge over the Main River in Germany, with a span of 124 feet, and was considered a major engineering breakthrough in bridge construction at the time. The Canadian bridge Pont de Quebec, Quebec City, **Figure 4**, which opened in 1919, after 30 years and two collapses, is the longest cantilever bridge in the world.

In a paper by Rajeshirke et al., India [5] the authors describe the use of balance cantilever bridges in India which are widely used in hilly regions where supporting from the bottom is difficult. The name Balance Cantilever Bridge is a construction methodology which balances out the cantilever portion and is one of the most effective methods of building bridges without the need of false work. Balanced cantilever bridges are used for special requirements like construction over traffic, short lead time compared to steel and use local labour and materials. Extradosed bridge is a unique type of bridge between Girder Bridge and cable-stayed bridge. As most of the literature covers either balance cantilever bridge or extradosed bridge, this paper introduces and attempts to summarize comparative study of balance cantilever and extra dose bridge with its span arrangement, span by depth ratio, and pre-stressing of steel [5].

3.5 Suspension bridge

Developed in the early 1800s suspension bridges were a marvel in bridge engineering and capable of spanning great lengths. The basic components of a suspension bridge are main cables, towers, and secure anchorages at both ends of the bridge. The deck carrying the dead load and vehicle traffic is hung from the suspension cables with vertical suspenders. The load carrying members are the main cables as tension members made of high-strength steel and are efficient in carrying loads. With this suspension cable configuration the dead weight of the bridge can be reduced making longer spans possible. Early suspension bridges had problems with vibrations and wind loading before the dynamics of wind loading on bridsges was understood. John Roebling was the first engineer to build suspension bridges designed for wind loading

Figure 5.
Pont de Quebec Opened 1919 987 m. CC BY-SA 3.0.

Figure 6.
John R Roebling Suspension Bridge Cincinnati, OH Opened 1867. CC BY-SA 3.0.

with the Roebling bridge in Cincinnati, Ohio, **Figure 6**, and the Brooklyn bridge in New York City.

In a paper by Arioglu [6] the author describes "suspension bridges as masterpieces of the engineering profession with conceptually clear cut 5-piece load-bearing systems which are highly hyperstatic and undergo large displacements under loads having nonlinear behavior and are sensitive to horizontal loads, such as wind loading. Suspension bridges are the most elegant, aesthetic and relatively economic structures of our civilization. Suspension bridge designs are based on mathematical models, using known patterns of physical behavior, but have many unknowns and uncertainties. This paper explores practical mathematical expressions obtained through regression analyses to predict key design parameters of long span suspension bridges such as main geometric dimensions, material quantities/qualities and dynamic properties for preliminary design calculations.

A large design parameter database matrix for 20 long span suspension bridges was collected to bring out heuristic approximations through regression analyses. These regression models are used to examine the design parameters of 1915 Çanakkale Bridge Project, which will break the longest span record with a main span length of 2023 m and the tallest tower record with 318 m (IP Point). It was observed that the dimensions, mass distributions and material qualities selected for the design of 1915 Çanakkale Bridge agree with the findings of this study." [6].

The key design parameters for regression models used by Arigulo on existing suspension bridges correlated well with the design parameters for the new 1915 Canakkale Bridge over the Dardanelles in Turkey. The bridge opened in March 2022 with a span of 3.7 km and is the longest suspension bridge in the world.

3.6 Truss bridge

The truss bridge is a load-bearing structure efficiently incorporating trusses in an array of triangular sections and has been around for centuries. Dynamic loads are accommodated by triangular elements which absorb tension and compression. The combination of tension and compression ensures the structure of the bridge is maintained and the decking area remains uncompromised even in relatively strong winds.

3.7 Tied arch bridge

Incorporating an arch structure supported by vertical ties between the arch and the deck, the tied arch bridge creates downward pressure from the arch structure to the deck of the bridge which translates into tension by the vertical ties. The tips of the arch structure are connected by a bottom chord. The deck strengthening chord connects the tips of each end of the acting like a bowstring which absorbs pressure.

4. Bridge components

Consideration is given to the three primary components of a bridge structure and their subcomponents of which all are subject to elements of deterioration and potential failure.

4.1 Foundation

A bridge foundation, for all types of bridges, consists of the following components:

Piles: The initial foundation of a bridge are piles, wood, steel, or concrete, driven into the ground to support the entire weight of a bridge. By distributing p Piles distribute weight and stresses applied by the bridge evenly through the ground making it stable and strong.

Caps: To provide additional load transferring capacity pile caps are placed on top of the pile foundation provide additional load transferring capacity to the piles and give maximum strength to the upper part of the bridge.

Bents: Forming the foundation for the substructure bents connect piles and caps.

4.2 Substructure

A bridge substructure consists of the follow components which transfer the bridge load forces to the foundation:

Abutments: Capable of withstanding high levels of horizontal force abutments are the vertical support at the ends of the bridge.

Piers: Providing support points for the bridge piers are mounted at the end of each span to reduce the effects of forces and vibrations.

Pier Caps: Acting as a space for the girders pier caps function to transfer loads on bearings from the superstructure components on the top.

4.3 Superstructure

The bridge superstructure.

Girders: Girders (or beams) join pile caps together and give support to the deck and can be over a single span spans joining all the bents, dependent on the length of the bridge. Girders usually have a truss design to improve stress and load resistibility, passing pressure to the foundation.

Bearings: Bearings are structural members capable of transferring loads from the deck to the substructure. These displace stresses and load to the piers through the girders to allow movement between parts of a bridge. The movement can be linear as well as torsional.

Trusses: Trusses are made by joining triangular components to divide loads and bending moments through the bridge. Some types are simple trusses, suspension, and also cantilever trusses. The truss network provides a surface for transportation which can be built as a deck truss, pony truss, or through truss. Each truss differs in how the traffic will move on the bridge.

Decks: Decks made of concrete or metal direct traffic load and include drainage systems, curbs, expansion components, sidewalks and approach slabs.

Barriers: Bridges have barriers on the sides for safety and protection of the decks.

Arches: A bridge with arches has a high degree of strength. Arches control the safety and load bearing ability of the bridge. The quantity of arches and materials used for construction is very important.

Spandrel: A space connecting the bridge pillars and deck beam is called the spandrel. There can be open or closed spandrels depending on the arch design.

5. Concrete bridges

Known for their longevity and low maintenance costs reinforced concrete (RC) bridges are designed to maintain their service life over long periods of time. But they deteriorate from the same elements as steel bridges: poor construction, and outdated designs for today's traffic loads. Subject to aggressive factors, such as over loading, vibration, extreme weather, freeze thaw cycles, chlorides in de-icing salts, plus air borne chlorides in marine environments the service life of the RC bridge is degraded. As the Federal Highway Administration has stated "Salt contamination is probably the most significant single contributor to bridge deterioration". These five primary elements are the contributing factors to the deterioration of RC bridges.

5.1 Elements of deterioration on concrete bridges

For reinforced concrete bridges there are two primary elements, or factors, that contribute to the deterioration of concrete structural members: salts and loads exceeding the original design criteria.

5.1.1 Deterioration from water and salt

The effect of salt on the decks and substructure of an RC bridge can be significant. The chloride ion as a major component of sodium chloride and calcium chloride is the most destructive element to an RC bridge in the corrosion deterioration process on the reinforcing bars, which expands and induces high tensile stresses in the surrounding concrete. Hairline cracks enlarge from freeze/thaw and traffic causing delamination and spalling of concrete. Water seepage (with salt) through faulty deck joints cause deterioration in abutment back walls, beam seats, pier caps, concrete pads, and end diaphragms.

5.1.2 Deterioration from overloading

The effect of heavier truck loads on today's bridges, which in many cases were designed over 50 years ago, produces an element of deterioration on the bridge deck and supporting components. Increased truck weights, volumes and speeds produce more cycles of larger stress ranges reducing the fatigue life of beams and girders. The longitudinal and vertical forces induced by heavy trucks tend to exacerbate the deterioration of the bearings, which are already weakened by salt-laden water and debris. An additional problem is many existing bridges lack sufficient grillage reinforcement which results in cracks and spalls on the abutments and piers near the bearings. If the bridge approach slabs are insufficient the impacts of heavier trucks hitting the bridge produce higher incidents of cracks and spalls of concrete header and back walls at the abutments [7].

6. Steel bridges

Widely used in different structural forms around the world steel bridges provide large span lengths for highway and railway bridges. With its strength, ductility, rapid construction, and compressive and tensile strengths of 370 N/sq. mm steel had advantages over other construction materials. It has higher strength in both tension and compression than concrete, and has strength to cost ratio and stiffness to weight ratio. Strength, ductility, toughness, weldability, weather resistance, chemical composition, shape, size, and surface characteristics are important properties of structural steel for designing and construction of steel bridges [7].

6.1 Elements of deterioration in steel bridges

Corrosion is the major element of deterioration in steel structures. Fatigue damage and brittle fracture are special problems for steel structures resulting from repetitive loadings over a long period of time. As Contreras-Nieto et al. [8] state in their paper as a significant number of steel bridges are approaching the end of their service life, understanding deterioration characteristics and rate of deterioration will help bridge stakeholders prioritize bridge maintenance, repairs, and rehabilitation. One prediction model uses data mining techniques to include logistic regression, decision trees, neural networks, gradient boosting, and support vector machine to the United States' national bridge inventory to estimate the probability of steel bridge superstructures reaching deficiency. The model uses data based on the defined scope of the research: design material (steel), type of design (stringer/multi-beam or girder), and deck type (cast-in-place concrete). The predictors of the model include age, average daily traffic, design load, maximum span length, owner, location, and structure length. The magnitude that these factors contribute to the likelihood of a steel bridge superstructure's deficiency was identified. Outcomes of the analysis afford bridge stakeholders the opportunity to better understand the factors that are correlated to steel bridge deterioration as well as provide a means to assess risks of superstructure deficiency for the sake of prioritizing bridge maintenance, repair, and rehabilitation [8]. The prediction model the author proposes considers a wide range of factors which correlate to a risk assessment on the rate of deterioration of the steel structure, with which the owners can determine maintenance and rehabilitation priorities.

6.2 Critical defects in steel bridges

The following are key factors causing deterioration of steel bridges:

a. Major cracks in girder flanges or in tension cords are critical.

b. Loss of section through corrosion in compression elements, or floor beam connection angles and rivets.

c. Fracture-critical member is one whose failure will result in a failure of the bridge structure, which have the material property of fracture toughness, and is dependent on material toughness.

d. Welds, holes, notches, loss of section, and pitting will affect a steel member's fatigue strength, in which welded members are more sensitive to fatigue induced cracks. Cover plate terminations, flange and butt splices, lateral bracing connections, stiffener end welds, are areas of concern for welded girders.

e. Lack of ductility, material toughness, stress conditions, drop in temperature will cause brittle fracture.

f. Out-of-plane bending distortion is a typical cause of fatigue-related failures.

g. An expansion bearing failure may occur caused by corrosion or pier or abutment movement subjecting main members to large tension or compression stresses. Temperature changes can also cause increases in stress levels [1].

7. Bridge failures

The following bridge failures are just a few of the many bridges that have collapsed in North America over the past decades, which unfortunately have taken many lives. The failures are contributed to many factors, all of which lead back to human error in the phases of the bridge design, construction, inspection, and maintenance.

7.1 Pittsburgh bridge collapse

On Jan 28, 2022, at 6:45 am a 52-year old steel-framed bridge, **Figure 7**, which spans a deep ravine in Frick Park, collapsed, sending six cars and a metro bus down with it. Fortunately no one was killed. Six people had minor injuries. According to the mayor, an inspection of the bridge was performed in September. The bridge has been described as in "poor" condition in previous reports from 2011 to 2017. A structural evaluation described the bridge meeting "minimum tolerable limits to be left in place as is."

Failure analysis: The root cause of the bridge failure was a single point of failure in a corroded structural element. The bridge was non-redundant and fracture critical, meaning a single point of failure will cause the bridge to collapse. Contributing factors were a failure in the inspection process to assess and accurately determine the structural capacity of the corroded steel elements and the decision to keep the bridge open despite a poor condition.

Figure 7.
Collapsed Pittsburg bridge January 2022. CC BY-SA 3.0.

7.2 Pedestrian bridge failure: Florida

A concrete pedestrian bridge under construction at the International University in Miami suddenly collapsed in March 2018. The bridge, which had not yet opened, killed five people. The 174 foot concrete structure, **Figure 8**, weighing 930 tons was non-redundant and fracture critical. Seven days before opening numerous wide and deep structural cracks developed, with cracks lengthening daily. Before the collapse the EOR stated there were no safety concerns.

Failure analysis: Root cause was a failure in a single structural element. Among the contributing factors was a failure by the Engineer of Record (EOR) and the Construction Engineer and Inspector (CEI) to recognize the severity of the developing cracks. In addition there was a lack of a clear line responsibility for enforcing safety and structural issues.

7.3 I-35 W Interstate Highway Bridge Collapse

In August 2007, in Minneapolis, MN, at 6 pm in rush hour traffic, the I-35 W bridge across the Mississippi river collapsed taking with it 111 vehicles, killing 13 people, and injuring 145.

The bridge, **Figure 9**, constructed in 1967, was non-redundant and fracture critical, meaning if one member failed the entire bridge could collapse. The cause of the bridge failure was a gusset plate which tore along a line of rivets. It was determined the gusset plate was too thin. The failure was triggered by the additional weight of construction equipment parked on the bridge and rush hour traffic. In the previous years the bridge was identified as having deficiencies, to include the gusset plates.

In a paper by Salem et al. [9] the authors analyzed the cause of the I-35 collapse. The I-35 Bridge over the Mississippi River in Minneapolis, Minnesota catastrophically failed during the evening rush hour on August 1, 2007, collapsing into the river. In the years prior to the collapse, several reports cited problems with the bridge structure.

The author's research analytically investigated the cause of the collapse using the Applied Element Method. The bridge was modeled using construction drawings, with relevant structural details and loadings. Structural details included the steel

Figure 8.
Collapsed pedestrian bridge Miami, FL. CC BY-SA 3.0.

Figure 9.
I-35 W Mississippi River Bridge collapse. CC BY-SA 3.0.

truss, gusset plates, concrete slabs, concrete piers, while structural loading included traffic and construction. AEM provided the cause of collapse of the I35-W Bridge. The cause of collapse was found to be the failure of the gusset plates at connections L11 and U10, which well agreed with the field investigations of the collapsed bridge. The under-designed thickness of the plates, their corrosion, and over loading due to traffic and construction loads at time of collapse were the reasons for the bridge collapse [9].

Failure analysis: The cause of the collapse was a single ½" gusset plate failing along a line of rivets. There were a number of contributing factors that came together leading to the bridge failure: design flaws, inadequate design review, inadequate inspection, MnDot policies not being followed, poor information flow, the organizational structure not addressing bridge conditions and safety. All these factors combined led to the bridge collapsing. It is noteworthy to note the AEM program used by the authors with the model of the bridge's construction drawings and structural details and loadings predicted the failure of the bridge exactly at gusset plate L11 and U10.

7.4 Hyatt Regency Hotel skywalk failure

The Hyatt Regency Hotel in Kansas City, Missouri, suffered a structural collapse in July 1981, in which two overhead walkways, **Figure 10**, failed under the loading of a large number of people. 114 people were killed and 216 injured. The primary cause of the failure was the induced vibrations from a number of people on the skywalks (overloaded) dancing to the rhythm of the music on the ground floor. It was the worst civil engineering failure in US history. With about 40 people on the second-level walkway, and another 20 on the fourth floor walkway, the fourth floor walkway gave way from a failed bolt connection, dropped onto the second floor walkway, where both plunged to the atrium floor.

Figure 10.
Fourth and second floor skywalks falling to the atrium. CC BY-SA 3.0.

Failure analysis: The root cause was the failure of the fourth floor skywalk suspension rod in a welded channel iron causing the skywalk to drop onto the second floor skywalk with both then dropping on the ground floor. Among the contributing factors was inadequate design in the skywalks, a failure in engineer review of shop drawings and field changes, a lack of oversight responsibility, and clear lines of authority starting with the Engineer of Record.

7.5 Silver Suspension Bridge

The Silver Bridge, **Figure 11**, at Point Pleasant WVA, crossing over to Gallipolis, Ohio, over the Ohio River, opened in 1928, **Figure 10**, and was the first bridge un the US to use the eye bar-link suspensions system. At 5 p.m. 15 December 1967, a single eye bar failed causing the bridge to collapse in seconds, killing 46.

Failure analysis: A cleavage fracture in lower limb of the eye of eye bar 330 was the cause of the bridge failure. The single-point failure caused entire bridge to collapse. The contributing factors were the fact the bridge had no redundancy, the cracked eye in the eye bar was not found during routine inspections, and the combined action of

Figure 11.
The Silver Bridge in 1928, as failed in December 1967. CC BY-SA 3.0.

stress corrosion and corrosion fatigue over 40 years. As result the US Congress passed a federal law requiring systematic inspection of all US bridges.

7.6 I-880 Cypress Freeway Collapse

The Loma Prieta earthquake in 1989 caused heavy damage in Santa Cruz County which collapsed the double-deck Cypress Street Viaduct, **Figure 12**, of Interstate 880 in West Oakland. The 6.9 magnitude earthquake caused 63 deaths and 375 injuries.

Moehle [10] states the viaduct was built-in the late 1950s on reclaimed marsh land the Cypress Street Viaduct was a double-deck freeway section made of non-ductile reinforced concrete. The Viaduct was designed as a two-tier multi-lane highway constructed of reinforced concrete upper and lower levels were connected by two-column bents in a combination of cast concrete and four pin (shear key) connections. The upper deck in some sections was not securely fastened to the lower deck, making this concrete susceptible to vibrations [10].

Yashinsky [11] indicates two major factors led to the collapse. The first was the geotechnical aspect of the central San Francisco Bay area. The second was the design of the concrete columns and bent caps and pin connections. Strong ground shaking in the marshland caused soil liquefaction. As the bridge vibrated during the earthquake, the pins connecting the upper level to the lower level also began to vibrate, causing the concrete surrounding the pins to crumble and break away. Without the presence of concrete under the support columns, the columns slid sideways under the weight of the upper deck and allowed a large portion of the upper deck to collapse [11].

Failure analysis: The root cause of the collapse of the Cypress Express Freeway was failure of pins connecting the upper and lower levels due to the strong ground shaking. Contributing factors were inadequate transverse reinforcement in the columns and deficient bent cap and pin connection designs and lack of compensation for the weak soil conditions [11].

Tests were performed by Monteiro et al. [12] on pieces of concrete extracted from the wreckage to assess structural integrity; many components of the Viaduct were found to be structurally sound. It was concluded that the concrete used had more than satisfactory strength. In addition, micro structural analysis of concrete

Figure 12.
Collapse of the I-880 freeway in the 1989 Loma Prieta earthquake. CC BY-SA 3.0.

samples taken from undamaged columns within the region of collapse showed that the concrete was produced and cast according to the proper procedures at the time of construction [12].

7.7 Pont de Quebec Collapse 1907

In August 1907 the Pont De Quebec, **Figure 13**, under construction on both sides of the St. Lawerence River, suddenly collapsed killing 75 workers and injuring 11.

Failure analysis: The root cause of the failure was an overweight structure for the bridge structural design. The primary contributing factor was an error in the design calculations causing the steel structure to collapse under its own weight. Other

Figure 13.
Pont de Quebec collapse in August 1907. CC BY-SA 3.0.

Figure 14.
Pont De Quebec in 2009 [13]. CC BY-SA 2.5.

contributing factors were lack of review by an independent bridge engineer and clear lines of project responsibility and authority.

Construction resumed in 1910. In 1916 the center span during assembly collapsed, killing 16 works. The cause was failure of a casting used in the hoisting of the center section. The bridge finally opened in 1919 (**Figure 14**).

8. Results and findings

This chapter presents the results and findings of seven research papers that focus on the causes of bridge failures and methods to analyze bridge structures. Presenting frequency of bridge failures Wardhana looks at causes from design, construction, material, and maintenance. Zhang considers structure type and material with an emphasis evaluating structural strength and stability in a bridge's early life to monitor the rate of fatigue in structural elements. Evaluating arch bridges Trajber considers structure type and service age proposing repair and rehabilitation to reduce the rate of deterioration in arch bridges. For cable-stayed bridges Padic analyses the number of design parameters needed to account for the interaction of load-bearing structural elements. Pajeskirke evaluates balanced cantilever bridges in India used for hilly terrain with little bottom support which require special construction requirements. Arioglu uses regression models to examine design parameters for suspension bridges with application to the new 1915 Canakkale Bridge in Turkey. In looking at the I-880 freeway collapse in California in 1989 Mochle points out the failures was the result of not identifying the weakest point in the bridge, specifically the connection pins between the two decks. Analyzing the I-35 bridge collapse in Minnesota Salem uses the Applied Element Method (AEM) to accurately identify the exact gusset plate that failed. The I-35 bridge had long standing deficiencies.

The seven bridge failures presented this paper highlight a variety of causes that led to structural failures. In each of the seven failures there are valuable lessons to be learned. These lessons should be used by designers to build stronger and better bridges with longer service lives. Bridges should not fail.

By looking at bridge failures and their causes Heggade, VP of Board of Management of Gammon India Ltd., presents in his paper the fact there are valuable lessons to be learned in failures. He notes a majority of bridge failures occur in service without external action, during construction and in false works. He indicates there have been an alarming trend of bridge failures in Asian countries and discusses aspects of learning lessons in bridge failures from the Indian context. The point he makes is that accurate documentation of bridge failures is necessary for improved bridge designs. Study of failure improves design concepts for robustness, extrapolation, and durability. The study of bridge failures is an invaluable source of information on bridge design limitations. Bridge design is a process of anticipation of failure. Heggade states bridge designers must learn from past bridge failures to improve deigns to prevent bridge failure [14].

In recommending improved methods to reduce bridge failures Zhang points out:

"Researchers need to strengthen their research on the stability and fatigue of steel bridges, as well as inspection and maintenance. Extreme loads such as flood, collision, and overload contribute to a large number of bridge failures because of the lack of extreme loads data and design theory defects. It is critical for bridges to have sufficient redundancy and capacity protection measures to reduce the probability of bridge failure due to extreme loads. Previous statistical methods and classification methods for the characteristics and causes of bridge failures lack unified standards, and a more scientific method needs to be established [8].

The key point Zhang makes is that unified standards with better scientific methods are needed to classify the characteristics and causes of bridge failures. More research is needed to significantly improve risks reduction of bridge structural failure.

One of the key findings to come out of the research for this paper is the need to better understand how and why bridge failures occur and to apply the lessons learned in failures to design and build better bridges, of all types, that will not fail.

9. Preventing future bridge failures: inspection and maintenance

Bridge inspection and maintenance are the two most important activities for an existing bridge to preserve its function and service life and to provide public safety. The American Society of Civil Engineers reports that one out of every nine bridges in the US structurally deficient. Age and deteriorated bridge conditions are a contributing factor to many recent bridge failures.

9.1 Inspection

As Silano and Henderson [7] states in his book "Bridge inspection and Rehabilitation" the primary purpose of bridge inspections is to ensure public safety. The secondary purpose is to preserve the remaining life of bridge structures through the early detection and addressing of deficiencies. Federal law governs the requirements of the Bridge Inspection Program. The United States Code (23 U.S.C. 151) requires the Secretary of Transportation, in consultation with State transportation departments, to establish national bridge inspection standards for the proper safety inspection and evaluation of all highway bridges. These requirements are spelled out in the Code of Federal Regulations (Part 650, Subpart C) and govern the National Bridge Inspection Standards (NBIS) through purpose, applicability, definition of terms, qualification of personnel, inspection frequencies, inspection procedures, inventory procedures, and supporting references. Federal Highway Administration (FHWA) has developed 23 Metrics for the Oversight of the National Bridge Inspection Program. These metrics are a risk-based assessment of the performance of state bridge inspection programs and compliance with the NBIS. Each year, bridge Inspection programs are audited by the FHWA for compliance on these metrics. And yet bridge failures still occur [7].

9.2 Maintenance

With proper inspection and identification of maintenance requirements to preserve the integrity of bridge structural members deterioration of bridge components to the point of failure can be prevented. Corrosion, one of the leading causes of section loss in steel members and concrete reinforcement, leads to strength degradation and increases the risk of failure. Timely maintenance can prevent bridge deterioration and potential failure. Lack of bridge maintenance is the most preventable of all bridge failure causes.

9.3 Non-destructive testing

The use of non-destructive testing (NDT) on concrete and steel bridge components is useful in determining material condition. Of the many NDT methods available for the bridge inspector, visual inspection is one of the most effective. To test for voids and de-laminations in concrete the impact-echo method is effective in detecting substrate de-laminations. This method was applied with a small mobile impact machine to detect de-bonding of CFRP plates on the bridges in Macedonia [15]. NDT impact methods used in the periodic inspections of bridges provide significant data on the bridge condition [16].

10. Conclusion

Bridges deteriorate and bridges fail. The challenge for bridge owners is how to reduce the rate of bridge deterioration and to prevent bridge failures. Bridge failures are rare but they do occur. The goal of this chapter is to understand the nature of elements that cause deterioration in concrete and steel bridges and the effects these elements have on bridge structural components. The chapter presents an overview of the seven basic types of bridges used world-wide in varying configurations and lengths to meet requirements for specific locations. Components of a bridge, from foundation to superstructure are discussed, of which all bridge components are subject to deterioration in some form. The two primary elements of deterioration are water with deicing chemicals on concrete and steel, and heavy vehicle traffic. To preserve the structural integrity and service life of bridges comprehensive inspection, maintenance, and strong funding programs are required. Inspection emphasis must be placed bridges that are non-redundant and fracture-critical to prevent future failures.

Seven bridge failures and seven research papers are presented in this paper. As became evident in researching bridge failures, all bridge failures have a common characteristic: human error involving flawed designs, a lack of design review and construction oversight, lack of clear lines of authority, coupled with inadequate inspection and maintenance over the service life of the bridge. Bridge failures are preventable. By using the lessons learned in each bridge failure designers, engineers, and inspectors can prevent future bridge failures. Bridge failures are preventable.

Conflict of interest

The author declares no conflict of interest.

Author details

Kenneth C. Crawford
Institute of Bridge Reinforcement and Rehabilitation, Bloomington, IN, USA

*Address all correspondence to: ken.cawford@ibrr.org

References

[1] Wardhana K, Hadipriono F. Analysis of Recent Bridge Failures in the United States. Journal of Performance of Constructed Facilities. 2003;**17**(3). DOI: 10.1061/(ASCE)0887-3828(2003)17:3(144)

[2] Zhang G, Liu Y, Jiang L, Yang J. Causes and statistical characteristics of bridge failures: A review. Journal of Traffic and Transportation Engineering (English Edition). 2017;**9**(12):388-406. DOI: 10.1016/j.jtte.2021.12.003

[3] Trajber D, Penava F, Rinaudo LA. Construction types and typologies of existing masonry arch bridges in Croatia. IOP Conference Series Materials Science and Engineering. 2021;**1209**(1):012063. DOI: 10.1088/1757-899X/1209/1/012063

[4] Radic J, Ivankovic A, Kindij A. Analysis and shaping of cable stay bridges: An example of Jarun Bridge. GRADEVINAR (Builder). 2009;**61-9**:837-851, UDK 624.5.012.46.001.3

[5] Rajeshirke S, Suryawanshi Y, Khadake N. Comparative study of balance cantilever bridge and extradosed bridge. Article, Research Gate. 2022;**10**(V):3365-3375. DOI: 10.22214/ijraset.2022.43126

[6] Arioglu E. Importance of "Heuristics" in suspension bridge engineering and 1915 Çanakkale Bridge. Developments in International Bridge Engineering, Selected Papers from Istanbul Bridge Conference. Heidelberg, Germany: Springer Verlag; 2021

[7] Silano L, Henderson A. Bridge Inspection and Rehabilitation, A Practical Guide. New York, NY: John Wiley & Sons; 1993. ISBN 0-471-53263-2

[8] Contreras-Nieto C, Shan Y, Lewis P. Characterization of steel bridge superstructure deterioration through data mining techniques. Journal of Performance of Constructed Facilities, ASCE. 2018;**32**(5)

[9] Salem H, Helmy H. Numerical investigation of collapse of the Minnesota I-35W bridge. Elsevier, Engineering Structures. 2014;**59**(2014):635-645

[10] Moehle J. Preliminary Observations on the Performance of Concrete Freeway Structures. Berkeley: National Information Service for Earthquake Engineering, Univ of California; 1999. Available from: www.eerc.berkeley.edu/loma_prieta/moehle.html

[11] Yashinsky M. "Cypress Street Viaduct", US Geological Survey Professional Paper, No. 1552-8. Library of Congress catalog-card No. 92-32287. 1998. pp. 19-26

[12] Monteiro P, Asselanis J, MacCracken W. Investigation of the microstructure and mechanical properties of the structural materials of the I-880 double-deck viaduct. ACI Materials Journal. 1991;**88**(3):288-293. ISSN 0889-325X

[13] Sébastien Savard—Own work, CC BY-SA 2.5. Available from: https://commons.wikimedia.org/w/index.php?curid=836550

[14] Heggade V. Lesson to be Learnt from Bridge Failures. The Bridge and Structural Engineer, IABSE. 2014;**44**(4):4-28

[15] Crawford K. NDT evaluation of long-term bond durability on CFRP-structural

systems applied to RC highway bridges. International Journal of Advanced Structural Engineering, Springer. 2016;**8**(2):161-168

[16] Sansalone M, Street W. The impact-echo method. NDTnet. 1998;**3**(2):3-5

Chapter 15

Remote Assessment of the Serviceability of Infrastructural Assets

Ikwulono D. Unobe and Andrew D. Sorensen

Abstract

The continued serviceability of structures and infrastructural assets over time is a very important component of design and construction. Traditionally, structural health monitoring techniques utilize sensors and field inspections to capture distress and defects in structural members to allow for repairs and retrofitting prior to catastrophic collapse. However, these processes are usually quite expensive (financially and time wise), making it difficult to assess all assets. Rather, it necessary to identify specific assets requiring field investigations. To do this, an ability to remotely compute the real time serviceability of these structures and periodically update their condition to determine the criticality of field inspections is important. Several methods and processes have been proposed for this purpose with respect to different loading conditions and type of structures. This chapter will offer insight into remote monitoring of structures and infrastructure over time as they sustain damage from varied loading conditions.

Keywords: infrastructure, remote monitoring, reliability analysis, Bayesian updating, infrastructural decay, structural health monitoring, wind loads

1. Introduction

Structures are primarily designed to have the capacity to withstand certain loads they can be expected to bear. Over their design lives, structural members can be expected to deteriorate, with an attendant decrease in their capacities. This is generally as a result of exposure to environmental factors, aging of the constituent materials, as well as the residual effects of loading conditions. Such degradation of structures occurs in a number of ways including fatigue, corrosion, cracking, and scour. Thus, maintaining infrastructure to ensure its continued ability to fulfill design criteria is very important for the safety and security of a society. To this end, accurate knowledge of the in-service state of infrastructural assets cannot be trifled with. This brings to the fore, the pertinence of periodic inspection and maintenance for these assets, especially those designed to last for a long time, to ensure they continue to meet specified design requirements.

The process of inspecting or monitoring structures with a view to identifying and assessing damage to them is referred to as structural health monitoring (SHM). Principally, the inspection aspect is carried out by visual means. However, such visual inspections have a number of shortcomings including subjectivity of the inspecting personnel, the ability to only capture surface damage in visible locations, and the limitation of the inspection to damage without a congruent monitoring of loads applied to the structure. In addition, the sheer volume of infrastructural assets requiring such inspections and maintenance precludes the possibility of thorough inspections being carried out regularly on most, increasing the possibility of critical information getting missed during cursory examinations. For example, the Federal Highway Administration specifies biennial routine inspections for every bridge within its purview. These inspections are primarily designated to be carried out via visual inspection, with more advanced equipment and methods only deployed for more in-depth inspection if deemed necessary. However, with about 600,000 bridges included in this inspection regime, rigorous visual inspection of each bridge becomes a very difficult task to achieve within the projected time frame, leading to increased risks of failure.

With the unreliability of such conventional lifetime assessment methods [1, 2], SHM has become a very important tool for assessing the lifetime of a structure. To curtail the stated shortcomings in the inspection and maintenance process, it became necessary to expand the original scope of SHM frameworks. In redefining the SHM process, Ref. [3] asserted that a monitoring system should include four operations namely acquisition, validation, analysis and management. Ideally, a single SHM system would collect information on both loads and system response to the loading. In addition to monitoring duties, an ideal SHM system would also incorporate some prognostic methods that will allow for the damage levels to be evaluated and the in-situ health of the structures to be determined. This will allow for an analysis of the present and future performance of the infrastructural assets. Integrating a variety of disciplines including material science, non-destructive evaluation, fracture and fatigue mechanics, structural dynamics and structural design, SHM frameworks can be designed to collect information on deflections and strains, system behavior, thresholds of systems and members, input values for lifetime assessment, and maintenance planning [1]. To this end, SHM frameworks originally designed for capturing and assessing the initiation and propagation of damage, have been expanded to collect other information related to the performance of the structure [1, 4]. To prevent damage to structures, the monitoring or data acquisition stage of SHM is carried out principally using non-destructive evaluation (NDE) techniques. These techniques allow for the evaluation of the integrity of materials and structures without compromising their continued abilities to meet design criteria. Used widely for detecting and characterizing defects and damage in infrastructure, NDE techniques have been extensively researched and offer a useful means of collecting information for SHM to adequately characterize the in-situ health of infrastructure.

A holistic monitoring system must be able to provide sufficient information for users to make decisions on the continued serviceability of the structure, and on maintenance, repair and replacement regimens. With the plethora of tools developed for capturing damage and structural response, the determination of the presence of damage has become a relatively trivial task. However, the extrapolation of the effect of the damage on the serviceability of the structure remains a relatively complicated undertaking. Currently, most systems deployed for SHM of infrastructural assets are essentially either sensor arrays or other equipment used to identify the existence of damage or distress in the structures under investigation, without requisite provision for quantifying the damage and defining the residual capacities of the structures. Thus, while providing good data on the loading of a system, and the damage incurred, these systems do not provide information on its continued serviceability, making manual inspections in the SHM process irreplaceable. To this end, SHM systems remain a precursor to physical inspections, giving inspectors an idea of the state of infrastructure and a basis for the determination of inspection and maintenance schedules. This brings to the fore the importance of accurately analyzing the data collected from the SHM systems.

Two approaches are used for the analysis and interpretation of the data collected, and primarily differ in the use or avoidance of a physics-based model for analyzing the behavior of the structure [5]. Model based approaches involve updates to models that capture damage and eroding capacity with respect to applied loads to reflect changes in the structural parameters observed from collected data [6–8]. In analyzing the strengths and weaknesses of the approaches, Ref. [5] determined that rather than being considered alternatives, both systems should be considered to be complementary, depending on the needs and requirements of the SHM system. Although well established, these methods are developed using idealizations of the structural behavior, without taking into considerations uncertainties in materials, geometry and loading conditions in analyzing the data collected. These uncertainties could lead to discrepancies between projected structural behavior and the actual behavior, possibly creating incorrect assumptions on the state of a structure. To this end, such deterministic methods remain at best approximations of the condition of a structure and must be used with caution. To overcome this shortcoming, a probabilistic process that accounts for these uncertainties, thus allowing for a more realistic estimation of the state of the structure, and its reliability can be applied. This inclusion of reliability methods in SHM processes, enhances the ability of monitoring systems and components in real time, and allows for the introduction of predefined alert levels to trigger specified actions once a value dips below a critical reliability index [5].

Beyond the use of reliability methods, Bayesian updating processes have become a popular means for updating the state of structures in SHM frameworks. These processes, utilize the data collected alongside prior knowledge on the performance of the structure to make inferences on its current state and future performance. The incorporation of Bayesian updating processes, moves SHM frameworks from being monitoring systems to becoming more holistic systems, inculcating both monitoring and analysis into making decisions on the condition and future performance of infrastructural assets.

Some steps are important for such SHM frameworks. These are obtaining data on the loading and response of structures, characterizing damage or distress from this information, analyzing the information, and making conclusions based on the characterizations. These steps are carried out to get a good grasp of the state of the structure, prior to full on-site inspections. However, with the cost of permanent sensors, it is quite unfeasible to instrument all infrastructural assets. The alternative, periodic inspections are also impractical due to the number and geographical spread of these assets. To counteract these challenges, this chapter proposes a trade-off between both. This alternative involves the use of existing and regularly updated data such as wind speeds, traffic information and ground motion from seismic events to extrapolate the condition of infrastructural assets exposed to these conditions, and update their conditions, allowing for the optimization of an inspection regimen. To this end, the objectives of this chapter include:

1. The identification of loading scenarios with readily available data and the extrapolation of such data to site specific conditions for the locations of infrastructural assets.

2. The development of a procedure for the use of these loading conditions to determine the in-service state of the structures.

3. The determination of a methodology that will allow for periodic updating of the state of these assets using the collected information and prior knowledge of the performance of the assets.

To meet these objectives, the rest of the chapter is designed to begin with a background and overview of the SHM process, and then focuses on a methodology designed to meet each objective. A case study example of traffic signal structures is included to demonstrate the use of the proposed methodology.

2. Overview of structural health monitoring

Designed principally to determine and assess structural integrity, structural health monitoring has continued to evolve with improvements in the technology and processes used in achieving the said goals. Some fundamental bases of SHM include an assumption that all materials have inherent flaws, and the necessity of at least two system states in order to assess damage. As an important tool for assessing the condition and lifetime performance of a structure, Ref. [9] opined that systems which only include sensors deployed on structures without a definition or classification system for the damage cannot truly be classified as SHM systems. The study stated that a true SHM system would include a quantifiable and pre-established definition of damage to be detected by the sensors. Also, in the development and deployment of a SHM system, a classification process for the identification of damage and assessment of its extent needs to be defined [9]. To improve the practice of SHM for infrastructure, Ref. [3] proposed a condition-based assessment framework for the management of bridges. This process, would ideally provide information on damage to the structure and erosion of structural resistance, as well as the probability of the structure's performance falling below a set standard and an estimation of the remaining useful service life. An SHM system encompassing these parts would arm inspectors with adequate information on the service condition of these infrastructural assets, allowing them to make decisions on the repairs/retrofitting, while drastically cutting down on the need for periodic manual inspections. There are five principal steps in a SHM process. These are detection, localization, classification, assessment and prediction [9, 10]. While the first two steps involve utilizing sensors and other equipment for monitoring via non-destructive evaluation of the structure under investigation, the last 3 steps involve analyses of the data collected from the monitoring process. Thus, it can be said that an optimal SHM process consists of monitoring of structural behavior, and analyzing this data for a proper prognostication of structural health.

2.1 System monitoring

Monitoring of structures for SHM involves use of instrumentation and processes for the detection and localization of damage. With the inability of sensors to measure damage, the data collection at the system monitoring stage is qualitative in nature, giving useful information on the presence and type of damage present in a structure, but not offering quantitative information on the extent of the damage and the remaining useful life of the structure or structural member. Most sensor systems deployed for the purpose of SHM consist principally of this monitoring process, without going further to quantify the damage and predict remaining useful life of the structures [9]. Although only one part of a holistic SHM process, the importance of proper deployment of sensors and instrumentation towards damage detection and monitoring cannot be overestimated. With sensing systems, there is a trade-off between sensitivity to damage of a sensing system and its noise rejection capability, and also the size of damage detectable is inversely proportional to the frequency range of excitation [9]. Generally, the length and time associated with the initiation and evolution of damage dictates the properties of the sensing system to be used. Non-destructive evaluation techniques have been widely investigated for this purpose. These techniques can be deployed in an online manner for continuous monitoring, or in an offline manner for inspection purposes. NDE methods researched and deployed for SHM include vibration-based methods [11–13], optical based methods [14], radiography [15], ultrasonic testing [16, 17], acoustic emission [18, 19], electromagnetic methods [20–22], magnetic particle inspection [23, 24] and thermographic methods [25–28]. These methods are usually used to capture damage in either continuous monitoring schemes or in inspection regimes. Several types of damage are usually the targets of such non-destructive evaluation processes. Ref. [13] in a follow up of a study by Ref. [29], determined that the material and geometric changes that can be characterized as damage by these systems include cracks, corrosion, buckling, creep, fastener loosening and loss of preload, debonding, delamination, microstructural degradation, and pull-out.

2.2 Prognostication of structural health

Beyond identifying the existence and type of damage, it is vital for SHM processes to adequately characterize the damage to determine the continued serviceability of the structure in the presence of these defects/damages. Determining the continued health of the structure and its constituent members involves analysis of the captured damage, and its effect on structural capacity. This requires the identification/development of damage models that accurately represent the initiation and propagation of the identified damage, and the degradation of capacity due to the damage propagation. Ref. [13] opined that a typical SHM scheme for development of damage models includes six principal steps including identification of the damage mechanism, identification of structural parameters affected, modeling of structure, sensitivity studies on the model, modeling of the damage mechanism, and sensitivity studies of variations of different parameters.

There are four possible levels of analyzing data collected from a SHM system to characterize the health of the structure. These are inferences using the raw data, analyzing the data to detect damage, deterioration or changes in structural behavior, localizing and quantifying the damage and/or changes, and predicting future performance based on current condition [5]. The first two levels do not usually require very rigorous levels of analysis, beyond quite straightforward determination of a deviation from the original state of the structure as an indication of damage. The last two levels on the other hand usually require an extensive examination of the data in a bid to determine the condition of the structure, the level of damage and distress, its residual capacity and the future performance that can be expected of it. To determine these levels, reliability analysis (level 3) and Bayesian updating (level 4) are used to make inferences from the data collected.

2.2.1 Reliability analysis

A very important aspect of structural health monitoring involves the use of the damage incurred by, and capacity of the structure to estimate its remaining useful life i.e. the residual capacity for specific loading scenarios. The determination of deterioration levels and changes in structural capacity, require both a characterization of damage as well as a proper determination of the effect of the damage on the structural capacity. Several studies have offered methods into using damage models to estimate residual capacities for structures and their members. These include estimation of remaining fatigue life [30, 31], post impact capacity [32], residual capacity post corrosion initiation [33], etc. In these studies, the computation of the residual capacities is done in either a deterministic or a probabilistic manner. Deterministic estimations of residual capacity are quite straightforward, and relatively uncomplex. However, with the presence of uncertainties in both loads, material properties and geometry of structural members, such deterministic estimations may not offer an accurate estimate of the damage and residual capacity of a structure.

To curtail the disadvantages of deterministic computations, probabilistic methods have been developed which consider the uncertainties associated with the different parameters involved in the load and resistance of structures. Two types of uncertainties are commonly encountered namely aleatory and epistemic uncertainties [34]. Aleatory uncertainty refers to uncertainties that cannot be reduced or minimized such as uncertainties associated with material or geometric properties. Epistemic uncertainties on the other hand refer to uncertainties which can be minimized as more knowledge is gained of the system and more care is taken in modeling it to closer reflect its actual state. Uncertainties from sensor measurements can be said to be epistemic, and can be reduced with the development or acquisition of more precise instrumentation.

Structural reliability in utilizing a probabilistic approach to define the performance of a structure, takes cognizance of uncertainties relating to different aspects of design and construction, thus improving the accuracy of analytically deduced performance of the structure [35, 36]. Fundamentally, reliability analysis sets out to determine the likelihood of a structure's performance failing to live up to its design criteria [37]. This analysis principally revolves around the definition and evaluation of a limit state. Limit states are commonly designed around safety, serviceability or durability criteria and define the boundary for acceptable performance and failure [3]. These limit states are evaluated to determine the likelihood that a structure or structural member fails to meet a specific design criterion (probability of failure). This probability of a system failing to meet the specific performance criterion is defined generically as shown in Eq. (1) below [38].

$$P_f = Pr[g(\boldsymbol{x}) < 0] \tag{1}$$

where: P_f is the probability of failure, $g(x)$ is a performance or limit state function and $\boldsymbol{x}$ is a vector of all the random variables included in the limit state function.

In reliability analysis, the reliability of a structure is quantified using a reliability index. This index, is a measure of structural reliability and captures the inherent influence of parameter uncertainties [39]. In a SHM framework, this reliability index becomes a quantified measure of the structure's likely performance under the loading scenario in question, and allows for more informed decision making on the structure. Used as a defining parameter for condition assessment, a reliability index can be defined as a decision criterion for structural performance, with the dropping of a structure's performance below this limit indicative of a need for immediate inspection and possible remedial actions. Inculcating a reliability analysis into the SHM framework will thus bridge the gap between capturing damage and distress, and extrapolating the effects of these captured damages on the performance of the structure.

These probabilistic methods have been used to good effect in several studies to characterize the expected behavior of structures under expected loading over the lifetimes of these structures [33, 40–43]. However, some of these studies assume the initial state of the structure is undamaged and do not update the probabilistic model to account for damage and eroding capacity during its design life. Others that do account for eroding capacity utilize specific models for certain time dependent loads that cannot be extrapolated to other loading types. In addition, although useful for estimating the performance of a structure at the set point in time, reliability analysis on its own does not give a good indication of future performance. Model updating i.e. the updating of the models describing a structure and its performance with new information collected from the structure is needed to do this. A commonly used model updating method is Bayesian updating.

2.2.2 Bayesian updating

Structural model updating involves the adjustment of a theoretical model to reflect the responses garnered from the actual structure, and thus improve the ability of the model to characterize the behavior and response of the structure to applied loads. For a model updating procedure to be useful in practice, it must be able to handle noisy data, relatively small datasets, errors in the model, incorporate existing knowledge about the structural system performance, and be insensitive to distributions of model parameters [44]. Bayesian techniques are particularly advantageous for model updating as they meet these criteria, thus making them ideal for use in SHM.

Used in many fields for the updating prior probabilistic models with observed data, Bayesian updating provides a consistent framework for introducing new information into existing probabilistic models towards improving their accuracy. This technique, by balancing prior information with observed data, allows for the proper estimation of posterior distributions of uncertain parameters, ensuring that logically consistent inferences can be drawn, and used in prognostic models.

In SHM, Bayesian techniques allow information gleaned from inspections as well as monitoring regimes to be combined and used in better predicting future performance of structures [41]. Using a Bayesian framework, Ref. [45] proposed a process for model updating to reflect the changing state of structures as new information on their state is gleaned from sensors attached to them. In following other model updating procedures, this process makes three base assumptions. These are the existence of variations within the model parameters, the understanding that the models only approximate actual systems and their behaviors, and the knowledge that the model and its corresponding system may be more sensitive to some parameters than to others [45]. This approach to model updating has some peculiar advantages including its explicit consideration of uncertainties, as well as the ability to incorporate both prior information and newly obtained data into the prediction process [41]. This

Bayesian updating process requires knowledge of a number of compositional parts including quantifiable damage level of the structure, or a damage model that allows this quantified damage to be computed, a capacity model, and a relationship between the capacity and damage, defining the erosion of capacity with increasing damage.

To utilize this updating procedure, two approaches are generally considered. The first involves the implementation of a monitoring scheme to assess a quantifiable damage level, and using this in a reliability model to compute a probability of failure. The second method is the use of monitoring schemes to monitor the load exposure of the structure and then use this in a damage model to estimate the damage done prior to the determination of updated probabilities of failure from a reliability model. Ideally, both these methods require some level of monitoring to collect information on the structure under investigation. As such, the first step in designing a Bayesian updating scheme for the health of in-service structures is the determination of data collection on loads and/or system response to the loads.

2.3 Remote assessment of serviceability state

Data collection for use in a SHM Bayesian updating framework usually involves the use of set in place sensors or the periodic inspection of the structures using mobile equipment/tools. While permanent sensors to collect data on infrastructural assets is an appealing idea, the sheer cost and logistical challenge of installing these sensors and maintaining them for all infrastructural assets limits the possibility of the use, especially for relatively low-cost infrastructural assets. Similarly, periodic inspections while useful are curtailed by the aforementioned logistical and practical challenges. These shortcomings underscore the challenges of implementing such updating schemes. To this end, a remote monitoring regime utilizing information which can be collected without the installation of onsite monitoring tools, and used to assess the state of the infrastructure would be very beneficial. Such a regimen, can be used in optimizing inspection intervals, thus reducing the cost of inspection and maintenance while also minimizing risk of abrupt failures.

The development of such remote monitoring and assessment methods, requires some very specific information. It is necessary to first identify possible types of damage/loading which can be remotely assessed without onsite sensors or equipment. Two possible loading scenarios which can be monitored in such a manner are wind loads and seismic loads. Wind loads are determined from wind speeds, and stations capturing this phenomenon are quite common around towns and cities. Seismic loads are captured from seismological stations and are quite similar for relatively large areas. With the proliferation of locations with equipment collecting data on these load types, it is quite possible to collect data from various sources and aggregate this data into loading history for infrastructural assets at a site of interest, and then use this information to obtain the state of the infrastructure. Periodically carrying out this process, timely information can be obtained and the Bayesian updating process used to update the condition of the asset in question.

Some studies have offered ideas and proposed paths towards such remote assessment of infrastructure. Ref. [31] touted the possibility of aggregating wind speeds collected from stations around a site of interest into the wind speeds for the location, and using these wind speeds in analyzing the damage to traffic ancillary structures. Ref. [30] took this idea further in developing a framework using historic wind speed data gleaned from different locations into making predictions on the remaining useful fatigue lives of these traffic structures. Ref. [46] similarly utilized historic wind data to determine the fatigue state of high mast illumination poles. Ref. [47] presented a framework for rapid assessment of seismic damage to bridges. This study combined probabilistic analysis with a machine learning algorithm to predict likely damage to bridges in the event of a seismic event, so as to optimize decision making on what to do about the said bridges after an earthquake. These studies aimed to help optimize inspection and maintenance regimes by offering a means of obtaining an idea of the state of these infrastructural assets and their remaining useful lives prior to scheduling inspections and repair/retrofitting actions. While offering cogent processes for the remote assessment of infrastructure, these studies stopped short of using a Bayesian updating process and thus while giving an estimate of the state, cannot be used to continually update the state of the structure over time.

Incorporating a Bayesian updating process into reliability-based decision analysis process for bridges, Ref. [48], determined that including prior information on the performance of the bridges had a telling effect on the resulting reliability analyses. Further buttressing the utility of a Bayesian updating process which incorporates such information in analyzing current and future performance [49], offered a process for estimating the remaining useful life of a structure after developing fatigue cracks using a Bayesian updating process. This study identified parameters leading to damage growth, and utilized simulations and Bayesian inference to identify unknown parameters that will allow for an accurate estimation of the fatigue growth rate.

In developing a framework for remote asset management, some criteria need to be met. These include the consideration of uncertainties in material and model parameters, the ability to leverage historic loading information in determining the condition of the structures, and beyond that, the ability to predict future performance based on the prior knowledge of the state, and updated information collected about the structure. To design a Bayesian updating framework for remote assessment, this study proposes laid out in the next section.

3. Methodology

The methodology for the remote asset management is very similar to that for any SHM process, and can be divided into two parts namely data collection and data analysis. The data collection process is relatively straightforward, and begins with an identification of the requisite information needed for the assessment. This data comes in form of historic information on the loading patterns, collected from monitoring stations within the vicinity of the location of interest. Cleaning the data and aggregating the information into a usable form specific to the location of interest can then be carryout using processes specific to the type of data. Data analysis on the other hand requires extensive knowledge about geometry of the structure, material properties and model parameters as detailed in the following sections. The overall process is as follows:

1. Determine the load and damage mechanism to be investigated.

2. Identify monitoring stations collecting the needed data in the vicinity of the site in question.

3. Collect the data and aggregate into site specific data.

4. Compute the damage using probabilistic methods.

5. Periodically update the reliability using Bayesian methods and timely information from monitoring stations.

6. Schedule inspections once the reliability falls below a predetermined threshold.

7. Update the reliability models with results from inspections and/or maintenance processes.

Details on each step in the process are laid out in the following sections.

3.1 Data analysis

In SHM, two levels of analyzing information collected on the state of a structure require extensive examination of said data. These are the levels of quantifying damage/changes to the structure and that of predicting future performance based on the latest information on condition [5]. Reliability analysis can be used for the first level, and Bayesian updating off the results from the reliability analyses used for the second.

Carrying out a reliability analysis of the structure in question requires a number of steps. To begin with, a limit state equation including both resistance and damage models needs to be ascertained. These models should ideally be models that include measurable parameters, which can be captured via a site inspection. As such, when a structure is flagged for inspection, measurements made during said inspection can be used in eventually updating the condition of the structure for future estimation of its condition.

After determination of a limit state, a method of solving the probabilistic problem needs to be decided upon. Popular methods for solving such problems include first order reliability method (FORM), second order reliability method (SORM), Monte Carlo simulations, Markov Chain Monte Carlo simulations, Hasofer-Lind procedure, and Rackwitz- Fiessler procedure. These methods each has its pros and cons and a determination of which would be ideal for the particular set of circumstances is needed. The process for the reliability analysis is as follows:

- Identify a limit state equation including both capacity and demand models appropriate to the scenario under investigation.

- Identify all nondeterminate variables in the limit state, their distributions and parameters.

- Compute the reliability index or probability of failure using these parameters and the limit state function in one of the methods mentioned above.

The relationship between the probability of failure and reliability index can be described using the normal cumulative function as shown in Eq. (2) [35, 38].

$$P_f = \Phi(-\beta) \qquad (2)$$

where Φ is the standard normal cumulative function.

The reliability analysis giving a probability of failure and a reliability index for the structure offers insight into the reliability parameters correlating to the point-in-time condition of the structure i.e. a Level 3 type assessment of the state of the structure.

Bayesian updating using the prior information on historic loadings, as well as fresh information from either the reliability analysis or from inspections can then be used to update the reliability models for a prediction of the possible future performance of the structure given the point-in-time knowledge of its condition. Based on the Bayes' theorem of conditional probability, the underlying idea can be used to update a quantified characteristic state of a structure, using the Bayesian framework as shown in Eq. (3) [48].

$$p_f(b|a) = \frac{p_f(a+b) - p_f(a)}{1 - p_f(a)} \tag{3}$$

where $p_f(b|a)$ is the probability of failure in b subsequent years given that it has survived a number of years, and $p_f(a+b)$ and $p_f(a)$ are the probabilities of failure in time a + b and a respectively.

Updating the probability of failure using timely information on loading, and /or from inspections, would allow for continuous monitoring of the condition, and prediction of future performance, offering a cheap and quick way to remotely obtain insights into the condition of infrastructural assets.

3.2 Case study

To demonstrate the laid-out methodology, a case study is presented. This illustrative example involves two traffic signal structures placed in different orientations at the same location. Selected from the cases presented in Ref. [30], these represent structures at a location which showed significant damage from wind forces in the aforementioned study. Installed in 1997, these traffic signal structures are in a location with significant wind loads a explained in Ref. [30]. To this end, the analysis carried out in the study, showed that each is expected to have degraded significantly, making them good candidates to test the remote assessment strategy laid out in this study. Both traffic signal structures are cantilevered structures, with the mast arm extending from a single pole, which also has a luminaire post over it.

3.2.1 Loading model

The first step in the methodology involved collecting data pertinent to the structure under investigation. For the traffic signal structure, the principal type of load affecting its performance relates to the wind forces acting on it. To this end, wind speeds were collected from different weather monitoring stations in the vicinity of the traffic structures. A process laid out in Ref. [30] for cleaning the data to get rid of outliers, incomplete data and erroneous readings was then used to obtain a dataset that representative of the historic wind speeds in the general area. These were then converted into hourly wind data using the Durst curve. Next, these wind data were aggregated into site specific wind data using the process described in Ref. [31]. The equation used in obtaining the wind data specific to the site is as shown in Eq. (4).

$$S_d = \frac{\sum_{i=1}^{n} \frac{S_{d,i}}{R_i}}{\sum_{i=1}^{n} \frac{1}{R_i}} \tag{4}$$

where S_d is the wind parameter for a specific time period d, n is the number of weather stations used in the interpolation process, $S_{d,i}$ is the wind parameter for the time period d at the weather station i, and R_i is the distance of weather station i from the site of interest.

The synthesized wind data collected and interpolated for the location as described above led to the computation of approximate historic wind information for the site, and thus the wind forces the structures located therein are expected to have borne over their service lives. With the historic wind forces acting on the structure collected, the next step involved the determination of stresses from these forces at critical locations, and the response of the structure to these loadings.

Assuming that the cyclic wind forces on the structure will lead to fatigue at certain critical locations on the structure, the stresses at identified critical locations due to the wind forces were computed. The base of the mast arm and the base of the pole were selected as critical fatigue locations as a number of studies have pinpointed these locations to be fatigue critical given the concentration of stresses there. The deterioration of the connections at these locations were then analyzed and used as a defining parameter for the service states of the traffic signal structures.

3.2.2 Deterioration model

To ensure ease in updating using information gleaned from site inspections, it is imperative that the structural degradation model used includes measurable degradation parameters. Prior studies on wind fatigue degradation of similar structures made use of the Miner's rule for cumulative fatigue damage in analyzing the fatigue damage. Although a valid process for estimating fatigue damage, this process does not give observable parameters, and would not be updateable using results from site inspections. To this end, a fracture mechanics approach for crack propagation is used in this study instead, and a limit state function defined related to the crack propagation through the weld at both critical locations as shown in Eq. (5) [50].

$$g(X,t) = \int_{a_i}^{a_f} \frac{da}{\left(Y(a)\sqrt{\pi a}\right)^B} - CS_R^B N \leq 0 \tag{5}$$

where a_i is the initial crack length, a_f is the crack size associated with failure, Y(a) is a geometry function accounting for shape of specimen and mode of failure, C is a material property, B is an equivalent damage material property, S_R is the equivalent stress range, and N is the number of stress cycles.

The limit state equation was then evaluated using the statistical parameters for the random variables shown in **Table 1**. This process was used to compute the annual reliability and probabilities of failure for the traffic signal structures. These can be expressed as the point in time probabilities of failure which do not consider the previous year's probability of failure. Next, an updated probability of failure for each year is computed using a Bayesian updating process as expressed in Eq. (3).

Parameter	Mean	Coefficient of Variation	Distribution
Thickness of Pole	0.18	0.25	Normal
Initial crack length	0.004–0.01	1	Normal
Stress	Variable	0.1	Normal
C	1.294×10^{-12}	0.5	Lognormal

Table 1.
Random variables used in the limit state equation.

3.2.3 Results

Cumulative probabilities of failure are obtained for the 25 year span the traffic structures have been in service. For the reliability analyses, failure is deemed to have occurred if the crack in the weld extends to the thickness of the tubular pole or mast arm. Assuming prior knowledge of the existing level of deterioration of the structure via a knowledge of an existing crack and the corresponding lengths, **Figures 1** and **2** show the annual reliabilities of the traffic structure, as a function of the service age. The influence of time on the reliabilities can be seen with the continual degradation of the reliability indices over time, irrespective of the initial size of the crack. However,

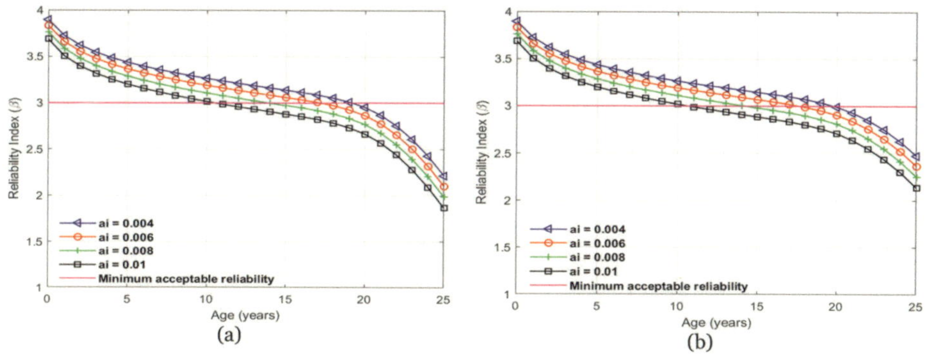

Figure 1.
Annual point-in-time reliabilities for pole to baseplate connection for (a) traffic structure 1, and (b) traffic structure 2.

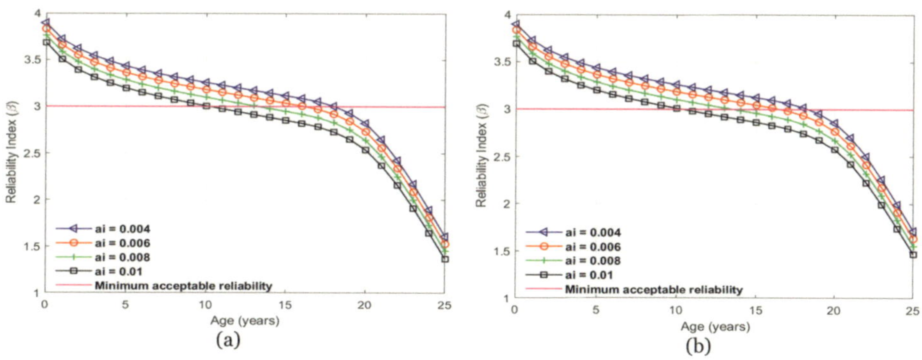

Figure 2.
Annual point-in-time reliabilities for mast arm to baseplate connection for (a) traffic structure 1, and (b) traffic structure 2.

the initial size of the defect (ai) also has a telling effect on the structures and the time until they require inspections and/or maintenance. For example, for the pole to base-plate connection o traffic structure 2, assuming a reliability index of 3 is the determined point at which an inspection becomes necessary, the traffic structure would be due for inspection in year 16 assuming an initial crack size of 0.08 to 0.1 inches, but would only be due for inspection in year 19 for an initial crack size of 0.04 inches. However, post inspection, these point-in-time reliabilities are not updated with the results of the inspection/possible maintenance, and thus they will no longer represent ground truth.

Annual reliability indices, computed with an inclusion of the influence of prior knowledge of the performance of the structure in the preceding years is shown in **Figures 3** and **4**. These indices are compared with those computed using the cumulative stress but without an inclusion of the previous performance shown in **Figures 1** and **2** for an initial crack size e of 0.01 inches.

The results for the updated reliabilities show the effect of prior knowledge on the reliability of a structure. The reliability and probability of failure for each year includes the prior knowledge that the structure did not fail in the previous year. From the results in **Figures 3** and **4**, it can be observed that the annual point-in-time reliability indices are significantly less conservative than the Bayesian updated reliabilities. Similar to the conclusions drawn in Ref. [48], not taking prior performance

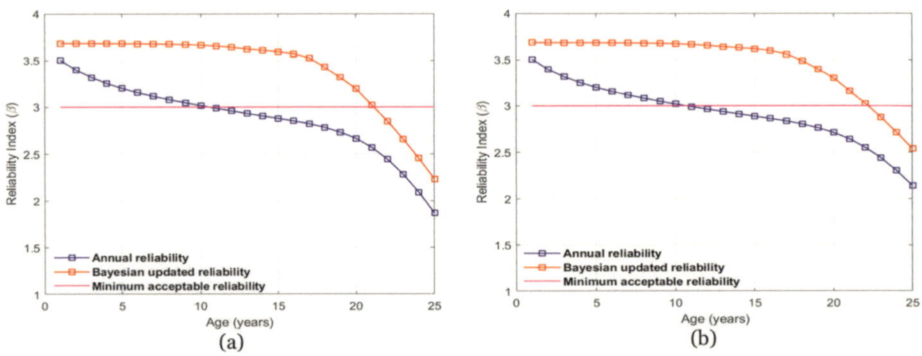

Figure 3.
Comparison of Bayesian updated reliabilities to annual point-in-time reliabilities for pole to baseplate connection for (a) traffic structure 1, and (b) traffic structure 2.

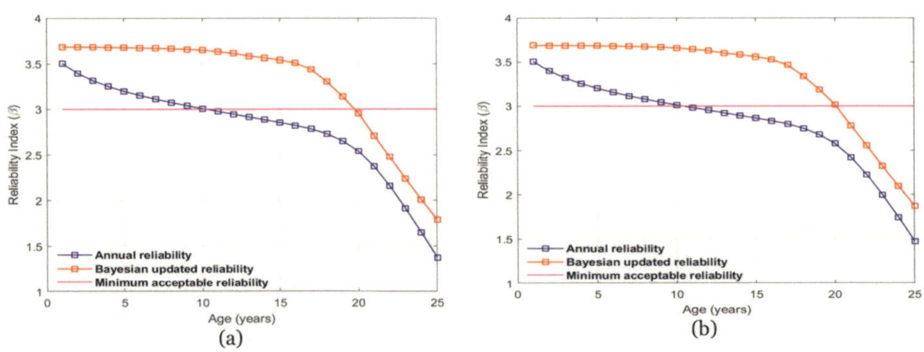

Figure 4.
Comparison of Bayesian updated reliabilities to annual point-in-time reliabilities for mast arm to baseplate connection for (a) traffic structure 1, and (b) traffic structure 2.

into account results in conservative reliability indices which may not be truly reflective of the performance of the structure. For example, at 25 years, the reliability index of traffic structure 1 is 1.87 with a corresponding probability of failure of 0.03. Comparatively, the Bayesian updated reliability index at this time is 2.23 and the probability of failure is 0.013. Essentially, this means that while the point-in-time reliability predicts that the likelihood of the structure failing in the next year as 3%, the Bayesian updated reliability gives a less conservative estimate of 1.3% probability of the structure failing. This seemingly less conservative result is because known information about the prior performance of the structure (i.e. not failing in prior years), is used in the Bayesian updated reliability but is not used in the point-in-time reliability estimate. Thus, while giving seemingly unconservative results, the Bayesian updating method does offer a realistic insight into the condition of the structure, considering its known performance. In addition, this method offers the flexibility of incorporating post inspection information, into an updated reliability for the structure.

4. Future work

This study set out to offer a framework for remote asset monitoring and assessment. Although a framework is laid out, there is still significant work to be done for widespread use.

A critical aspect of this framework involves the use of easily accessible data pertaining to the loading conditions related to the location of the structure. Wind data is used in the case study presented. However, it is important to identify data related to other types of loading conditions that can be obtained conveniently and updated regularly. It should also be possible to extrapolate these data to reflect the conditions of the site of interest, and use them in damage/deterioration models for the structure.

Damage and capacity deterioration models which can be used with such easily accessible data and which also capture measurable damage need to be identified for different loading scenarios pertinent to infrastructural assets. In addition to utilizing collected data, these models need to characterize damage in a way that can be physically measured in order to allow for updates to be made to the model from field inspections and maintenance work carried out.

Although offering a handy way to remotely assess infrastructural assets and update their conditions, the damage and deterioration models used in this study as well as in other studies on using reliability analyses are commonly based on idealizations of the structural systems. More field data is needed to determine the correlation between the structural systems and these models, in order to improve their predictive capabilities.

5. Conclusions

In a bid to optimize the SHM process, this study set out to offer a framework for the remote monitoring of infrastructural assets prior to scheduling field inspections and maintenance programs. Based on a Bayesian updating process, a process of obtaining remotely accessible data, extrapolating site-specific conditions from this data, and computing the time dependent performance indices for a structure at a specific location was laid out. Using a pair of traffic signal structures as a case study, the Bayesian updated reliabilities and the point-in-time reliability indices were computed. The point-in-time reliability indices offered more conservative results, due to the indices not considering the prior performance of the structure. The Bayesian updated reliability indices on the other hand accounted for the past performance of the structure and it's continued serviceability. Beyond determining its current serviceability, the Bayesian updating process also provides room for including results from field inspections and/or maintenance work in future performance indices, which the point-in-time reliability does not, further buttressing its value in remote assessment frameworks.

Author details

Ikwulono D. Unobe* and Andrew D. Sorensen
Department of Civil and Environmental Engineering, Utah State University, Logan, UT, USA

*Address all correspondence to: david.unobe@usu.edu

References

[1] Peil U. Civil infrastructure load models for structural health monitoring. Encyclopedia of Structural Health Monitoring. 2009;7:131-157

[2] Schütz W. Fatigue life prediction by calculation: Facts and fantasies. Structural Safety and Reliability. 1994;**1**: 1125-1131

[3] Karbhari VM. Design principles for civil structures. Encyclopedia of Structural Health Monitoring. 2009;**84**: 1388-1397

[4] Boller C. Structural health monitoring—An introduction and definitions. Encyclopedia of Structural Health Monitoring. 2009;**1**:1-23

[5] Catbas FN. Structural health monitoring: Applications and data analysis. In: Structural Health Monitoring of Civil Infrastructure Systems. Cambridge, UK: Elsevier; 2009. pp. 1-39

[6] Doebling S. Damage Detection and Model Refinement Using Elemental Stiffness Perturbations with Constrained Connectivity. Salt Lake City, UT, USA: Adaptive Structures Forum; 1996. p. 1307

[7] Friswell MI, Penny JET. Is damage location using vibration measurements practical. Euromech 365 International Workshop: Damas. 1997;**97**:351-362

[8] Zimmerman DC, Simmermacher T. Model correlation using multiple static load and vibration tests. AIAA Journal. 1995;**33**:2182-2188

[9] Farrar CR, Worden K, Dulieu-Barton J. Principles of structural degradation monitoring. Encyclopedia of Structural Health Monitoring. 2009;**82**:1350-1370

[10] Rytter A. Vibrational Based Inspection of Civil Engineering Structures. Aalborg Denmark: Aalborg University; 1993

[11] Adams RD, Flitcroft JE, Short D, Walton D. Vibration Testing as a Nondestructive Test Tool for Composite Materials. West Conshohocken, PA, USA: ASTM International; 1975

[12] Cawley P, Adams RD. The location of defects in structures from measurements of natural frequencies. Journal of Strain Analysis for Engineering Design. 1979;**14**:49-57

[13] Sundararaman S. Static damage phenomena and models. Encyclopedia of Structural Health Monitoring. 2009;**8**: 158-188

[14] Mackerle J. Finite-element modelling of non-destructive material evaluation, an addendum: A bibliography (1997–2003). Modelling and Simulation in Materials Science and Engineering. 2004;**12**:799

[15] Bossi RH, Iddings FA, Wheeler GC. Nondestructive Testing Handbook, Vol. 4, Radiographic Testing. USA: American Society for Nondestructive Testing. Inc; 2002

[16] Krautkrämer J, Krautkrämer H. Ultrasonic Testing of Materials. Berlin, Germany: Springer Science & Business Media; 2013

[17] Hardt DE, Katz JM. Ultrasonic measurement of weld penetration. Welding Journal. 1984;**63**:273s-281s

[18] Prosser WH, Jackson KE, Kellas S, Smith BT, McKeon J, Friedman A. Advanced waveform-based acoustic

emission detection of matrix cracking in composites. Materials Evaluation. 1995; **53**:1052-1058

[19] Wevers M. Listening to the sound of materials: Acoustic emission for the analysis of material behaviour. Ndt & E International. 1997;**30**:99-106

[20] Rao BPC, Jayakumar T, Raj B. Electromagnetic NDE techniques for materials characterization. In: Ultrasonic and Advanced Methods for Nondestructive Testing and Material Characterization. Hackensack, NJ, USA: World Scientific; 2007. pp. 247-277

[21] Grimberg R, Premel D, Savin A, le Bihan Y, Placko D. Eddy current holography evaluation of delamination in carbon-epoxy composites. Insight. 2001;**43**:260-264

[22] Zoughi R. Microwave Non-destructive Testing and Evaluation Principles. Vol. 4. Berlin, Germany: Springer Science & Business Media; 2000

[23] Papaelias M, Cheng L, Kogia M, Mohimi A, Kappatos V, Selcuk C, et al. Inspection and structural health monitoring techniques for concentrated solar power plants. Renewable Energy. 2016;**85**:1178-1191

[24] Lovejoy MJ. Magnetic Particle Inspection: A Practical Guide. Berlin, Germany: Springer Science & Business Media; 1993

[25] Bagavathiappan S, Lahiri BB, Saravanan T, Philip J, Jayakumar T. Infrared thermography for condition monitoring–a review. Infrared Physics & Technology. 2013;**60**:35-55

[26] Meola C, Carlomagno GM, Giorleo L. Geometrical limitations to detection of defects in composites by means of infrared thermography. Journal of Nondestructive Evaluation. 2004;**23**: 125-132. DOI: 10.1007/s10921-004-0819-z

[27] Ibarra-Castanedo C, Genest M, Piau J-M, Guibert S, Bendada A, Maldague XP, et al. Active Infrared Thermography Techniques for the Nondestructive Testing of Materials. Ultrasonic and Advanced Methods for Nondestructive Testing and Material Characterization. Hackensack, NJ: World Scientific; 2007. pp. 325-348

[28] Maldague X, Ziadi A, Klein M. Double pulse infrared thermography. Ndt & E International. 2004;**37**:559-564

[29] Adams D. Health Monitoring of Structural Materials and Components: Methods with Applications. New York, NY, USA: John Wiley & Sons; 2007

[30] Unobe ID, Bean B, Sorensen AD. Creating an asset management plan for traffic signal structures through interactive explorations of wind induced fatigue damage. Structure and Infrastructure Engineering. 2022; **18**:1-14

[31] Diekfuss JA. Reliability-Based Fatigue Assessment of Mast-Arm Sign Support Structures. Milwaukee, WI, USA: Marquette University; 2013

[32] Thomas RJ, Steel K, Sorensen AD. Reliability analysis of circular reinforced concrete columns subject to sequential vehicular impact and blast loading. Engineering Structures. 2018;**168**: 838-851

[33] Liu M, Frangopol DM. Time-dependent bridge network reliability: Novel approach. Journal of Structural Engineering. 2005;**131**:329-337

[34] Melchers RE, Beck AT. Structural Reliability Analysis and Prediction. New York, NY, USA: John Wiley & Sons; 2018

[35] Nowak AS, Collins KR. Reliability of Structures. Boca Raton, FL, USA: CRC Press; 2012

[36] Jiang Z, Hu W, Dong W, Gao Z, Ren Z. Structural reliability analysis of wind turbines: A review. Energies (Basel). 2017;**10**:2099

[37] Dang C, Valdebenito MA, Faes MGR, Wei P, Beer M. Structural reliability analysis: A Bayesian perspective. Structural Safety. 2022;**99**: 102259

[38] Ayyub BM, McCuen RH. Probability, Statistics, and Reliability for Engineers and Scientists. Boca Raton, FL, USA: CRC Press; 2016

[39] Der Kiureghian A. Analysis of structural reliability under parameter uncertainties. Probabilistic Engineering Mechanics. 2008;**23**:351-358. DOI: 10.1016/j.probengmech.2007.10.011

[40] Akgül F, Frangopol DM. Rating and reliability of existing bridges in a network. Journal of Bridge Engineering. 2003;**8**:383-393

[41] Enright MP, Frangopol DM. Condition prediction of deteriorating concrete bridges using Bayesian updating. Journal of Structural Engineering. 1999;**125**:1118-1125

[42] Saad T, Fu CC. Determining remaining strength capacity of deteriorating RC bridge substructures. Journal of Performance of Constructed Facilities. 2015;**29**:04014122

[43] Unobe ID, Sorensen AD. Multi-hazard analysis of a wind turbine concrete foundation under wind fatigue and seismic loadings. Structural Safety. 2015;**57**:57. DOI: 10.1016/j.strusafe.2015.07.003

[44] Beck JL, Katafygiotis LS. Updating models and their uncertainties. I: Bayesian statistical framework. Journal of Engineering Mechanics-Proceedings of the ASCE. 1998;**124**:455-462

[45] Vanik MW. A Bayesian Probabilistic Approach to Structural Health Monitoring. Pasadena, CA, USA: California Institute of Technology; 1997

[46] Dawood M, Goyal R, Dhonde H, Bradberry T. Fatigue life assessment of cracked high-mast illumination poles. Journal of Performance of Constructed Facilities. 2014;**28**:311-320

[47] Mangalathu S, Hwang S-H, Choi E, Jeon J-S. Rapid seismic damage evaluation of bridge portfolios using machine learning techniques. Engineering Structures. 2019;**201**:109785

[48] Stewart MG, Val D, v. Role of load history in reliability-based decision analysis of aging bridges. Journal of Structural Engineering. 1999;**125**:776-783

[49] Coppe A, Pais MJ, Haftka RT, Kim NH. Using a simple crack growth model in predicting remaining useful life. Journal of Aircraft. 2012;**49**: 1965-1973

[50] Chung H-Y, Manuel L, Frank KH. Optimal Inspection of Fracture-Critical Steel Trapezoidal Girders: A Summary. Austin, TX: Center for Transportation Research, University of Texas at Austin; 2004